Evolution Through Genetic Exchange

Michael L. Arnold

Department of Genetics, University of Georgia, Athens, Georgia, USA

OXFORD

UNIVERSITY PRESS

This book has been printed digitally and produced in a standard specification
in order to ensure its continuing availability

OXFORD
UNIVERSITY PRESS

Great Clarendon Street, Oxford OX2 6DP
United Kingdom

Oxford University Press is a department of the University of Oxford.
It furthers the University's objective of excellence in research, scholarship,
and education by publishing worldwide. Oxford is a registered trade mark of
Oxford University Press in the UK and in certain other countries

© Oxford University Press 2006

The moral rights of the author have been asserted

Reprinted 2012

British Library Cataloguing in Publication Data
Data available

Library of Congress Cataloging in Publication Data
Data available

ISBN 978-0-19-922903-1

Printed and bound by CPI Group (UK) Ltd, Croydon, CR0 4YY

Some books—if I may use a comparison I have used elsewhere—must be regarded more as we regard those cathedrals where work of many different periods is mixed and produces a total effect, admirable indeed but never foreseen nor intended by any one of the successive builders. It is misleading to think of . . . an author in our modern sense and throw all the earlier work into the category of 'sources'. He is merely the last builder, doing a few demolitions here and adding a few features there.

(C.S. Lewis, *The Discarded Image—An Introduction to Medieval and Renaissance Literature*, 1964, p. 210)

To Frances, Brian, and Jenny.
You make it possible for me to attempt great things.

Preface

As the title suggests, this book concerns the evolutionary role played by genetic exchange. In some cases this exchange can be termed natural hybridization; for other examples it is more apropos to refer to the process as lateral or horizontal gene transfer, and yet for other instances the distinction between these processes is blurred. One goal of this work then is to exemplify the similarities—with regard to fundamental evolutionary outcomes—of genetic exchange, whatever its underlying mechanism. I am unaware of another synthesis that has highlighted these similarities.

One reason I have been unable to identify such a synthesis (other than the fact that I might have overlooked it) is that the processes of natural hybridization and lateral gene transfer are seen as non-overlapping in their mechanism and outcome. Undoubtedly some of the avenues for genetic exchange—such as introgression through sexual recombination versus lateral gene transfer mediated by transposable elements—are based on definably different molecular mechanisms. However, even such widely different mechanisms may result in similar effects on adaptations (new or transferred), genome evolution, population genetics, and the evolutionary/ecological trajectory of organisms.

A second reason, somewhat related to the first, is an organism-biased viewpoint among some evolutionary biologists. This viewpoint can be summed up by a quote from a colleague at a recent congress on evolutionary biology—'Lateral gene transfer occurs in microorganisms and has nothing to do with animals. In contrast, natural hybridization affects the latter, but not the former.' I answered this colleague in the following manner. First, I reminded him that genetic exchange *sensu lato*

could produce the same, important evolutionary consequences, notwithstanding the underlying mechanism causing the exchange. For example, this is seen in the evolution of novel adaptations—involving lateral gene transfer—leading to the flea-borne, deadly, causative agent of plague from a rarely fatal, orally transmitted, bacterial species. In terms of evolutionary outcomes, this is quite similar to the adaptations accrued from natural hybridization between annual sunflower species resulting in the formation of several new species. Furthermore, I pointed out that natural hybridization *and* lateral gene transfer have impacted both 'higher' and 'lower' organisms. Consider, on the one hand, evidence for introgressive hybridization between the lineages leading to humans and chimpanzees, and the impact of retroviral-like elements on these same lineages, and on the other hand the effect of natural hybridization and lateral gene transfer on species of trypanosomes or yeast.

The chapter structure of this book reflects my goal of presenting the conceptual framework indicated above. This conceptual framework can be summarized with a metaphor. Thus, instead of a simple bifurcating tree of life, I believe that a better metaphor for describing evolutionary diversification is a web of life. I also believe that this is not a novel viewpoint. The recognition of divergence accompanied by ongoing, or at least episodic, genetic exchange was championed in the last century (e.g. see the work of Ledyard Stebbins and Edgar Anderson) and is also reflected in a growing number of contemporary, evolutionary publications. As with Darwin's descent with modification, a number of groups/workers in evolutionary biology have arrived at the same conclusion nearly simultaneously. I cover the work by many of the groups contributing to this paradigm shift, but I will apologize beforehand for inevitable, and totally accidental, oversights.

Chapter 1 includes a brief history of pre-Darwinian, evolutionary studies concerning genetic exchange. I then turn my attention to various organismal systems to exemplify some ways in which post-Modern Synthesis research has been pursued to test the evolutionary role of genetic exchange. I have chosen examples of well-developed evolutionary model systems including plants, animals, bacteria, and viruses. I have chosen these as exemplars because they reflect broadly based, in-depth studies of the possible evolutionary consequences from natural hybridization and/or lateral gene transfer. Some cases (e.g. viral lineages) also afford the opportunity to point to the uncertainty in categorizing them as either natural hybridization or lateral gene transfer. This reflects a tension that I hope to maintain throughout this book. The tension results from the observation that these processes of genetic exchange blend together somewhat in terms of mechanism of occurrence and evolutionary outcome.

Chapter 2 introduces the topic of species concepts and the study of genetic exchange. I will use only four (biological, phylogenetic, cohesion, and prokaryotic) of the many definitions to illustrate how our concept of (i) what a species is and (ii) how it originates influences what evolutionary importance (or lack there of) we ascribe to genetic exchange. I end this chapter by suggesting how we might use species concepts—often the bane of evolutionists interested in gene exchange—to afford a clearer understanding of the importance of natural hybridization and lateral gene transfer.

In Chapter 3, I discuss methodologies to test the hypothesis of genetic exchange. The goal of this chapter is to indicate the diversity of approaches that facilitate such tests. I begin by examining conflicting conclusions, drawn by different investigators, concerning the importance of gene flow in the genus *Quercus*. I use this to indicate the importance of utilizing a comparative approach when investigating possible instances of genetic exchange. I then discuss five methodologies to test for such exchange: (i) trans-generational hybrid zone analyses; (ii) estimates of phylogenetic discordance;

(iii) analyses involving gene genealogies and models of speciation; (iv) estimations of intragenomic divergence; and (v) the application of nested clade analysis. I review examples that highlight the specific utility of each of these approaches.

In Chapter 4, I turn my attention to the barriers that limit genetic exchange. The conceptual framework for this chapter assumes that specific barriers are part of a multi-tiered process. In sexually reproducing organisms these barriers can be easily placed into the categories of pre- and postzygotic. However, even for those organisms thought to spend most of their life history reproducing asexually (e.g. bacterial species) limits to genetic exchange can be typified with this concept. Various ecological, behavioral, gametic, viability, and fertility barriers can thus isolate both 'asexual' and 'sexual' species. The model that I adopt in this chapter is one in which multiple, life-history-stage-specific characteristics must be overcome for gene exchange to occur.

Chapter 5 illustrates possibly the most significant fact highlighted in my first book, *Natural Hybridization and Evolution* (1997, Oxford University Press): hybrid genotypes, like any other set of genotypes, demonstrate a range of fitness estimates. In keeping with the subject of this present work, I demonstrate this point with examples involving natural hybridization, viral recombination, and lateral gene transfer. For all classes the conclusion is the same, some hybrid/recombinant genotypes have lower, some the same, and others higher fitness estimates relative to their progenitors.

Chapter 6 reflects a fundamentally important outcome of genetic exchange, that of gene duplication. As throughout this book, I consider instances where the genetic exchange, in this case resulting in genomic duplications, has derived from sexual reproduction as well as other forms of recombination. I consider how genetic-exchange-induced duplications are seen as affecting genome evolution, gene function, adaptations and radiations of entire clades. In keeping with the tenor of the entire book, I emphasize evolutionarily creative outcomes that derive from the duplication of the genomic components of microorganisms, plants, and animals.

Chapter 7 includes examples of the effects of gene exchange on the formation of new evolutionary lineages. In some cases this can be viewed as the origin of new species, in others the organisms formed from genetic exchange are given subspecific classifications. In some instances, the genetic exchange event is seen to be the basal event for entire evolutionary assemblages. Regardless of the evolutionary timing of such events and the taxonomic identity of the products, the importance remains in the outcome: that is, novelty has been produced, both in terms of new organisms and new adaptations. As in other chapters, I will use examples from a wide range of microorganismic and multicellular lineages. Furthermore, I will highlight cases of diversification that can be placed into the categories of polyploid, homoploid, ecological, and recombinational speciation. Yet, as mentioned above, I will not limit my discussion to only those cases that involve the origin of single lineages recognized as 'species'. Rather, I will include an array of exemplars that reflect the diversity of outcomes that include the formation of novel evolutionary lineages and clades.

Topics in Chapter 8 include those associated with positive and negative effects of genetic exchange on the fate of endangered flora and fauna. In particular, I consider the role that such gene flow may have (i) in replenishing populations with limited genetic variability or (ii) in causing genetic assimilation of rare forms by more numerous, related taxa. I will argue that if evolutionary diversification is indeed a web-like process, then we should not give weight to whether members of one evolutionary lineage exchange genes with another when attempting to determine a value for conservation. Instead, we should ask the question of whether genetic exchange can help or hinder the conservation of manageable units (i.e. taxa). I also examine the hypothesis that invasive species sometimes originate through introgressive hybridization.

Chapter 9 reviews findings relating to the effect of genetic exchange on the evolution of the human lineage and lineages of organisms with which we interact (e.g. disease vectors, food sources). I consider fossil and genetic data that suggest gene flow between various archaic taxa, and between archaic taxa and anatomically modern *Homo* (for this section I borrow heavily from Arnold and Meyer 2006). I demonstrate the disastrous effects of natural hybridization and lateral gene transfer in disease and disease-vector evolution. However, I also consider the highly beneficial results from the same processes in producing food products, drugs, and even clothing. Chapter 9 concludes with a discussion demonstrating that *Homo sapiens* has played an active role in producing new strands of the web of life. Our species has accomplished (and continues to accomplish) this through the genetic modification of microorganisms, plants, and animals. In the past, we were limited in this exercise to methods involving crosses between related organisms— through introgressive hybridization. Most recently, however, we have begun a program of lateral gene transfer, through 'genetic modification', to produce hybrids carrying genes from more and more distantly related forms.

In Chapter 10 I begin by reflecting upon the major theme of this book: that genetic exchange is pervasive across all biological lineages. I discuss the implications of this regarding the tree-of-life and web-of-life concepts. I include a discussion of research directions that will benefit our understanding of the role of genetic exchange in evolution. Some of these, including the use of genomic information to discern web processes, are gaining momentum with the appearance of many new data-sets. Others, such as studies that investigate the role of ecological setting on the outcome of genetic exchange, are rare, yet they represent another Golden Fleece because of their potential to yield new insights of major importance. I conclude Chapter 10 with a quote from Darwin that (as is usual for his writings) encapsulates the theme I try to communicate.

There is a glossary of definitions at the back of the book. All terms listed in the Glossary are italicized in the text where they are first used.

Many undergraduate, graduate, and postgraduate scholars, along with evolutionary biologist colleagues at the University of Georgia and elsewhere, have provided crucial data and concepts that form the basis of this book. I owe a huge debt to Ed Larson—my colleague and friend—who spent countless hours with me in discussion and writing sessions that led to our *Wilson Quarterly* article in which we summarized the concept of the web of life. Wyatt Anderson, the late Marjorie Asmussen,

John Avise, Jim Hamrick, Jessie Kissinger, Rodney Mauricio, Daniel Promislow, and John Wares from the Department of Genetics, University of Georgia, have given much of their limited and valuable time to discussions of the ideas presented in this book. Over the many years of our friendships, Dan Howard and Loren Rieseberg have given encouragement and guidance in the research paths taken by my group. Much of their guidance has resulted in the ideas and data collected in this book. The following graduate students and postdoctoral associates in the Department of Genetics, University of Georgia, read, and helped revise, portions of this book: Scott Cornman, Eleanor Kuntz, Noland Martin, Monica Poelchau, Jeff Ross-Ibarra, and Scott Small. Paul Harvey and Peter Holland of the University of Oxford graciously expended much time and effort that resulted in my Visiting Research Fellowship at Merton College, Oxford. Ian Sherman, Abbie Headon, Kerstin Demata, and Stefanie Gehrig provided invaluable editorial support at all stages of this project. During the writing of this book, I have been supported financially by a grant from the National Science Foundation (DEB-0345123). Finally, I take full responsibility for any mistakes or omissions in this text.

As with my first book, I must direct most of my thanks and praise to my wife Frances and my children—now adults—Brian and Jenny. Without their continuous love and support, without their understanding and patience, without their encouragement and faith in my ability to do what I thought impossible—that is, to write books—this project would not have even been contemplated. As with *Natural Hybridization and Evolution*, I dedicate this book to them.

Contents

CHAPTER 1

History of investigations

The large number of cases of introgression [reported led to the complaint] that we 'were seeing hybrids under every bush'. The truth of the matter was that we had learned under exactly which bush to look.

(Anderson 1951)

. . . hybridization between species or near-species can provide the genetic variation necessary for natural selection to produce a new adaptive norm.

(Lewontin and Birch 1966)

On a phylogenetic time scale, lateral transfer is one cause of reticulate evolution that can result in inconsistencies in taxonomic relationships.

(Kidwell 1993)

Horizontal transfer, even at very low levels, produces a mosaic chromosome comprised of genes of differing ancestries and durations in the genome.

(Lawrence and Ochman 1998)

This tree-of-life notion of evolution attained near-iconic status in the mid-20th century with the modern neo-Darwinian synthesis in biology. But over the past 15 years, new discoveries have led many evolutionary biologists to conclude that the concept is seriously misleading and, in the case of some evolutionary developments, just plain wrong. Evolution, they say, is better seen as a tangled web.

(Arnold and Larson 2004)

1.1 Pre-Darwin, Darwin, the Modern Synthesis and genetic exchange: development of a paradigm

The role of genetic exchange in evolution has been of intense interest and debate for at least a century. However, the accumulation of data sets, the consideration of new models, and the resurrection of old models has led to a paradigm shift, reflected by an appreciation of the importance of gene transfer. In spite of this, arguments emphasizing an unimportant role for genetic exchange are still reflected in recent evolutionary literature (e.g. Coyne and Orr 2004). The goal of this text is to highlight numerous examples that illustrate the widespread and significant impact from the various mechanisms of genetic exchange on organismal evolution (i.e. molecular to phenotypic).

Considerations of the relative importance of genetic exchange began with conceptualizations and studies concerning the evolutionary role of *natural hybridization* (see the Glossary for a definition). Much later came the recognition of the additional pervasiveness of *lateral* or *horizontal gene transfer* (Katz 2002; Zhaxybayeva *et al.* 2004; Beiko *et al.* 2005; Gogarten and Townsend 2005; Oborník and Green 2005; Simonson *et al.* 2005; Sørensen *et al.* 2005; Shutt and Gray 2006). Although mythological 'hybrids' are legion in Greek, Egyptian, and Hindu writings and art, scientific investigations into the process and products of natural hybridization are not seen clearly until the mid-eighteenth century and the work of Linnaeus. Linnaeus believed in a Special Creation model for the origin of species. Surprisingly, he also held that species could arise through hybridization (i.e. *hybrid speciation*) between previously created forms. In this regard, he wrote in his *Disquisitio de Sexu Plantarum* (1760; as cited by Grant 1981, p. 245), 'It is impossible to doubt that there are new species produced by hybrid generation.' Just over 100 years later, the German botanist Joseph Kölreuter studied the consequences of experimental hybridization in plants. Kölreuter (as referenced by Darwin 1859, pp. 246–247) found that hybrids from heterospecific crosses were often sterile, a result that cast doubt on the possibility of *hybrid species* formation. As with Kölreuter, Darwin seemed most impressed by the fact that heterospecific crosses were difficult to form and that the offspring from such crosses (i.e. 'mongrels') were generally highly infertile. Such observations led Darwin to conclude that the most likely outcome of natural hybridization would be the rare production of hybrid offspring with uniformly low fitness, and thus reflected an evolutionary dead-end (Darwin 1859, pp. 276–277).

The view of genetic exchange (in particular, natural hybridization) as an evolutionarily important process came to be championed by workers such as the Dutch botanist Johannes Lotsy in the period between the publication of *On the Origin of Species* and the neo-Darwinian synthesis. Unlike Kölreuter and Darwin, Lotsy (1931) envisioned natural hybridization as the primary mechanism for evolutionary change. Specifically, Lotsy (1931) hypothesized that new taxa originated from the interbreeding of individuals from different syngameons ('... an habitually interbreeding community ...', Lotsy 1931, p. 3), and not from the gradual accumulation of heritable differences. In contrast, the advent of the neo-Darwinian synthesis, or Modern Synthesis ushered, in a conceptual framework unfavorable to an appreciation for the role of *reticulate evolution*. As the concepts of the Modern Synthesis took root and spread—particularly through the writings of Ernst Mayr and Theodosius Dobzhansky—the viewpoint that evolutionary diversification resulted almost entirely through allopatric divergence (i.e. in the absence of gene flow) became entrenched. Furthermore, if it did occur, genetic exchange (once again in regard to

natural hybridization) was seen as an epiphenomenon of 'normal' evolutionary processes. Mayr voiced the viewpoint of many evolutionary zoologists in the following ways: 'Successful hybridization is indeed a rare phenomenon among animals' and '... available evidence contradicts the assumption that hybridization plays a major evolutionary role...' (Mayr 1963, p. 133).

Yet, simultaneously with the establishment of the paradigm that gene exchange was of little significance for animals, workers in botanical systems were finding evidence for its widespread and important role in diversification and adaptation. Edgar Anderson and Ledyard Stebbins, in particular, formulated arguments concerning the effects of genetic exchange and then demonstrated experimentally the effect of natural hybridization on the evolutionary trajectory of numerous plant assemblages. Anderson highlighted, for example, the key importance of environmental disturbance as a cause of *introgressive hybridization* (Anderson 1948; see also de Kort *et al.* 2002a, b for an example of introgressive hybridization due to habitat modification). Anderson (1949) also predicted future results demonstrating the role of *adaptive trait transfer* in producing introgressed forms capable of invading novel habitats (e.g. Figure 1.1; Lewontin and Birch 1966).

Though possibly not a part of his consideration, Anderson also presaged the understanding of how such adaptations might form in microorganisms through the processes of lateral gene transfer, viral recombination, etc. Indeed, the occurrence of new adaptations through genetic exchange between microorganisms became evident at this time (i.e. the 1950s) with the discovery that bacteria could acquire multiple antibiotic resistances through lateral gene transfer (Ochman *et al.* 2000). However, it would take approximately 40 years, and the advent of multiple-gene genealogies, before it was understood that genetic exchange among microorganisms was rampant, and of fundamental evolutionary importance (Boucher *et al.* 2003; Olendzenski *et al.* 2004; Ochman *et al.* 2005; Simonson *et al.* 2005). Like studies of eukaryotes, genetic exchange involving members of the other domains of life was detectable by non-concordance between phylogenies constructed using different genes. This non-concordance indicated that a large

Figure 1.1   Distribution of the Queensland fruit fly, *Dacus tryoni* (now *Bactrocera tryoni*). This species has extended its range during the past 150 years to include locations east of the dashed line. This range expansion is hypothesized to be the result of the introgression of genes from *Dacus neohumeralis* (now *Bactrocera neohumeralis*) controlling adaptations to the colder climates (from Lewontin and Birch 1966).

fraction of the genomes of microorganisms had been acquired laterally from other taxa (see Doolittle 1999 for a review). Yet, between the discovery of the transfer of antibiotic resistance and the collection of genomic data in the 1980s and 1990s, the evolutionary impact of genetic exchange among microorganisms was thought to be slight (Simonson *et al.* 2005).

Like Anderson, Ledyard Stebbins emphasized the role of natural hybridization in affecting the evolution of plant species. In 1954, in a classic paper entitled 'Natural hybridization as an evolutionary stimulus', Stebbins, with Anderson, argued that natural hybridization was of widespread and fundamental importance for the formation of novel adaptations and new evolutionary lineages (Anderson and Stebbins 1954). Furthermore, Stebbins proposed that *allopolyploid species* made up a large proportion of flowering plants (Stebbins 1947). It may seem somewhat surprising then that evolutionary biologists in general were seen to endorse a negative outlook on the evolutionary role of genetic exchange, especially given that Stebbins was recognized as one of the architects of the Modern Synthesis. Yet, as the synthesis hardened this was the scientific atmosphere.

I have argued elsewhere (e.g. Arnold 1997, p. 13) that this negative viewpoint has as much to do with sociology and philosophy as it does with biology. Thus I believe that two philosophical issues have resulted in the conclusion that natural hybridization is inherently deleterious. The first of these is the fact that genetic exchange may prevent systematic/taxonomic treatments from being sharply defined. It is much more satisfying to have multiple, resolved, and concordant phylogenies. Reticulate events may derail these analyses by leading to non-concordant phylogenetic hypotheses (e.g. Rieseberg *et al.* 1990b; but see McDade 1992). The second factor affecting negative viewpoints toward the role of genetic exchange can be couched in terms of violations of species integrity. Paterson (1985) reflected a similar conclusion when stating, 'In English, notice how approbative are words such as "pure", "purebred", "thoroughbred" and how pejorative are those like . . ."hybrid". Such cultural biases . . . might well predispose the unwary to favour ideas like that of "isolating mechanisms" with the role of "protecting the integrity of species" '. Natural hybridization is seen as a breach in this integrity.

However, the above does not explain fully why the Modern Synthesis was a watershed for the reaffirmation of the opinion voiced by Kölreuter and Darwin, especially in the face of so much evidence of the evolutionary role played by genetic exchange. I believe that one explanation for this turn of events lies in the profligacy of publications by Mayr and Dobzhansky. Their emphasis on allopatric divergence, with little or no role for

diversification concurrent with genetic exchange, was the most widely publicized and well-known paradigm. As a result, I would suggest that there has been an attempt by many to explain the evolutionary pattern of all organisms with a narrowly based, zoological paradigm. I will argue in later chapters that the effects from web processes (i.e. genetic exchange due to introgression, viral recombination, and lateral gene transfer) are replete in animal species as well, in this case mediated by introgressive hybridization. However, until recently this has not been a well-appreciated fact, as reflected by the following two quotes penned by evolutionary zoologists: 'Botanists recognize the importance of introgressive hybridization in evolution. Our results . . . indicate that zoologists must do the same' (Dowling and DeMarais 1993) and 'Today, DNA sequence data and other molecular methods are beginning to show that limited invasions of the genome are widespread, with potentially important consequences in evolutionary biology, speciation, biodiversity, and conservation' (Mallet 2005).

At the end of the time period encapsulating the Modern Synthesis (*c.*1950) the tenets that species arise in allopatry, through the gradual accumulation of mutations, and that the evolutionary history of species complexes does not include the exchange of genetic material were established. In the ensuing decades, this viewpoint, though not well supported by data from plants and microorganisms—nor, as recognized later, by already existing data from many animal groups (Arnold 1997)—was adopted as fact. This in turn had (and continues to have) an effect on how subsequent analyses were designed and what conclusions were drawn (e.g. see Barton and Hewitt 1985 and Coyne and Orr 2004 for examples of such approaches in studies of genetic exchange). For example, a recent study of the two sister species, *Drosophila yakuba* and *Drosophila santomea*, found that the mitochondrial DNA of the former species has replaced that of the latter through introgressive hybridization (Llopart *et al.* 2005). Furthermore, this same analysis detected 'significant introgression' for two of 28 (7%) of the nuclear loci studied. In spite of their findings of such high levels of introgression, Llopart *et al.* (2005) adopted the classical argument that '. . . gene flow has not been extensive . . .'.

In contrast to the above, studies of some species complexes have indeed been couched in terms of a web-of-life rather than a tree-of-life metaphor (see Arnold 1997 for a review), while others were designed with an appreciation of both metaphors. The following examples illustrate the type of study that has informed an understanding of the dual processes of genetic exchange and evolutionary diversification. I use these examples for two purposes. The first purpose is to exemplify the scientific development of four species complexes—including plants, bacteria, animals, and viruses—as model systems for studying genetic exchange during the post-Modern Synthesis period. The second purpose is to demonstrate the similarity in outcomes, in terms of genetic structure and evolutionary trajectory, resulting from genetic exchange mediated by such diverse mechanisms as sexual reproduction, lateral gene transfer, and viral recombination. For the microorganisms, the literature reviewed is very recent. This reflects the general lack of an appreciation for the extent of genetic exchange (in the form of lateral gene transfer and viral recombination) that has occurred among microorganisms until relatively recently (the late 1980s).

1.2   Post-Modern Synthesis: case studies of genetic exchange

1.2.1   Louisiana irises

Taxonomic uncertainty fuels evolutionary clarification
In 1931, a paper appeared in the *Contributions of the New York Botanical Gardens*. Entitled 'Botanical interpretation of the Iridaceous plants of the Gulf States', and authored by J.K. Small and E.J. Alexander, this paper was the stimulus for an entire scientific cottage industry reflected in studies by research groups now spanning seven decades (e.g. Viosca 1935; Foster 1937; Riley 1938, 1939, 1942, 1943a, b; Randolph *et al.* 1961, 1967; Bennett and Grace 1990; Arnold *et al.* 1990a; Cruzan and Arnold 1993; Carney *et al.* 1994; Burke *et al.* 1998a, b; Johnston *et al.* 2004; Bouck *et al.* 2005; Martin *et al.* 2005). The focus of Small and Alexander's (1931) study was the plant group commonly known as the Louisiana irises. Ironically, Small and Alexander's main conclusion—that there existed over 80 species

of Louisiana iris within the confines of Louisiana—was demonstrably wrong. Instead, Small and Alexander had discovered, and named in some detail, products of introgressive hybridization between three morphologically, developmentally, reproductively, and ecologically definable species (Viosca 1935; Foster 1937; Riley 1938). Although Small and Alexander's taxonomic work was flawed, their studies resulted in recognition of the Louisiana irises as an excellent example of the evolutionary effects arising from web processes. Specifically, these species occupy a fundamentally important scientific and historical niche as the exemplar of Edgar Anderson for the process of introgressive hybridization (Anderson 1949).

In rapid succession, Viosca (1935), Foster (1937), and Riley (1938, 1939, 1942, 1943a, b) responded to Small and Alexander's (1931) taxonomic treatment with overwhelming evidence that the majority of 'species' named by them were actually recombinant (i.e. hybrid) genotypes (e.g. Figure 1.2; Riley 1938). The data generated by Viosca, Foster, and Riley demonstrated that three species, *Iris fulva*, *Iris brevicaulis*, and *Iris hexagona*, had hybridized naturally to form the 70 or more remaining forms described by Small and Alexander (1931). Comparisons of these species showed differences in pollinators, ecological setting, and flowering seasons. However, when in sympatry, *I. fulva*, *I. brevicaulis* and *I. hexagona* formed *hybrid zones* (Viosca 1935; Riley 1938). This then was the explanation for the many forms recognized by Small and Alexander (Figure 1.2).

Edgar Anderson published his classic *Introgressive Hybridization* (Anderson 1949) subsequent to the initial flurry of research regarding the number of Louisiana iris species. In the first chapter of his book, Anderson chose the morphological characters (Figure 1.2) described by Riley (1938) to illustrate the process of introgressive hybridization (i.e. 'introgression'; Anderson and Hubricht 1938). In an earlier paper Anderson (1948) had used these same species to exemplify the effect of human disturbance on introgression. In both of these treatments, Anderson proposed a significant role for hybridization in the evolutionary history of Louisiana iris species. In particular, he concluded that natural hybridization had resulted in introgression. Anderson held that the general

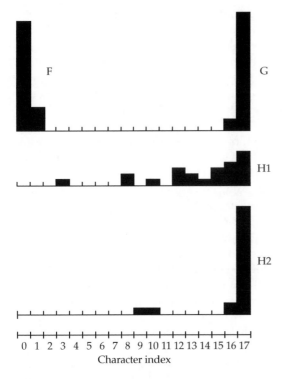

Figure 1.2 Character index distribution for populations of Louisiana irises. Each population (i.e. F, G, H1, H2) demonstrates admixtures of characters from *Iris fulva* and *Iris hexagona* var. *giganticaerulea*. A character index score of 0 indicates *I. fulva* individuals, a score of 17 typifies *I. hexagona* var. *giganticaerulea* plants and intermediate scores reflect hybrid genotypes. The hybrid plants possess morphologies typical of plants listed as different species by Small and Alexander (1931) (from Riley 1938).

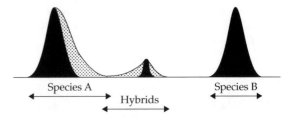

Figure 1.3 Expected morphological distributions for parental species, F₁, and later-generation hybrid individuals. The filled regions indicate parental and F₁ morphologies, while the stippled areas indicate later-generation hybrid phenotypes. The arrows beneath the distributions indicate the taxonomic descriptions of individuals possessing the various morphologies (from Anderson and Hubricht 1938).

significance of introgression was to greatly increase the '. . . variation in the participating species . . .' and thus '. . . far outweigh the immediate effects of gene mutation' (Anderson 1949, pp. 61–62). Anderson's viewpoint concerning the evolutionary importance of natural hybridization in the Louisiana irises (and other species; Anderson 1949) rested on his conclusion that introgression had impacted the gene pools of the hybridizing species.

On the second page of *Introgressive Hybridization* Anderson stated, 'Riley, Foster, Viosca, and Anderson are in virtual agreement concerning the following account . . .'. This 'account' referred to the description of introgression in the Louisiana irises. One name omitted was that of L.F. Randolph. Yet, data from a cytological study by Randolph (1934) were cited in the same paragraph containing

the affirmation of agreement among the other authors. I have suggested previously that, at the time of publication of Anderson's book, Randolph may already have held a contrary viewpoint concerning the importance of introgressive hybridization among these species. Regardless, Randolph's publication, entitled 'Negative evidence of introgression affecting the stability of Louisiana iris species' (Randolph *et al.* 1967), indicated clearly his opposition to the conclusions drawn by Anderson (Arnold 1994). Randolph's paper reported morphological—with limited pollen fertility—data for populations of *I. fulva*, *I. brevicaulis*, *I. hexagona*, and their *natural hybrids*. In contrast to Anderson (1949), the results from the morphological and pollen fertility studies led Randolph *et al.* (1967) to conclude that there was only localized hybridization and introgression between all of the Louisiana iris species. Randolph *et al.* (1967) thus concluded that natural hybridization among the Louisiana irises was not a 'typical example' of an extensive, and evolutionarily important, role for introgressive hybridization.

One explanation for the opposing conclusions of Anderson and Randolph lay in the lack of resolving power afforded by the data sets used by both research groups. Anderson and Hubricht (1938) had discussed such a lack of power for detecting introgression when using quantitative (e.g. morphological) traits (Figure 1.3). They emphasized that the effect of repeated backcrossing between hybrids and parental taxa was to transfer a small proportion of genetic material from one species into another (i.e. introgression). Such transfer might

Figure 1.4 Proportion of *I. fulva* (filled) and *I. hexagona* (unfilled) genetic markers in 10 Louisiana iris populations. Each pie diagram represents a separate population sample. *I. fulva* populations 1–3 and *I. hexagona* populations 4–7 show no evidence of introgression. However, the morphologically *I. fulva*-like population 9 and *I. hexagona*-like population 10 reflect the effect of introgressive hybridization, as does the contemporary hybrid zone (population 8; from Arnold *et al*. 1990a).

result in evolutionarily novel and important genotypes (Anderson and Hubricht 1938; Anderson 1949; Anderson and Stebbins 1954; Lewontin and Birch 1966; Chiba 2005). However, the '. . . wider spread of a few genes . . . might well be imperceptible . . .' (Anderson 1949, p. 102). Thus for Anderson and Randolph, the available methodologies made it problematic to test whether introgression had affected greatly the evolution of the Louisiana irises. The test of the alternate viewpoints would await the availability of discrete molecular markers.

The evolutionary genetics of introgression and hybrid speciation
The joining of the necessary genetic markers with a research program interested in the evolutionary role of genetic exchange occurred during the late 1980s. These initial studies supported Anderson's contention (Anderson 1948, 1949) that introgression had indeed impacted greatly this species complex (Figure 1.4; Arnold *et al*. 1990a). We were thus able to find genetic footprints of past and ongoing

introgression between each of the three species, *I. fulva*, *I. brevicaulis* and *I. hexagona* (Figure 1.4; Arnold *et al*. 1990a, b; Nason *et al*. 1992).

The initial genetic surveys led to further analyses by which we dissected the effects of the two avenues for gene exchange in plants (i.e. seed and pollen) in causing the observed introgression. Either mode was seen as possible/probable for the Louisiana iris species; all species possess buoyant seeds that float in the bayous and marshes of their native habitats and have pollen vectors (bumblebees and hummingbirds) capable of long-distance flights between allopatric populations (Viosca 1935; Emms and Arnold 2000; Wesselingh and Arnold 2000). Inferring seed movement was made possible through assays of the maternally transmitted chloroplast DNA (cpDNA; Arnold *et al*. 1991; Cruzan *et al*. 1993). We were therefore able to use the cpDNA as a seed parent-specific (i.e. maternal) marker, and the biparentally inherited, nuclear markers as monitors of pollen flow. Used in concert, data from these marker systems supported the hypothesis of introgression being initiated first by direct pollen flow between the species, rather than by seed dispersal, followed by pollen exchange (Figure 1.5; Arnold *et al*. 1991, 1992; Arnold 1993; Arnold and Bennett 1993).

As discussed above, Randolph discounted the effects from introgressive hybridization among the Louisiana iris species. However, he did posit a hybrid origin for a species he designated, *Iris nelsonii*. Randolph (1966) demonstrated a naturalist's eye when he described *I. nelsonii*, on the basis of limited habitat, cytological, and morphological evidence, as a tri-hybrid product of *I. fulva*, *I. brevicaulis*, and *I. hexagona*. When our group tested this hypothesis using species-specific nuclear (ribosomal DNA, restriction fragment length polymorphisms (RFLPs), randomly amplified polymorphic DNAs (RAPDs), and allozymes) and cytoplasmic (cpDNA) markers, Randolph's insightful description of the derivation of *I. nelsonii* was supported (Arnold *et al*. 1991; Arnold 1993).

Pre-zygotic barriers to F_1 formation
A failed experiment led to a series of investigations into the cascade of barriers that limit reproduction among *I. fulva*, *I. brevicaulis* and *I. hexagona*. In an

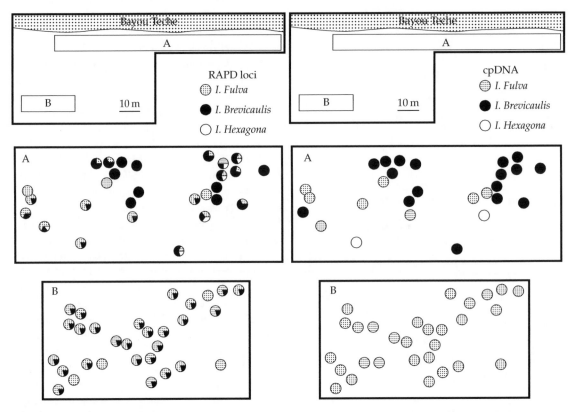

Figure 1.5 Distribution of species-specific nuclear (RAPD) and chloroplast (cpDNA) markers in a Louisiana iris hybrid zone involving *I. fulva*, *I. brevicaulis*, and *I. hexagona*. Each pie diagram represents a single plant. Though nuclear markers introgress freely, the cpDNA markers are restricted in their distribution as indicated by the lack of cpDNA introgression between the A and B portions of this zone, even though the distance between these subpopulations is only approx. 30 m. This indicates the restricted amount of seed-mediated relative to pollen-mediated introgression (from Arnold 1993).

effort to simulate the evolution of a hybrid zone, a mixed population of *I. fulva* and *I. hexagona* was established by introducing 200 individuals of the latter species into a natural population of *I. fulva*. The experimental design incorporated the natural transfer of pollen by bumblebees and humming-birds within and between the two species. Surprisingly, the expected high-frequency production of F_1 hybrids was not obtained. Over the 3-year study we instead observed only 0.03 and 0.74% F_1 seeds from *I. fulva* and *I. hexagona* fruits, respectively. Furthermore, no F_1 or later-generation hybrid plants became established in this mixed population (Hodges *et al.* 1996). In contrast, numerous (Arnold *et al.* 1990a, b, 1992; Nason *et al.* 1992; Arnold 1993; Cruzan and Arnold 1993, 1994; Johnston *et al.* 2001)

analyses of natural hybrid zones involving various combinations of *I. fulva*, *I. brevicaulis*, and *I. hexagona* detected hundreds of genotypes consistent with advanced-generation (i.e. BC_2, BC_3, F_2, F_3, etc), but not F_1-generation hybrids. These combined results suggested the need for a series of studies designed to test for pre- and post-pollination reproductive barriers to F_1 formation. I will discuss the results of these analyses in Chapter 4 of this volume in the context of illustrating reproductive isolation as a multi-staged process. In short, we found that both pollinator preferences and *gamete competition* affected the production of both F_1 and later-generation hybrid individuals (e.g. Carney *et al.* 1994; Emms and Arnold 2000; Wesselingh and Arnold 2000).

Hybrid zone evolution

The discovery of Louisiana iris hybrid zones, replete with later-generation hybrid genotypes, suggested that post-F_1 hybrid plants should be formed at significantly higher frequencies relative to the F_1 generation (Arnold 1994, 2000). The discovery of a high frequency of adult plants in natural hybrid zones demonstrating later-generation hybrid genotypes is consistent with this hypothesis. We also chose to test experimentally whether the formation of first-generation backcross progeny (i.e. BC_1) and second filial generation (i.e. F_2) hybrid progeny demonstrated a relaxation of the constraints apparent for the F_1 generation. To do this, artificially produced F_1 plants were introduced into an *I. fulva/I. hexagona* mixed population (Hodges *et al.* 1996) monitored previously for the frequency of F_1 formation (Arnold *et al.* 1993). In all, 100 F_1 rhizomes were planted in this mixed population (Hodges *et al.* 1996).

There was an enormous disparity in the number of parental and F_1 flowers produced during this flowering season (i.e. >1500 parental compared with 34 F_1). It thus seemed unlikely that the F_1 individuals would act as the pollen parent for more seeds on the parental flowers than would the more numerous, alternate parental individuals. However, this was exactly what Hodges *et al.* (1996) discovered. The frequencies of F_1 offspring formed in *I. hexagona* and *I. fulva* fruits were 0.74 and 0.03%, respectively. In contrast, the percentage of BC_1 progeny formed on *I. hexagona* and *I. fulva* were 6.9 and 1.7%, respectively. The extremely rare F_1 flowers, relative to parental flowers, caused an order of magnitude or more increase in hybrid production in the fruits of the two parental species (Hodges *et al.* 1996). This finding was consistent with the hypothesis that the formation and establishment of F_1 hybrids between the Louisiana iris species was the bottleneck for hybridization, introgression, and hybrid speciation. However, once this rare event occurred, the F_1-generation plants functioned as a substantial bridge for further genetic exchange and evolutionary diversification (Arnold 2000).

In Louisiana irises, there appears to be a relaxation of the barrier for genetic exchange subsequent to the F_1 generation. However, we have collected numerous data with regard to mating patterns and hybrid zone structure that detected factors that limited which later-generation hybrid genotypes contributed to further evolutionary innovations. In particular, the Louisiana iris parental and hybrid populations are genetically structured by widespread asexual reproduction (i.e. from rhizomatous spread; Burke *et al.* 2000a; Cornman *et al.* 2004). This asexual reproduction results in copies of identical genotypes distributed in natural iris populations. Such clonal reproduction could affect hybrid zone evolution if sexual reproduction occurred most often between genotypically similar or identical individuals. Cruzan and Arnold (1994), Hodges *et al.* (1996), and Cornman *et al.* (2004) did indeed find evidence for a higher than expected frequency of matings between similar hybrid genotypes. In addition, evidence of post-zygotic selection, resulting in hybrid zones with fewer and more similar hybrid genotypes than expected, was also discovered (Cruzan and Arnold 1994; Cornman *et al.* 2004).

Taken together, the above studies suggest that (i) the F_1 generation is the most restrictive sieve for the establishment of genetic exchange among *I. fulva*, *I. brevicaulis*, and *I. hexagona* and (ii) F_2, BC_1, and later-generation hybrids are formed at significantly higher frequencies than F_1 progeny, but that (iii) genetic structuring caused by clonal reproduction and post-zygotic selection, leading to a higher frequency of recruitment of similar genotypes, limits the genotypic complexity of hybrid zones.

Hybrid fitness

It has been argued for decades that natural hybridization between divergent populations of animals is an evolutionary deadend . . . A common rationale for this conclusion is the view that the outcome of hybridization episodes is governed by selection against hybrids independent of the environment. Although some plant biologists (e.g. Anderson 1949; Stebbins 1959) have been more open to environment-dependent models of hybrid zone evolution, process-oriented studies of plant hybrid populations have been few. This has resulted in a dearth in our understanding of patterns of hybrid fitness and the importance of environment-dependent selection in plants (Arnold 1997; p. 113).

This statement is now only partially correct. First, though some authors—zoologists in particular—still argue for its lack of importance in the evolutionary history of many, if not most, organisms (e.g. Schemske 2000; Coyne and Orr 2004), more and more researchers have discovered patterns that suggest a major role for genetic exchange in evolutionary diversification (e.g. see Arnold 2004a, b; Arnold et al. 2004; Seehausen 2004; Mallet 2005 for reviews). In addition, we now have extensive, process-oriented studies of natural hybridization in plants. In particular there are numerous studies that provide estimates of the fitness of hybrids as it relates to introgression and hybrid speciation (e.g. Schweitzer et al. 2002; Fritz et al. 2003; Rieseberg et al. 2003). However, the above quote reflects well the need for more of such data sets in order to test (i) the appropriateness of various hybrid zone models, (ii) the role of natural hybridization in the evolution of plant lineages, and (iii) the impact of web-like processes in the evolution of organisms in general (Arnold and Larson 2004).

With regard to analyses involving the Louisiana iris species complex, hybrid fitness has been estimated in a number of ways. The motivation for these studies was the recognition of the part played by hybrid fitness estimates (or assumptions) in assessments of the evolutionary importance of genetic exchange mediated by natural hybridization (Arnold 1992, 1997; Arnold and Hodges 1995). Though no single study has produced measurements of lifetime fitness for Louisiana irises—something difficult, or impossible, to accomplish for long-lived genotypes that can potentially maintain themselves indefinitely through asexual reproduction (e.g. Burke et al. 2000a; Cornman et al. 2004)—fitness estimates have been generated for multiple life-history stages and for numerous parental and hybrid genotypes. These experiments have detected the following: (i) differential seed viability correlated with various hybrid and parental classes (Cruzan and Arnold 1994); (ii) habitat-dependent and -independent fitness values for hybrid and parental species' genotypes (Emms and Arnold 1997; Burke et al. 1998a, b; Emms and Arnold 2000;

Wesselingh and Arnold 2000; Johnston et al. 2001, 2003; Cornman et al. 2004; Bouck et al. 2005; Martin et al. 2005); and (iii) high relative fitness due to heterosis and transgressive segregation for certain hybrid genotypes (Burke et al. 1998a; Johnston et al. 2004). Overall these investigations into the relative fitness of Louisiana iris genotypes demonstrate that different hybrid and parental genotypes can demonstrate a range of fitness values (e.g. Figure 1.6; Johnston et al. 2003). In some instances relative fitness is environment-dependent, and in other instances it is not. A fair proportion of hybrid genotypes demonstrate high relative fitness regardless of environment while other classes of hybrid genotypes possess relatively low fitness, also regardless of the environmental setting. Unlike earlier (Darwin 1859; Mayr 1963) and more recent (Coyne and Orr 2004) conceptualizations, our studies thus indicate that many hybrid genotypes demonstrate high fitness when compared to their parents.

Consistent with the findings from the studies of Louisiana irises (Randolph 1966; Arnold 1993), there are numerous examples of lineages receiving genetic material through sexual reproduction, other forms of recombination (e.g. that found in viruses), or horizontal transfer, resulting in the expression of novel adaptations (Figure 1.7). Particularly striking

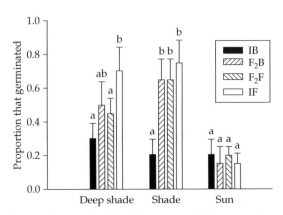

Figure 1.6 Proportion of seeds that germinated in four genotypic classes of Louisiana iris (IB, *I. brevicaulis*; IF, *I. fulva*; two F_2 hybrid crosses, F_2F, F_2 plants formed using F_1 plants with *I. fulva* cytoplasm; F_2B, F_2 plants formed using F_1 plants with *I. brevicaulis* cytoplasm). All data shown were from a single soil moisture treatment (i.e. moist). Letters indicate results from means comparisons within each light treatment (from Johnston et al. 2003).

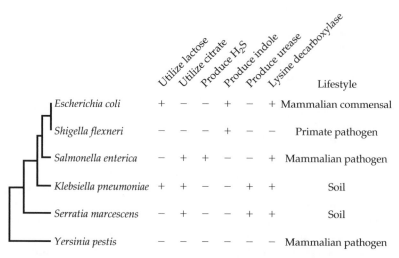

	Utilize lactose	Utilize citrate	Produce H$_2$S	Produce indole	Produce urease	Lysine decarboxylase	Lifestyle
Escherichia coli	+	−	−	+	−	+	Mammalian commensal
Shigella flexneri	−	−	−	+	−	−	Primate pathogen
Salmonella enterica	−	+	+	−	−	+	Mammalian pathogen
Klebsiella pneumoniae	+	+	−	−	+	+	Soil
Serratia marcescens	−	+	−	−	+	+	Soil
Yersinia pestis	−	−	−	−	−	−	Mammalian pathogen

Figure 1.7 Phylogenetic relationships and characteristic traits of various bacterial species. + indicates the presence and − the absence of a trait in more than 85% of the strains of a given species. Genes contributing to the species-specific characteristics have been acquired through horizontal transfer (from Ochman *et al.* 2000).

examples of acquisition of novel adaptations can be seen in human pathogens, such as plague. Furthermore, this organism is an excellent example of the debate over the relative importance of genetic exchange versus other processes, like point mutations and gene silencing, in the development of adaptations necessary for the invasion of new habitats. In the case of pathogens, the expression of the novel adaptations causes a novel set of pathologies in host species.

1.2.2 Plague

Yersinia pestis: *microevolution and origin*
Alexandre Yersin is credited with isolating the causative agent for plague (his namesake, *Yersinia pestis*) in 1894 (Perry and Fetherston 1997). As the announcement of the discovery was being made, the third pandemic of plague, this time originating in China and spreading along shipping routes that intersected Hong Kong, was underway (Achtman *et al.* 1999). The death toll from plague, at least during recorded history, has been estimated at approx. 200 million (Perry and Fetherston 1997). A large proportion of these fatalities occurred during the three pandemics that swept through different portions of the known world: (i) the Justinian plague from 541 to 544 AD in the Mediterranean basin, Mediterranean Europe, and the Middle East;

(ii) the European Black Death from 1347 to 1351 AD (followed by epidemic cycles until the nineteenth century); and (iii) the pandemic begun in the Yünnan province of China around 1855 that subsequently spread via steamship routes to Africa, Australia, Europe, Hawaii, India, Japan, the Middle East, the Philippines, North America, and South America (Perry and Fetherston 1997). Significantly, *Y. pestis* is now well defined as a derivative of the rarely lethal, enteric bacterial species *Yersinia pseudotuberculosis* (Achtman *et al.* 1999, 2004; Parkhill *et al.* 2001; Chain *et al.* 2004). In the context of this book, the question we need to ask is, did genetic transfer play a role in the evolution of this pathogen? More specifically, did genetic transfer result in the origin of adaptations necessary for the ecological shift from a gastrointestinal bacterial species, transferred through contaminated food and water, into a 'systemic invasive infectious' disease-causing pathogen (Parkhill *et al.* 2001) transferred either subcutaneously by an insect intermediate or through the air by infected humans? These questions are now answerable using data from studies of both the population genetic structure and genomic constitution of *Y. pestis* and its congeners (Achtman *et al.* 1999, 2004; Parkhill *et al.* 2001; Chain *et al.* 2004).

Achtman and his colleagues carried out two genetic surveys (Achtman *et al.* 1999, 2004) to define microevolutionary patterns within *Y. pestis*. In the first of these they estimated the population genetic structure of three *Yersinia* species (*Y. pestis*, *Y. pseudotuberculosis*, *Y. enterocolitica*) by collecting (i) partial sequence data from six genes (*dmsA*, *glnA*, *manB*, *thrA*, *tmk*, *trpE*) and (ii) RFLP data using the insertion element IS*100* as a probe. The gene sequence information was collected for worldwide samples of *Y. pestis* (36 strains), *Y. pseudotuberculosis* (12 strains), and *Y. enterocolitica* (13 strains). The RFLP data were collected from 49 strains, representing three biotypes (i.e. biovars) of *Y. pestis*. Though no differences were detected in their virulence or pathology in humans or other animal hosts, these biovars had been defined previously by their ability to convert nitrate to nitrite and to ferment glycerol (Perry and Fetherston 1997). The RFLP data allowed an assessment of microevolutionary patterns, independent of phenotype, within *Y. pestis* (Achtman *et al.* 1999). The conclusions from this first analysis were summarized by the title of the paper containing these data, '*Yersinia pestis*, the cause of plague, is a recently emerged clone of *Yersinia pseudotuberculosis*' (Achtman *et al.* 1999). It was thus found that the plague bacteria differed little in sequence variation from either of its congeners and, in particular, was highly similar to *Y. pseudotuberculosis*. This study did however detect molecular variation that clustered the three biovars into separate clades.

The second analysis of *Y. pestis* population genetic structure also defined microevolutionary/population genetic structure within and among the strains sampled (Achtman *et al.* 2004). However, this analysis did not find reciprocal monophyly for the various biovars. Instead geographic origin of the samples was a better predictor of phylogenetic patterning. This study did define the time of origin for the *Y. pestis* lineage as being *c.*10 000–13 000 years before present (YBP), with a cladogenetic event occurring some 6500 YBP giving rise to strains more commonly associated with human populations (Achtman *et al.* 2004).

Yersinia pestis: lateral transfer, gene loss, and the evolution of virulence

Two analyses, both using a genomic approach, illustrate the diverse array of genetic processes that have contributed to the evolution of *Y. pestis*. Conclusions

from the first of these, a study by Parkhill *et al.* (2001), emphasized the role of genetic exchange in giving rise to the evolutionary trajectory resulting in plague. First, these authors detected sequence footprints (i.e. insertion sequence element perfect repeats and anomalous GC base-composition biases) in the genome of *Y. pestis* indicating numerous recombination events (Parkhill *et al.* 2001). Second, they discovered evidence for widespread gene inactivation in the *Y. pestis* genome, relative to its sister taxon *Y. pseudotuberculosis*. The large-scale gene inactivation was indicated by the presence of approx. 150 pseudogenes. Specifically, they detected numerous genes thought to be associated with the ancestral, enteric-bacterial habitat that had been preferentially silenced (Parkhill *et al.* 2001). Third, these authors argued that *Y. pestis* acted as the *recipient* of DNA from multiple *donors*, including bacteria and viruses (Parkhill *et al.* 2001). The data assembled in this analysis suggested that *Y. pestis* had developed adaptations as a result of both lateral exchange *and* gene silencing.

In contrast to Parkhill *et al.* (2001), the conclusions drawn from a more recent study (Chain *et al.* 2004) emphasized the role of gene loss and modification in the evolutionary pathway leading to the plague bacterium. These latter authors thus concluded that 'Extensive insertion sequence-mediated genome rearrangements and reductive evolution through massive gene loss . . . appear to be more important than acquisition of genes in the evolution of *Y. pestis*' (Chain *et al.* 2004). Yet, even with this emphasis, these authors detected a role for lateral transfer in producing 32 *Y. pestis*-specific chromosomal genes as well as two *Y. pestis*-specific plasmids. Significantly, one of these unique plasmids carries a gene that encodes phospholipase D (Hinnebusch *et al.* 2002; Hinchliffe *et al.* 2003). This gene product has been shown to be necessary for the viability of the plague bacteria in the midgut of its vector, the rat flea *Xenopsylla cheopsis* (Hinnebusch *et al.* 2002). Its function is believed to involve protection of the bacterium from digestion by a blood plasma cytotoxic digestion product found in the gut of *X. cheopsis*. Put into context, the plague bacterium has, since the divergence of the *Y. pestis* and *Y. pseudotuberculosis* lineages *c.*10 000–13 000 YBP,

(i) acquired 32 chromosomal genes and two plasmids containing genes of key function for the bacterium's life history and (ii) lost, through deletion or silencing, approx. 470 genes present in *Y. pseudotuberculosis* (Parkhill *et al.* 2001; Chain *et al.* 2004).

Taken together, the results of the various studies involving *Y. pestis* may be best reflected by the following quote: 'The evidence of ongoing genome fluidity, expansion and decay suggests *Y. pestis* is a pathogen that has undergone large-scale genetic flux and provides a unique insight into the ways in which new and highly virulent pathogens evolve' (Parkhill *et al.* 2001). A multitude of genetic events, including extensive genetic exchange, followed by natural selection, have molded the evolutionary trajectory of this organism.

This same conclusion has also been drawn from over three decades of studies involving a very different species group, the Darwin's finches (genus *Geospiza*). Obviously, like the Louisiana irises, the genetic exchange among species of *Geospiza* results from sexual reproduction and introgression, rather than lateral gene transfer. Yet, like the plague bacterium and Louisiana irises, studies into this group of organisms did not begin with an assumption of genetic exchange-derived evolution. Instead, the realization that a web-of-life metaphor described better the pattern of evolutionary diversification in this model system developed as data were collected from observations of natural populations and through experiments. As such, the evolutionary analyses into Darwin's finch species reflect another excellent example of how studies of genetic exchange have evolved as more and more data came to light.

1.2.3 Darwin's finches

Hybridization and the origin of species variability?

The most curious fact is the perfect gradation in the size of the beaks in the different species of *Geospiza*, from one as large as that of a hawfinch to that of a chaffinch, and . . . even to that of a warbler . . . Seeing this gradation and diversity of structure in one small, intimately related group of birds, one might really fancy that from an original paucity of birds in this archipelago, one species had been taken and

modified for different ends (Darwin 1845, pp. 401–402).

With this statement, Darwin may have been describing accurately the morphologically detectable consequence of repeated episodes of hybridization and selection among the birds named in his honor, the Darwin's finch species. However, as indicated above (section 1.1), Darwin's conclusions concerning the evolutionarily unimportant role of the formation of 'mongrels' did not predispose him to conclude that genetic exchange contributed to their gradated variation. Instead, he emphasized the role of natural selection alone in refining the variation and through this process '. . . from an original paucity of birds in this archipelago, one species had been taken and modified for different ends'. Intriguingly, his further observation that '. . . instead of there being only one intermediate species . . . there are no less than six species with insensibly graduated beaks' (Darwin 1845, p. 402) is the expected outcome from the formation of the F_1 individuals followed by either (i) their breeding with one another to form F_2 individuals or (ii) backcrossing with one or both of the parental species (Anderson and Hubricht 1938; Anderson 1949).

In contrast to Darwin's emphasis on the singular role of natural selection in the diversification of these finches, Lowe some 90 years later presented a paper entitled, 'The finches of the Galápagos in relation to Darwin's conception of species' (Lowe 1936). In this paper, Lowe suggested the following: '. . . it is difficult to resist the conclusion that in the Finches of the Galápagos we are faced with a *swarm of hybridization segregates* which remind us . . . of the "plant" swarms described by Cockayne and Lotsy in New Zealand forests as the result of natural crossings . . . I think it was William Bateson who always maintained that the Finches . . . could only be explained on the assumption that they were segregates of a cross between ancestral forms . . .' (Lowe 1936, pp. 320–321). Yet, the significance of Lowe's work may not have resided mainly in his conclusions, but rather in its catalytic effect in stimulating the investigations of Lack. Thus was born the '. . . the first modern treatment of the finches . . .' (Grant 1993) culminating with Lack's 1947 classic, *Darwin's Finches*.

Unlike Lowe, Lack's conclusions followed the Darwinian paradigm of negating a major role for genetic exchange in affecting the evolution of Darwin's finch species. Though Lack recognized that Lowe and others felt the variability in the finch species was likely a result of '. . . an unusually large amount of interbreeding between the species . . .' (Lack 1947, p. 95) he explained the overall high variability and the intermediacy of certain specimens in the following way. 'Some forms . . . are intermediate in appearance between two species, but in most cases this is probably due to the intermediate nature of their ecological requirements and not to a hybrid origin . . . To conclude, it seems probable that hybridization has not played an important part in the origin of new forms of Darwin's finches' (Lack 1947, p. 100).

Since Lack's seminal studies, the role of genetic exchange through introgression has been considered as a possible evolutionary mechanism impacting the diversification of the Darwin's finches. However, in reviewing these various studies Grant (1993) concluded, 'Since 1947, and prior to the study reported here, hybridization in the Galápagos has been neither neglected nor satisfactorily demonstrated.' Until the work of Peter and Rosemary Grant and their colleagues, the evolutionary role of genetic exchange among the *Geospiza* species was, at best, speculative. Thus, their earliest description of rare hybridization events leading to introgression between these species (Grant and Grant 1992) presaged the important evolutionary consequences of genetic exchange—not only for the Darwin's finches, but also for the bird clade in general (Grant and Grant 1992).

Environmental shifts and the relative fitness of hybrids
The fluctuation, over time, in the relative impact of genetic exchange on the evolution of Darwin's finches is an excellent example of its characteristic pattern of episodic importance. Furthermore, the frequency and strength of effect from genetic exchange in *Geospiza* also exemplify the importance of interactions between novel, hybrid genotypes and the environment. From 1976 to 1982 hybridization between *Geospiza fortis* and *Geospiza fuliginosa* produced 32 fledglings and that between

G. fortis and *Geospiza scandens* produced one (Grant and Grant 1993). Of these hybrid fledglings, only two of the *G. fortis* × *G. fuliginosa* F_1 individuals survived to breed (even then only reproducing after 1983). The single *G. fortis* × *G. scandens* F_1 survived less than 1 year without reproducing (Grant and Grant 1993). In contrast, between 1983 and 1991 hybrid birds not only survived, but also exceeded the value (i.e. 1.0) of fledglings necessary to replace themselves by a factor of 1.3 (Figure 1.8). During this same time *G. fortis* and *G. fuliginosa* did not produce sufficient fledglings to maintain their group sizes (Grant and Grant 1993). Specifically, 'In the period 1983 to 1991 finches bred in 6 of 9 years. Those that hybridized were at no obvious disadvantage. They bred as many times as conspecific pairs and produced clutches of similar size . . . *G. fortis* × *G. fuliginosa* pairs (3.9 eggs) . . . *G. fortis* pairs (3.9) . . . *G. fuliginosa* pairs (4.0)', *G. scandens* × *G. fortis* pairs (3.7) and *G. scandens* pairs (3.7; Grant and Grant 1992).

The observed transition in fitness estimates, particularly for hybrid offspring, generates the obvious question concerning causality. In this regard, the main causal factor was identified to be the El Niño of 1982–1983 that produced approx. 1400 mm of rainfall (Grant and Grant 1993). As a comparison, from 1976 to 1982 the maximum rainfall in any given year was approx. 150 mm. This extraordinary climatic fluctuation resulted in an equally extraordinary changeover in ecological setting for the Darwin's finch individuals. Although the El Niño event in 1982–1983 was ecologically revolutionary, the environmental perturbations did not stop with this year. Indeed, another extremely high rainfall

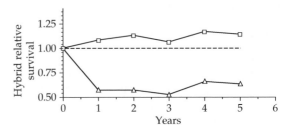

Figure 1.8   Fitness of F_1 hybrids between *G. fortis* and *G. fuliginosa* expressed in terms of survivorship relative to the survival of *G. fortis* before (1976–1981; △) and after (1983–1987; □) the 1982–1983 El Niño event (from Grant and Grant 1993).

total of approx. 600 mm accumulated in 1987. In contrast, rainfall in the years 1985 and 1988 was negligible (Grant and Grant 1993). The ecology of the Galápagos islands and the habitats for the finch species and hybrids were placed in upheaval by this series of climatic disturbances. Grant and Grant (1993) explained the changeover in this way: 'Most Galápagos islands . . . support drought-adapted communities of plants. Effects of abundant rainfall . . . on these communities were profound . . . Changes in plant communities caused changes in the granivorous finch populations . . . The evidence for selection for small beak size in the *G. fortis* population . . . is consistent with the survival advantage experienced by hybrids at the same time . . . the two evolutionary changes were apparently caused by the common factor of a change in food supply.' The high fitness of hybrids and smaller classes of *G. fortis* individuals was seen to be a direct result of the decrease in abundance of large, hard seeds and concomitant increase in the abundance of small seeds (Figure 1.9; Grant and Grant 1993, 1996).

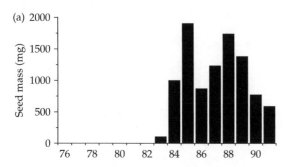

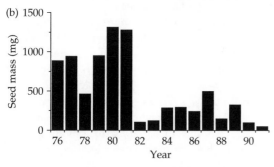

Figure 1.9 Changes in seed abundance before and after the El Niño event of 1982–1983. (a) Small seed abundance and (b) large seed abundance (from Grant and Grant 1993).

Genetic and phenotypic convergence caused by introgression and natural selection

The differentiation of the ground finch species based on morphological data is not reflected in . . . DNA sequence phylogenies . . . We suggest that the absence of species-specific lineages can be attributed to ongoing hybridization involving all six species of *Geospiza*.

(Freeland and Boag 1999)

The footprints of past and current hybridization in the Darwin's finch species are apparent in the degree of overlap in their genetic constitution. This is the conclusion reached by Freeland and Boag (1999) based upon sequence data from both the mitochondrial DNA control region and the nuclear internal transcribed spacer 1 region. Each of these genomic components is expected to accumulate mutations rapidly and therefore be appropriate for assessing relationships among recently diverged species. However, in place of species-specific genotypes among the *Geospiza* taxa, Freeland and Boag (1999) discovered incongruencies between (i) species identifications of individuals from genetic markers on the one hand or morphological characters on the other, and (ii) mitochondrial and nuclear markers in individual finches (Figure 1.10; Freeland and Boag 1999). Consistent with these results were findings from a recent investigation of the genetic identity between sympatric populations of species pairs compared with the genetic identity between allopatric populations (Grant *et al.* 2005). This latter study discovered a pattern in which species were more similar genetically to a sympatric relative than to allopatric populations of that same relative. Like those of Freeland and Boag (1999), these results were best explained by introgressive hybridization between sympatric populations of different species (Grant *et al.* 2005).

The above studies indicate the large degree to which introgressive hybridization has impacted the *Geospiza* species. However, specific examples of ongoing hybridization indicate a complex pattern of effects of introgression on different species. For example, Grant *et al.* (2004, 2005) discovered the following with regard to introgression of genetic

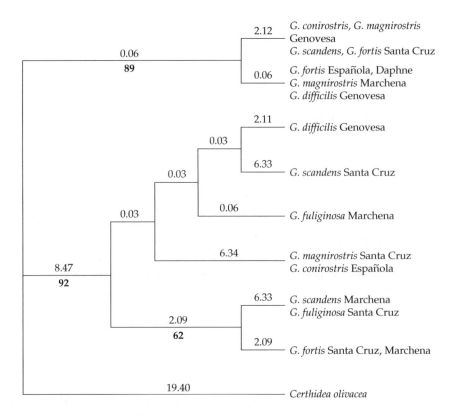

Figure 1.10 Phylogenetic tree of *Geospiza* species based upon nuclear ribosomal internal transcribed spacer sequences. Collection locality is noted after the species name. Branch lengths and bootstrap values are written above and below the branches, respectively (from Freeland and Boag 1999).

markers between *G. fortis* and *G. scandens* on the Galápagos island of Daphne Major: (i) there has been an increase in overall genetic heterozygosity in *G. scandens*, but not in *G. fortis*; (ii) alleles found in only *G. fortis* prior to the 1982–1983 El Niño are now found in high frequency in *G. scandens*; (iii) known hybrids (F_1 and BC_1 individuals) increased significantly in *G. scandens* samples between 1982 and 2002, but did not increase in *G. fortis* samples; and (iv) the two species converged genetically, but the convergence was explained by an increase of similarity of the *G. scandens* samples to *G. fortis*, and not vice versa (Figure 1.11; Grant *et al.* 2004). The asymmetrical nature of the effects from introgressive hybridization between these two species can also be seen in comparisons of genetic variation and morphological character change over the same 30-year period (Figure 1.12; Grant *et al.*

2005). Indeed, the best-known morphological characters of this model system—i.e. beak size and shape—also appear to have been impacted by recent, and presumably past, genetic exchange. Specifically, Grant *et al.* (2004) found that *G. scandens* became significantly more like *G. fortis* in terms of beak characteristics. Overall then, *G. scandens* was affected much more by introgression as reflected in its being drawn more toward the *G. fortis* genetic and morphological type than *G. fortis* to the *G. scandens* constitution (Grant *et al.* 2004, 2005). This is a pattern that is very reminiscent of Darwin's observation of infinitesimally graduated beak shapes between the various species.

As discussed above, the change in the fitness of hybrids subsequent to the El Niño event of 1982–1983 was caused by a change in the environment. In the

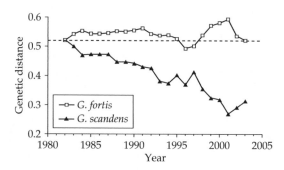

Figure 1.11 Genetic convergence of G. fortis and G. scandens. G. scandens became increasingly similar to the 1982 G. fortis sample, while G. fortis retained its initial genetic distinctness. In the 1982 sample, G. scandens individuals demonstrated an average genetic distance of > 0.5 compared to G. fortis samples. After 30 years, the genetic distance between the two species had dropped to approx. 0.3 (from Grant et al. 2004).

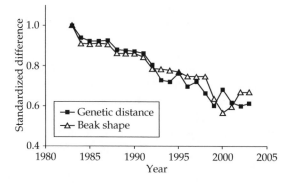

Figure 1.12 Convergence in overall genetic identity and beak shape of G. fortis and G. scandens from Daphne Major island (from Grant et al. 2005).

same way, the introgression and convergence of these two species may be largely a result of natural selection favoring the G. fortis phenotype and thus hybrids/G. scandens individuals that approach a G. fortis type (Grant et al. 2004, 2005). However, demography and behavior, also affected by differential survivorship during the environmental fluctuations (i.e. flood and drought years), most likely played a role as well (Grant and Grant 2002). The greater proportion of introgression from G. fortis into G. scandens appears to have been, at least initially, affected by more severe competition for

mates among G. fortis than G. scandens females. The environmental fluctuations thus reduced the survivorship of G. scandens females, leaving the surviving females with a surplus of conspecific males with which to pair. This led to (i) G. scandens females mating with conspecific males, but (ii) a fraction of the G. fortis females pairing with the excess of G. scandens males. Backcrossing of the F_1 progeny formed in the mixed nests occurred with G. scandens females because mate choice is affected largely by imprinting on the paternal (in this case, G. scandens) song (Grant and Grant 2002).

Introgressive hybridization as a cause of adaptive radiation?

Several reports have suggested a role for introgressive hybridization in the *adaptive radiation* of Darwin's finches. As discussed above, Lowe (1936) argued that the pattern of morphological variation among the finch species was the result of hybridization. Likewise, Freeland and Boag (1999) concluded that 'Hybridization has apparently played a role in the adaptive radiation of Darwin's finches.' Furthermore, Seehausen (2004) used the genetic and morphological data for the finches to propose them as an example of adaptive radiation from an initial 'hybrid swarm' stage. Indeed, radiation within this species complex may have been afforded somewhat because 'Hybridization may enhance fitness to different degrees by counteracting the effects of inbreeding depression, by other additive and nonadditive genetic effects, and by producing phenotypes well suited to exploit particular ecological conditions' (Grant et al. 2003). However, Grant et al. (2005) sound a cautionary note concerning the role of hybridization in the adaptive radiation of Darwin's finches when they state that '... we doubt that hybridization was necessary for any part of the adaptive radiation of Darwin's finches proposed under the hybrid swarm hypothesis ...'

Regardless of the impact of introgression on the adaptive radiation within *Geospiza*, its effect on adaptive evolution cannot be doubted. Introgressive hybridization has affected the evolutionary trajectory of at least certain species (e.g. G. scandens on Daphne Major) through the transfer of genes that allow an increase in fitness in changing environments (Grant *et al.* 2004). As Petren *et al.* (2005) concluded, 'Gene flow does not constrain phenotypic divergence [in Darwin's finches], but may augment genetic variation and facilitate evolution due to natural selection.' This conclusion is very similar to that reached in cases of genetic exchange in other organismal complexes, including that of our next example, influenza viruses.

1.2.4 Influenza

Genetic reassortment, natural selection, and the origin of influenza strains

Like the Darwin's finches, plague bacteria, and Louisiana irises, influenza viral types evolve through a combination of genetic exchange and natural selection (Ghedin *et al.* 2005). In particular, 'The introduction and subsequent spread in the human population of influenza A viruses with a novel hemagglutinin (HA) or a novel HA and neuraminidase (NA) subtype results from a sudden and major change in virus antigenicity... Pandemic strains contain new HA or NA genes derived from animal influenza A viruses. Influenza A viruses of 15 recognized HA subtypes and 9 NA subtypes are known to circulate in birds and other animals, creating a reservoir of influenza A virus genes available for genetic reassortment with circulating human strains...' (Subbarao *et al.* 1998). The NA and HA mutations often arise and accumulate in a reservoir outside the human host. Recombination among these different viral lineages may then lead to a viral pathogen that is transferable, and highly pathogenic, to humans (Subbarao *et al.* 1998).

Before and after recombination events among the influenza strains, natural selection continues to favor antigenic shifts that allow '... escape from immunity induced by prior infection or vaccination' (Smith *et al.* 2004). For example, genetic exchange between a human and a swine HA gene, producing a recombinant form, has been postulated as a key step in the evolution of the variant that caused the 1918 Spanish Flu pandemic (Gibbs *et al.* 2001; but see Worobey *et al.* 2002; Taubenberger *et al.* 2005). In addition, it is likely that selection for the specific sequence of the HA and NA genes present in the 1918 variant

contributed greatly to its heightened virulence (Kobasa *et al.* 2004; Taubenberger *et al.* 2005). Indeed, Kobasa *et al.* (2004) documented a change from a non-pathogenic to pathogenic form of influenza A when sequences complementary to the 1918 NA and HA variants were inserted. The synthetic viral strain caused a pathology characteristic of that found in victims during the 1918 pandemic—that is, infection of the entire lung and infiltration of inflammatory cells leading to severe hemorrhaging (Kobasa *et al.* 2004). Recombination may have created the unique combination of viral genes in the 1918 variant, but natural selection likely contributed to its virulence.

Humans are not the only species affected adversely by the web-like evolution of influenza strains. For example, outbreaks of swine influenza in the USA were first documented during the 1918 Spanish Flu pandemic (Zhou *et al.* 1999). Since that time, influenza A-type viral infections have risen to be the most common cause of respiratory disease in pigs (Zhou *et al.* 1999). Like human pathogens (and those of other species), the swine variants are compilations of genes from multiple sources (Table 1.1). The H3N2-type swine isolates contain genomic elements from the human H3N2 and the pig H1N1 viral lineages (Zhou *et al.* 1999). Other H3N2 swine isolates are tripartite reassortants, possessing avian-like, human-like, and pig-like genomic elements (Zhou *et al.* 1999; Richt *et al.* 2003).

A recent analysis also indicated that felid species were susceptible to infection by the avian variant of most concern as a potential cause of the next human pandemic, H5N1. Kuiken *et al.* (2004) demonstrated that this avian virus could infect cats both through contact between infected and uninfected animals and through uninfected cats feeding on birds infected with H5N1. These data indicated the risk of lethal infections in domestic cats and the risk that cats would act as a reservoir of viruses that could move from the infected felids into humans (Kuiken *et al.* 2004).

Genetic reassortment, natural selection, host shifts, and the origin of epidemics and pandemics

In any given year, influenza infections claim approx. 500 000 lives (Stöhr 2002). In addition,

Table 1.1 Two influenza A viral strains (SWNC98 and SWTX98) isolated from infected pigs during a North American outbreak of swine influenza. The diverse origins of the eight genes compared in this analysis demonstrate the mosaic nature of the viral genomes caused by reassortment among human, swine, and avian viral lineages (Zhou *et al.* 1999)

| Gene | Region compared (base pair position) | Viral isolate SWNC98 | | Viral isolate SWTX98 | |
		Putative lineage of origin	Per cent homology shared with lineage of origin	Putative lineage of origin	Per cent homology shared with lineage of origin
PB2	44–1600	Swine	93	Avian	93
PB1	1088–2236	Human	98	Human	99
PA	25–620	Swine	94	Avian	97
HA	78–1061	Human	98	Human	98
NP	34–1024	Swine	98	Swine	98
NA	26–1411	Human	99	Human	99
M	52–964	Swine	95	Swine	98
NS	22–866	Swine	97	Swine	98

highly virulent strains of influenza A kill, or at least cause humans to destroy, hundreds of millions of fowl (Normile and Enserink 2004). The evolutionary progression of the yearly influenza epidemics and pandemics includes a circuitous course through various hosts. Reservoirs in which these viral lineages are housed, mutate, and recombine with one another include species as diverse as domestic and wild ducks, chickens, tree sparrows, quail, egrets, geese, Grey herons, horses, pigs, and humans (Steinhauer and Skehel 2002; Li *et al.* 2004; Hulse-Post *et al.* 2005; Liu *et al.* 2005). The life history of the viral strains, which includes passing through multiple hosts, is apparently required for the process of viral genetic exchange. From this life history emerges genomes of pathogenic forms that are mosaics (i.e. from RNA recombination or gene segment reassortment; Steinhauer and Skehel 2002) of strains from multiple hosts.

As discussed above, a recombination event has been postulated as causal in the origin of the highly virulent 1918 pandemic strain (Gibbs *et al.* 2001). Notwithstanding the debate over the mosaic nature of the Spanish Flu genome (Gibbs *et al.* 2001, 2002; Worobey *et al.* 2002; Taubenberger *et al.* 2005), 'The viruses that caused the pandemics of 1957 and 1968 were derived from reassortment of avian and human strains ... In 1957, the HA, NA and PB1 gene segments came from an avian virus and the other segments were obtained from the human H1N1 viruses ... The H3N2 viruses that emerged in 1968 resulted from the replacement of HA and PB1 gene segments of circulating human H2N2 viruses with their counterparts from an avian source. The host species in which these reassortment events took place is not known ...'; however, 'Pigs have been proposed as a potential "mixing vessel" for reassortment ... and both avian and human strains can replicate in these hosts ...' (Steinhauer and Skehel 2002).

Presently, the influenza A viral lineage of most concern as the potential seed for the next human pandemic is the H5N1 avian influenza strain (Laver *et al.* 2000). Six of the 18 Hong Kong residents infected in 1997 with this 'new' virus (i.e. never isolated previously from humans) died. The steps taken to limit the epidemic included destruction of all the chickens—the carriers from which the humans contracted the disease—in Hong Kong (Laver *et al.* 2000). Of course, one of the major concerns stems from the likelihood that genes will enter a pathogen's genome, through recombination or other means, resulting in the transfer or *de novo* origin of adaptations. As stated above, this is a possible outcome when the greatly increased genotypic diversity provided by genetic exchange is acted upon by natural selection (Anderson 1949; Anderson and Stebbins 1954; Lewontin and Birch 1966; Arnold 1997; Kim and Rieseberg 1999; Ochman *et al.* 2000; Gogarten *et al.* 2002; Rieseberg *et al.* 2003; Martin *et al.* 2006). Of course, the effects from the adaptive evolution of the influenza A viral lineages, though beneficial to the virus, can be a devastatingly severe form of natural selection on their host populations, including those of humans (Webby and Webster 2003; Kuiken *et al.* 2004).

The epidemiological and evolutionary history of H5N1 viral lineages is instructive as an exemplar of how genetic exchange and natural selection can act in concert, resulting in the newest pandemic threat. In the case of H5N1, these processes have resulted in an organism with a mosaic genome that demonstrates radically different adaptations. Various influenza A genomes have contributed, and continue to contribute, to the evolution of H5N1 viruses. For example, the 'internal' genes (nonstructural protein, matrix protein, nucleoprotein, and three polymerase protein genes) present in H5N1 lineages that first infected humans in 1997 were donated by H9N2 viruses from quail (Guan *et al.* 1999). Genetic exchange resulting in the lethal 1997 variant actually involved an already circulating H5N1 lineage with both H9N2 and H6N1 lineage viruses (Guan *et al.* 1999). This occurred most likely as a result of the coexistence of the three viral lineages in the same host. The strain of H5N1 that caused human infections and deaths in 1997 has not been isolated since (Guan *et al.* 2002). However, the precursors of this strain were not eliminated by the control measures taken at the time (Guan *et al.* 2002; Li *et al.* 2004). Since 1997, genetic exchange among numerous aquatic bird and terrestrial poultry viral lineages has given rise to a plethora of H5N1 variants (Figure 1.13; Guan *et al.* 2002; Webby and Webster 2003; Li *et al.* 2004). In total, Li *et al.* (2004) reported the presence of 14 H5N1 lineages resulting from genetic exchange. Of these, one became predominant by 2004 (Figure 1.13). Therefore, the '. . . series of genetic reassortment events traceable to the precursor of the H5N1 viruses . . . gave rise to a dominant H5N1 genotype (Z) in chickens and

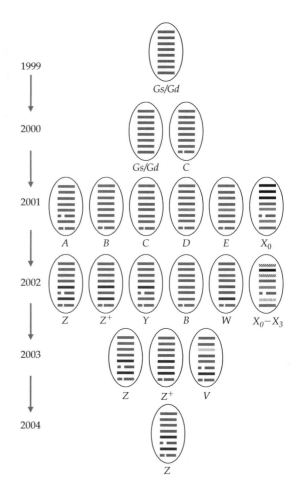

Figure 1.13 Genotypes of H5N1 influenza viral lineages demonstrating mosaicism caused by genetic exchange. The eight genes sampled were (horizontal bars starting at the top downwards): *PB2, PB1, PA, HA, NP, NA, M,* and *NS.* Each color represents a viral lineage (red indicates origin from the ancestral viral lineage Gs/Gd/1/96). Letters indicate different genotypes defined by gene phylogeny: a shared phylogenetic lineage was used as evidence of a shared ancestor. Genotypes A, B, and C were formed from genetic exchange between the ancestral Gs/Gd/1/96 and one or more aquatic avian viruses. Genotype D was formed when the *NP* gene from an H9N2 virus replaced the *NP* gene of genotype C. Genotype E arose through the replacement of the *NP* gene of genotype C with another *NP* avian virus gene. Further genetic exchange involving genotype E and additional aquatic avian influenza viruses resulted in the genotypes X_0–X_3. Genotype W differs from genotype B in its *PB2, NP,* and *M* genes. Genetic exchange of genotypes A and/or B with aquatic avian viral lineages gave rise to genotypes V, Y, Z, and Z^+. Genotype V may have resulted instead from reassortment of genotype Z with other aquatic avian viruses (from Li *et al.* 2004). See also Plate 1.

ducks that was responsible for the . . . highly pathogenic avian influenza H5N1 disease in poultry . . .' (Figure 1.13). Furthermore, the 2003/2004 outbreak was '. . . unprecedented in its geographical extent . . .' and 'ominous' in its transmission to humans (Li *et al.* 2004).

With regard to adaptive evolution, the observation that H5N1 can transfer directly between avian and human hosts is highly significant. Prior to the 1997 outbreak, infection of humans by avian influenza was postulated to require an intermediate host. In this regard, '. . . the receptor specificity of avian influenza viruses was thought to prevent their direct transmission to humans. Transmission from aquatic birds to humans was hypothesized to require infection of an intermediate host, such as the pig, that has both human-specific . . . and avian-specific . . . receptors' (Webby and Webster 2003). The H5N1 viral lineage that caused six fatalities out of 18 infections had obtained a novel adaptation through genetic exchange and natural selection allowing its direct spread from birds to humans. In addition, Li *et al.* (2004) argued that the replacement of older viral lineages by more recent reassortant forms (Figure 1.13) was an indication that the younger lineages possessed higher survivorship through the acquisition of novel adaptations.

The significance of genetic exchange among influenza A viral lineages for the acquisition of novel adaptations was also indicated by a study of the evolution of increased pathogenicity to mammals of H5N1 from 1999 to 2002 (Chen *et al.* 2004). In particular, these investigators found that the viruses occurring in domestic ducks between these years gained the ability to infect and kill mice without further adaptations within the mammalian host. As mentioned above, a key step in the evolution of pathogenicity to mammalian systems by an avian virus is the development of the ability to be transmitted from the avian reservoir to a mammalian host. Chen *et al.* (2004) used swine influenza epidemics as an analogy for how genetic exchange can accomplish the evolution of this adaptation. In particular, the H3N2 viruses isolated from a 1997 swine influenza outbreak were shown to be mosaics of human, pig, and

avian viral genomes (see previous section). In another case, the H1N1 swine influenza viruses from Europe were found to contain avian influenza genes that had been transferred via genetic exchange in 1979 (Chen *et al.* 2004). Comparisons of the data from the increase in lethality of H5N1 to mammalian species, as well as the data for the swine influenza epidemics, caused Chen *et al.* (2004) to conclude, 'Our results demonstrate that while circulating in domestic ducks, H5N1 viruses gradually acquired the characteristics that make them lethal in mice . . .' and also that '. . . avian influenza virus genes, when they are part of certain gene constellations, seem able to promote transmission among mammalian species.' The 'constellations' that evolved through numerous genetic exchanges between various avian and mammalian viruses were the basis of these new adaptations for transmissibility and virulence.

1.3 Summary and conclusions

Since the times of Linnaeus, Kölreuter, and Darwin the importance ascribed to genetic exchange in affecting the evolution of organisms has waxed and waned. In the past decade, numerous research groups—studying a wide array of species complexes—have reported findings that indicate an evolutionary pattern best described as a web rather than as a bifurcating tree. The research described above exemplifies the types of studies leading to the recognition of the web-of-life metaphor. Because these examples include viral, bacterial, plant, and animal species or variants, the molecular mechanisms underlying the genetic exchange are extremely diverse. They include RNA recombination, lateral gene transfer, and recombination during meiosis. In spite of the breadth of mechanisms, the outcomes for each of these exemplars are quite similar. Louisiana irises include a new tri-hybrid species possessing novel adaptations. Darwin's finch species have changed both genetically and phenotypically due to the action of natural selection acting upon introgressed genotypes leading to changes in suites of adaptations (e.g. beak shape). Plague bacteria arose as a consequence of lateral gene transfer and gene silencing resulting in an adaptive shift from a non-lethal gut bacterium. Finally, influenza A continues to be a moving target for avian and mammalian immune systems as recombinants between viruses housed in such diverse organisms as ducks, chickens, and humans recombine to form new mosaic genomes. The effects of genetic exchange on evolutionary pattern and process are pervasive, important, and long lasting.

CHAPTER 2

The role of species concepts

No one definition has as yet satisfied all naturalists; yet every naturalist knows vaguely what he means when he speaks of a species ... From these remarks it will be seen that I look at the term species, as one arbitrarily given for the sake of convenience to a set of individuals closely resembling each other ...

(Darwin 1859, pp. 44, 52)

... a species is a group of individuals fully fertile *inter se*, but barred from interbreeding with other similar groups by its physiological properties (producing either incompatibility of parents, or sterility of the hybrids, or both).

(Dobzhansky 1935)

... the most inclusive group of organisms having the potential for genetic and/or demographic exchangeability.

(Templeton 1989)

While the current data set is rather limited compared with the total breadth of prokaryotic diversity ... [it indicates that the species definition] ... could be as stringent as including only strains that show a > 99% [average nucleotide identity] or are less identical at the nucleotide level, but share an overlapping ecological niche ...

(Konstantinidis and Tiedje 2005)

A phylogenetic species is an irreducible ... cluster of organisms, diagnosably distinct from other such clusters, and within which there is a parental pattern of ancestry and descent ...

(Cracraft 1989)

2.1 Species concepts and understanding genetic exchange

The definitions for the various mechanisms of genetic exchange considered in this book do not depend upon the species concept applied. Furthermore, I would argue that holding to a given species concept does not preclude studying any of the mechanisms or outcomes of genetic exchange. Indeed, I would be so bold as to state that every study of genetic exchange, and indeed evolutionary biology, has at least an implicit assumption of how

to define the category of species. Even a cursory examination of the scientific literature describing natural hybridization and lateral gene transfer reveals a close association between these studies and debates concerning the nature of species and the process of speciation. For example, natural hybridization has been described alternatively as unimportant or of profound importance in plant and animal evolution based largely on the researcher's underlying definition of species. However, considerations of the evolutionary importance and consequences from any form of genetic exchange have been affected by the underlying species concept.

The relegation of genetic exchange into the category of evolutionary epiphenomena has been caused partially by the assumption that exchange between species is, by definition, impossible. For example, Willi Hennig is considered the architect of modern phylogenetic analysis and thus the founder of the *phylogenetic species concept*. However, in his classic work, *Phylogenetic Systematics* (Hennig 1966), he emphasized repeatedly an application of the *biological species concept* (Dobzhansky 1935, 1937; Mayr 1963). This led him to the following conclusion regarding the origin of hybrid species: 'Is not the species concept that the species includes all individuals that together are capable of producing completely fertile offspring, and must we not then consider groups whose individuals can produce new species by hybridization as partial groups of one species? If we speak of the origin of species by species hybridization, are we not guilty of circular reasoning between premise and conclusion?' (Hennig 1966, p. 208). In this way hybrid speciation, and indeed hybridization if it is defined as occurring only between species, is defined out of existence.

At the other end of the reproductive spectrum, Hennig (1966) considered whether asexual organisms posed a significant problem for species definition; that is, whether the requirement of the potential for bisexual reproduction in defining species was of major importance. He concluded that it was indeed crucial that species be defined by attributes of bi-sexual reproduction and thus be demonstrable members of a 'reproductive community' (Hennig 1966, p. 65). In any event, he did not feel that asexuality was an important consideration because of the scarcity of such lineages among animal taxa. Furthermore, as a general statement of the importance of non-sexually reproducing organisms, his statement that '... the question of the species concept in organisms without bisexual reproduction is no more than a relatively subordinate species problem of systematics' (Hennig 1966, p. 44) summed up well the viewpoint of many of the architects of the neo-Darwinian synthesis (e.g. Mayr 1963). Given that organisms that reproduce at least partially through asexual/clonal mechanisms make up a large proportion of overall biodiversity, and

that they often demonstrate the evolutionary effects of genetic exchange, their dismissal from evolutionary considerations reflects not so much the 'species problem', as a problem with species concepts. Hey *et al.* (2003) reflect, in the following manner, the tension created by the complexity contained within the term species and the application of any species concept: 'These two ideas—that species are categories that are created ... by the biologists who study them, and that species are objective, observable entities in nature—have long been in conflict.'

In this chapter I will consider four species concepts: biological, phylogenetic, *cohesion*, and *prokaryotic*. For the phylogenetic species concept, I will discuss two competing methods for defining species: history- and character-based phylogenetic species (Baum and Donoghue 1995). In the case of the prokaryotic species concept, I will also review two approaches for delimiting species: the phylo-phenetic species concept (Rosselló-Mora and Amann 2001) and the phylogenetic and ecological divergence concept (Cohan 2002). The discussion of the biological, phylogenetic, cohesion, and prokaryotic species concepts will illustrate (i) how the strict application of some concepts may limit investigations into the evolutionary role of genetic exchange, (ii) how, in contrast, a discerning application of any given species concept can facilitate such studies, and (iii) that we can largely avoid the tension described by Hey *et al.* (2003) in the same manner as that used to dismiss genetic exchange as an evolutionarily important process; that is, through an appropriate application of definitions. I will borrow heavily from *Natural Hybridization and Evolution* (Chapter 2, Arnold 1997) for the discussion of the biological, phylogenetic, and cohesion species concepts. I believe that my earlier descriptions remain useful for understanding the issues surrounding the application of these concepts to studies of genetic exchange.

2.2   Genetic exchange considered through four species concepts

2.2.1   Biological species concept

It is somewhat of a false dichotomy to represent the biological species concept as the only framework that is defined by the occurrence, or lack thereof, of genetic exchange. Most, if not all, concepts consider genetic exchange to be of primary concern for defining species (e.g. see Sites and Marshall 2004 for a discussion of several methodologies for delimiting species that assume reproductive isolation). As illustrated above, one has only to read *Phylogenetic Systematics* (Hennig 1966), the canonical treatment for the phylogenetic species concept, to see the importance given to defining reproductive communities as a litmus test for aiding in the description of species. It is more accurate to state that the biological species concept, unlike some others, emphasizes the development of barriers to genetic exchange as the paramount process by which species should be defined (Figure 2.1).

Mayr's (1942) definition that 'Species are groups of actually or potentially interbreeding natural populations, which are reproductively isolated from other such groups' was based on earlier descriptions by Dobzhansky (1935, 1937). This concept thus defines species based upon the presence of reproductive isolation, with the process of speciation equivalent to the development of barriers to reproduction. Logically, if one substitutes the process of genetic exchange *sensu lato* in the place of introgression through sexual recombination, 'reproductive isolation' can be extended to include organisms that do not reproduce sexually (e.g. Arnold 2004a). I do, however, realize that the substitution of 'genetic exchange' (inclusive of, for example, introgression, lateral gene transfer, and viral recombination) in place of strict sexual recombination incorporates many processes and organisms not included by workers like Mayr and Dobzhansky. Mayr (1963, p. 28) did, however, address the perceived difficulties presented for the biological species concept by asexually reproducing organisms. He stated, 'Various proposals have been made to resolve the difficulty that asexuality raises for the biological species concept. Some authors have gone so far as to abandon the biological species concept altogether . . . I can see nothing that would recommend this solution. It exaggerates the importance of asexuality . . .' Though Mayr was speaking mainly of animal taxa, he did make a most perceptive observation regarding 'asexual organisms' in general when he stated, 'Indeed

clandestine sexuality appears to be rather common among so-called asexual organisms' (Mayr 1963, p. 27). If broadened to include all the various forms of genetic exchange, Mayr's conception that truly asexual organisms are very rare, or non-existent, is supported quite well by subsequent studies (e.g. Gogarten *et al.* 2002).

As emphasized in section 2.1, a strict application of the biological species concept negates the possibility of species exchanging genes; reproductive isolation defines species and it is thus incorrect to speak of species exchanging genetic material. Within the framework of the biological species concept, the problem of genetic exchange between well-defined species can be addressed in two ways. These two approaches were formulated by Mayr (1942, 1963) and have been used repeatedly by subsequent workers. First, Mayr (1942, 1963) argued—from the standpoint of sexual reproduction and introgression between animal taxa—that if viable, fertile hybrids were produced then one should consider the hybridizing forms to be subspecies or semispecies. Second, he allowed that species might occasionally meet and mate, but that 'The majority of such hybrids are totally sterile . . . Even those hybrids that produce normal gametes in one or both sexes are nevertheless unsuccessful in most cases and do not participate in reproduction. Finally, when they do backcross to the parental species, they normally produce genotypes of inferior viability that are eliminated by natural selection' (Mayr 1963, p. 133). In this way genetic exchange, and in particular introgressive hybridization, is reduced to a rare vagary of biology, and is seen as not worth studying.

Following Harrison (1990), I have defined natural hybridization as successful matings in nature between individuals from two populations, or groups of populations, that are distinguishable on the basis of one or more heritable characters (Arnold 1997). Furthermore, I have incorporated the effect of genetic exchange through its various mechanisms under the description of reticulate evolution (i.e. a web-like set of phylogenetic relationships reflecting genetic exchange [through lateral transfer, viral recombination, introgressive hybridization, etc.] between diverging lineages). In the context of these definitions, the potential evolutionary role of genetic

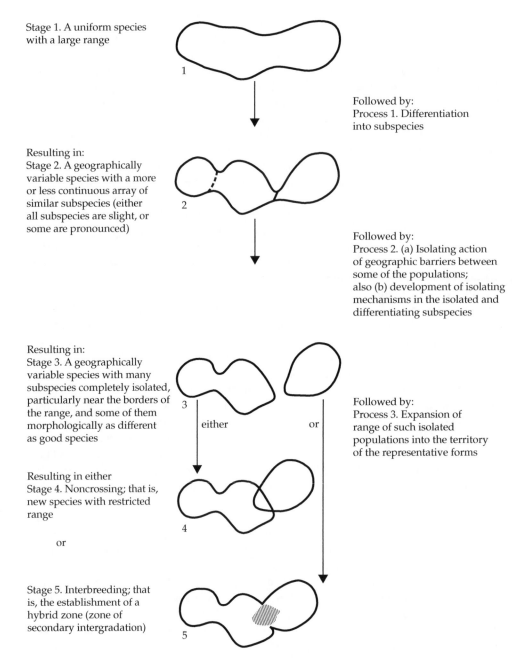

Stage 1. A uniform species
with a large range

Followed by:
Process 1. Differentiation
into subspecies

Resulting in:
Stage 2. A geographically
variable species with a more
or less continuous array of
similar subspecies (either
all subspecies are slight, or
some are pronounced)

Followed by:
Process 2. (a) Isolating action
of geographic barriers between
some of the populations;
also (b) development of isolating
mechanisms in the isolated and
differentiating subspecies

Resulting in:
Stage 3. A geographically
variable species with many
subspecies completely isolated,
particularly near the borders of
the range, and some of them
morphologically as different
as good species

either   or

Followed by:
Process 3. Expansion of
range of such isolated
populations into the territory
of the representative forms

Resulting in either
Stage 4. Noncrossing; that is,
new species with restricted
range

or

Stage 5. Interbreeding; that
is, the establishment of a
hybrid zone (zone of
secondary intergradation)

Figure 2.1 Stages and processes associated with allopatric speciation as envisioned by Mayr (1942, p. 160).

exchange is not diminished by the application of the biological species concept. Under these definitions, it does not represent a violation of the biological species concept for organisms from different evolutionary lineages to form viable, fertile hybrids, recombinants, or reassortants. For example, subspecies and semispecies are defined under the biological species concept in part by the presence of ongoing or potential gene flow (i.e. subspecies are made up of populations of a single species; Mayr

1963, p. 348). For the same reasons, given the definitions from this and my earlier work (Arnold 1997), evolutionary outcomes from genetic exchange such as adaptive trait introgression are also not explicitly discounted by the biological species concept. Yet, Mayr also argued that the results from crosses between even subspecies or semispecies are often maladaptive and transitory. He stated this viewpoint in the following manner: 'In natural populations there is usually severe selection against introgression. The failure of most zones of conspecific hybridization to broaden . . . shows that there is already a great deal of genetic unbalance between differentiated populations within a species' (Mayr 1963, p. 132). In contrast to these conclusions, a number of zoological examples have come to light in which natural hybridization is seen to be a frequent and, evolutionarily, important process (e.g. Dowling and DeMarais 1993; Grant *et al*. 2005).

One way in which to illustrate the importance of genetic exchange is to catalog its widespread occurrence in organisms as diverse as bacteria and mammals (Arnold 1997 and the present book). However, it is equally crucial to emphasize that the rarity of a particular event is not predictive of its potential evolutionary importance. If it were the case that rarity of occurrence predicted unimportance, then mutations that result in an increase in fitness—which of course form the basis for Darwinian or adaptive evolution—would be disregarded. In the same way, the frequent observation of lowered fertility and viability of F_1 hybrids, relative to conspecific individuals, could lead one to conclude that these progeny will play a minor role in the evolution of a given species complex. A portion of the following quote was used in section 1.1 to exemplify how minor of a role was ascribed to genetic exchange (in the context of this quote, introgressive hybridization) in the evolution of animal groups. However, it is an excellent example of the conclusion that rarity of occurrence equals lack of importance. 'The total weight of the available evidence contradicts the assumption that hybridization plays a major evolutionary role among higher animals. To begin with, hybrids are very rare among such animals, except in a few groups with external fertilization. The majority of such hybrids are totally sterile, even where they display 'hybrid vigor.' Even those hybrids that produce

normal gametes in one or both sexes are nevertheless unsuccessful in most cases and do not participate in reproduction. Finally, when they do backcross to the parental species, they normally produce genotypes of inferior viability that are eliminated by natural selection. Successful hybridization is indeed a rare phenomenon among animals' (Mayr 1963, p133). In contrast to this prediction, there are numerous instances in which, for example, the rarity of viable gametes has not prevented important evolutionary effects from F_1 and later-generation hybrid formation (Arnold *et al*. 1999). One example from the plant literature involves the annual sunflower species *Helianthus annuus* and *Helianthus petiolaris*. Experimental crosses involving these two species produce F_1 individuals that possess pollen fertilities ranging from 0 to 30% (mean approx. 14%; Heiser 1947). Additionally, experimentally produced F_2 and first backcross generation plants produce a maximum seed set of 1 and 2%, respectively (Heiser *et al*. 1969). Notwithstanding these extremely low levels of fertility and viability in the initial hybrid generations, natural hybridization between *H. annuus* and *H. petiolaris* has resulted in at least three stabilized hybrid species (Rieseberg 1991).

It can be seen from the above discussion that the application of the biological species concept has almost always led to the conclusion that genetic exchange, particularly natural hybridization, produces relatively unimportant evolutionary consequences. Yet, the definition of species in terms of reproductive barriers led Dobzhansky to view the process as extremely important. In this case, however, the ascription of importance was once again merely a reflection of the assumption of a uniformly maladaptive outcome from genetic exchange. Dobzhansky (1940, 1970) argued that the significant evolutionary effect from natural hybridization was to finalize the construction of prezygotic barriers to reproduction (i.e. 'reinforcement'; Blair 1955). Reinforcement was hypothesized to occur when individuals from previously allopatric, and genetically divergent, populations came back together spatially and mated, resulting in less fit hybrid offspring and thus selection favoring those individuals that mated with conspecifics (Dobzhansky 1940, 1970; Noor 1995; Hoskin *et al*.

2005; Lukhtanov *et al.* 2005). As seen through the window of the biological species concept, such instances reflect the role of natural hybridization as merely a means for finalizing speciation. Though of possibly large evolutionary effect for certain groups of organisms (see Servedio and Noor 2003 for review), reinforcement reflects once again an emphasis on hybridization/genetic exchange as being maladaptive.

Reinforcement is a model describing how natural hybridization can act as a mechanism by which reproductive isolation (and within the biological species concept the process of speciation) is completed. Under this model, natural hybridization plays a key, though secondary, role in 'biological' speciation. Similarly, the majority of zoological, and some botanical, treatments of hybridization have utilized this process as a tool for identifying how barriers to reproduction accumulate (e.g. Ramsey *et al.* 2003; Britton-Davidian *et al.* 2005). In many cases, the origin of barriers to reproduction has been inferred using hybrid zones as indicators of incipient speciation. The framework of the biological species concept suggests that speciation is an extended, gradual process occurring in an allopatric setting (Mayr 1942). As an extension of these assumptions, hybridizing taxa are considered to be at some genetically, evolutionarily, and ecologically intermediate point between conspecific and specific status. Studying hybrid zones is thus seen as an adjutant to viewing steps in early stages of speciation (i.e. before reproductive barriers are completed). The inferred utility of hybrid zones for studying 'biological speciation' is reflected in their description as a 'window' (Harrison 1990) through which an evolutionary biologist may peer, or a 'natural laboratory' (Hewitt 1988) in which to experiment. In the context of the biological species concept, hybrid zones allow the opportunity to determine the processes that have been causal in the origin of barriers to one mechanism for genetic exchange; that is, introgressive hybridization.

As stated at the outset of this section, the estimate of the strength of reproductive barriers as the litmus test for species and speciation is not unique to the biological species concept. However, the fact that this species concept has been widely applied has resulted in an emphasis on cessation of gene flow as the necessary and sufficient component for speciation. In a similar way, the strict application of the phylogenetic species concept is also problematic for considerations of evolutionary effects from genetic exchange. However, it is also the case that some work within the context of the phylogenetic species concept has taken into account the possibility of reticulate, rather than simply bifurcating, evolutionary networks (e.g. Baum and Donoghue 1995). As with the biological species concept, and indeed any species concept, applying the phylogenetic species concept does not need to limit studies that test the importance of reticulate evolution. Furthermore, if the application of any concept limits the study of evolutionary processes, then that species concept must be modified until it reflects biological reality.

2.2.2 Phylogenetic species concept

The origin of the phylogenetic species concept can be traced back at least to Hennig (1966), although Darwin (1859) also presented evolutionary diversification as a bifurcating tree in the only illustration found in *On the Origin of Species*. However, Hennig's treatment is the basis for the modern phylogenetic approach for studying evolutionary diversification. In the context of this discussion, Hennig's methodology can be placed within the history-based format of the phylogenetic species concept. Hennig (1966) argued that populations belonging to the same species must share a single, common ancestor (i.e. that they be monophyletic). Similarly, Cracraft (1989) defined species as '. . . an irreducible (basal) cluster of organisms, diagnosably distinct from other such clusters, and within which there is a parental pattern of ancestry and descent . . .' Inherent then in Hennig and Cracraft's (and indeed Darwin's) definitions of species is the aspect of evolutionary/phylogenetic history. The object of some variants of the phylogenetic species concept is thus the determination of the pattern of phylogenetic relationships as the basis for the definition of species. Use of the Hennigian concept to define species, and the process of speciation, relies upon the identification of ancestral and derived character states. Understanding the polarity of character evolution allows a resolution of the pattern of branching within phylogenies and demarcates monophyletic species 'boundaries' (i.e. 'irreducible clusters'; Cracraft 1989).

As indicated in section 2.1, the application of the requirement of monophyly eliminates the possibility of heterospecific hybridization in the origin of new species. Such an origin would be necessarily polyphyletic. Cracraft (1989) illustrated the viewpoint of Hennig and others when stating, 'In the majority of cases, phylogenetic species will be demonstrably monophyletic; they will never be nonmonophyletic, except through error.' Hennig (1966, p. 207) recognized that 'Special complications would arise if new species could also arise to a noteworthy extent by hybridization between species.' The complications of such events were, however, discounted by arguing that '. . . in all cases in which a 'polyphyletic origin of species' has been recognized, the species involved were so closely related that they could just as well be considered races of one species . . . We must consider a polyphyletic origin of . . . stages below species to be possible, through hybridization for example, but this does not touch the question of the monophyly of the higher taxa' (Hennig 1966, pp. 208–209). As discussed above, Hennig (1966) not only used monophyly in his species definition, but also the requirement that different species reflect separate (i.e. non-hybridizing) reproductive communities. Nixon and Wheeler (1990) outlined a similar viewpoint in the following way: 'With the strictest application of the phylogenetic species concept, species that show extensive intergradation will be treated as a single species'.

Hybrid speciation is only one of the many outcomes that may result from genetic exchange; other outcomes include introgression between the donor and recipient, genetic assimilation of one taxon by another, and reinforcement of barriers to reproduction (see Arnold 1992, 1997 for reviews). Furthermore, the definitions of hybridization and reticulate evolution adopted for this book result in the observation that genetic exchange can have similar effects whether occurring between or within species. Yet, the context for a majority of cases of reticulate evolution involves organisms placed in taxonomic categories of species or higher—at the extreme between taxa from different domains of life

(Boucher *et al.* 2003). Therefore, as additional genomic information becomes available, it may become more and more difficult to use the requirement of strict monophyly (i.e. all sequences in organisms from taxonomic groupings of species or above are from a single ancestor) in determining species boundaries (e.g. Pinceel *et al.* 2005).

It may seem counterintuitive that a 'phylogenetic' approach for defining species could lack a history-based assumption. It would instead seem logical that all such analyses, given that they have a phylogenetic/cladistic analysis as a goal, should instead be grouped within the history-based category. However, this is not the case for the character-based methodologies. A character-based approach thus '. . . does not rely on phylogenetic analyses in order to be implemented . . .' but rather defines '. . . species as the smallest aggregation . . . diagnosable by a unique combination of character states . . .' (Nixon and Wheeler 1990). It is illustrative that in their review of concepts for defining species, Sites and Marshall (2004) placed this approach in the category of 'nontree-based methods' for species delimitation.

Notwithstanding the significant theoretical and methodological differences between the history-based and character-based phylogenetic species concepts, the latter, like the former, assume that species are 'diagnosable'. In this case, diagnosability reflects unique character sets that are not found in other species. If genetic exchange occurs, those characters exchanged will not be used to define the species and thus the process itself (i.e. genetic exchange) cannot be detected. Once again definitional constraints limit genetic exchange to taxonomic groups below the species level. This, in turn, disallows consideration of exchanges between well-differentiated taxa, thus reflecting a large proportion of such cases.

In Chapter 1, I illustrated the similar outcomes from gene exchange in both eukaryotes and prokaryotes. In the next section we will see that, unlike the biological or phylogenetic species concept, the cohesion species concept has taken into account the role of gene exchange in explaining and predicting evolutionary diversification. I will continue to argue that this is necessary for a complete understanding of organismal evolution in general.

2.2.3 Cohesion species concept

The cohesion species concept defines species as '... the most inclusive group of organisms having the potential for genetic and/or demographic exchangeability' (Templeton 1989). Templeton (1989) argued that one of the main weaknesses of the biological species concept was its lack of applicability to either asexual organisms or to taxa belonging to *syngameons*. In Templeton's (1989) terminology the difficulty resulted from either 'too little' or 'too much sex'. Templeton (1989) proposed the cohesion species concept to address problems with the biological species concept as well as other species concepts. The cohesion species concept was thus designed to accommodate all of the microevolutionary processes thought to contribute to speciation (Templeton 1989). 'Genetic exchangeability' incorporated the process of gene flow and 'demographic or ecological exchangeability' the processes of genetic drift and natural selection (Templeton 1989). Specifically, '... genetic exchangeability ... is the ability to exchange genes during sexual reproduction' and '... demographic exchangeability occurs when all individuals in a population display exactly the same ranges and abilities of tolerance to all relevant ecological variables' (Templeton 1989). For the current discussion it is important to ask how the cohesion species concept can accommodate biological instances in which genetic exchange between divergent lineages is detected (i.e. when there is 'too much sex').

Unlike the biological, phylogenetic, and other species concepts, cohesion species are defined on the basis of independent evolutionary/ecological trajectories (Templeton, 1989, 2001). Among other implications from this concept, cohesion species may arise as products of a reticulation event between lineages defined as species (Templeton 1981, 2001). As importantly, cohesion species may participate in some degree of genetic exchange with other lineages and yet maintain their species status (Templeton 1981, 1989, 2001, 2004b). For example, syngameons (Figure 2.2) are a result of species having greater genetic than ecological exchangeability. In contrast to the biological or phylogenetic species concepts, the cohesion species concept does not require that these taxa be reduced to subspecific

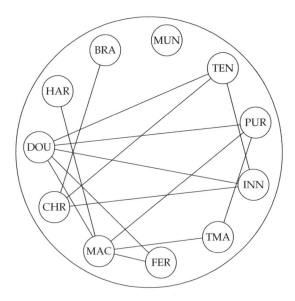

Figure 2.2 An illustration of a syngameon involving the Pacific Coast irises, series *Californicae*. Each small circle represents a separate species. The species designations are as follows: BRA, *Iris bracteata*; MUN, *Iris munzii*; TEN, *Iris tenax*; PUR, *Iris purdyi*; INN, *Iris innominata*; TMA, *Iris tenuissima*; FER, *Iris fernaldii*; MAC, *Iris macrosiphon*; CHR, *Iris chrysophylla*; DOU, *Iris douglasiana*; HAR, *Iris hartwegii*. A line connecting two circles indicates natural hybridization between those taxa (from Lenz 1959).

categories. Rather than viewing hybridization as a problem to be overcome for the process of divergent evolution to proceed, the application of the cohesion species concept indicates that it is more constructive and instructive to incorporate reticulate evolution into an evaluation of evolutionary process and pattern (Templeton 2004b).

Another strength of the cohesion species concept is that it allows the construction of hypotheses that are testable within the widely applied phylogeographic paradigm of Avise (2000b). By constructing such null hypotheses, it is possible to incorporate geography and distribution of population genetic variation into a genealogical analysis (Figure 2.3 Templeton 2001, 2004b). The rejection of the null hypotheses, (i) that geographically associated samples are from a single evolutionary lineage and (ii) that the lineages discovered by (i) are not exchangeable in terms of genetic or ecological characteristics, results in the inference that more than

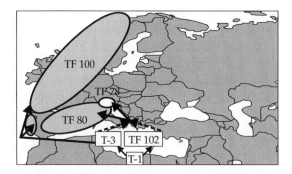

Figure 2.3 Phylogenetic and reticulate evolutionary patterns overlain on the geographic distribution of the brown trout species complex. The evolutionary patterns were inferred using DNA sequence data from the transferrin (TF) locus. Ovals indicate the approximate geographic locations of clades (as defined by the TF haplotypes, such as TF 100). Solid arrows indicate phylogenetic transitions between clades; dashed arrows indicate recombination events (from Templeton 2004b).

one evolutionary lineage (or species) is present in the sample (Templeton 2001). It could be argued that the phylogenetic species concept, but not the biological species concept (without the occurrence of sympatric populations), also allows such a rigorous identification of evolutionary lineages with separate trajectories. However, it is problematic to decide the weight to be given to the characters that identify a 'phylogenetic species' (Avise 2000a). In contrast, the cohesion species concept allows a statistical test of whether evolutionarily distinct lineages exist, and also whether reticulation, past or present, has impacted said lineages (Templeton 2001, 2004b). As such, the application of this paradigm reflects an encompassing, rather than exclusionary, approach for the deciphering of evolutionary pattern and process. In the same way, the prokaryotic species concept can be seen to embody the web-of-life metaphor, rather than a tree-of-life metaphor based on the biological or phylogenetic species concepts.

2.2.4 Prokaryotic species concept

The pervasive effects of lateral gene transfer in the evolutionary history of prokaryotic taxa are now well established (Boucher *et al.* 2003). This appreciation of the fundamental importance of genetic

exchange in the diversification of bacterial species has been factored into some systematic treatments. For example, Rosselló-Mora and Amann (2001) reflected the generality of problems inherent in all species concepts as well as the development, in their view, of a scientifically more robust prokaryotic species definition by stating, 'The species concept is a recurrent controversial issue that preoccupies philosophers as well as biologists of all disciplines. [The] prokaryotic species concept has its own history and results from a series of empirical improvements . . .'

With regard to the present discussion, one of the salient 'improvements' in bacterial systematic studies has been the addition of DNA sequence information (Doolittle 1999; Rosselló-Mora and Amann 2001; Cohan 2002; Stackebrandt *et al.* 2002; Konstantinidis and Tiedje 2005). Workers applying any of the prokaryotic species concepts utilize this type of data. However, it is of particular significance for studies that incorporate phylogenetic and/or phenetic approaches in defining bacterial species (e.g. Woese 1987; Doolittle 1999; Stackebrandt *et al.* 2002). Both the phylo-phenetic species concept (e.g. Rosselló-Mora and Amann 2001) and the phylogenetic and ecological divergence concept (e.g. Cohan 2002) utilize DNA sequence information in determining the taxonomic status of bacterial isolates. Furthermore, both frameworks incorporate a consideration of phenotypic/ecological distinctiveness. Finally, there is an *implicit* and *explicit* inclusion of phenetic similarity, estimated from DNA sequence information, for the phylogenetic and ecological concept and the phylo-phenetic species concept, respectively (Rosselló-Mora and Amann 2001; Cohan 2002). However, unlike the phylo-phenetic species concept, Cohan (2002) drew parallels with concepts applied to eukaryotes when he stated, '. . . A species is a group of organisms whose divergence is capped by a force of cohesion; divergence between different species is irreversible; and different species are ecologically distinct.' Furthermore, he argued that unlike commonly accepted systematic treatments for bacteria, '. . . these universal properties are held not by the named species . . . but by ecotypes . . .' and 'A named species is thus more like a genus than a species' (Figure 2.4; Cohan 2002).

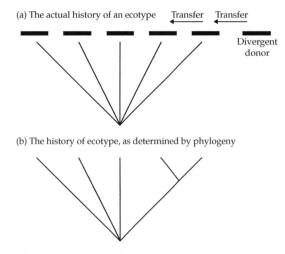

(a) The actual history of an ecotype

(b) The history of ecotype, as determined by phylogeny

Figure 2.4 The effect of lateral gene transfer on phylogeny reconstruction within a bacterial ecotype. (a) The true evolutionary history/phylogeny. (b) The inferred history/phylogeny illustrating that lateral gene transfer following diversification has resulted in an incorrect inference of a recent, shared ancestor for the two lineages on the extreme right (from Cohan 2002).

Consistent with both of the iterations of the prokaryotic species concept, overall genetic similarity, coupled with phenotypic traits associated with ecological setting, has been emphasized as a basis for systematic treatments of prokaryotes (Stackebrandt *et al.* 2002; Gevers *et al.* 2005; Konstantinidis and Tiedje 2005). Thus, 'A bacterial species is essentially considered to be a collection of strains that are characterized by at least one diagnostic phenotypic trait and whose purified DNA molecules show 70% or higher reassociation values . . .' (Konstantinidis and Tiedje 2005). The reference to reassociation values was in regard to the yardstick that strains from the same species would demonstrate at least 70% DNA-DNA hybridization during reassociation kinetics experiments (Konstantinidis and Tiedje 2005).

With the advent of direct DNA sequencing, there was a need to correlate sequence data with the 70% reassociation standard. Without such a correlation, groups in which sequence data were available for some taxa, but not all, could not be treated exhaustively using this systematic methodology (Stackebrandt *et al.* 2002). However, even given the availability of such a correlation (e.g. Rosselló-Mora

and Amann 2001), the phenotypes of the bacterial taxa of interest are also taken into account when defining species. However, the effect of horizontal gene transfer on phenotype is known to be widespread. Indeed, the profound effect of the inclusion of laterally transferred adaptations on evolutionary treatments is easily discernible in prokaryotic systems (Doolittle *et al.* 2003). Thus genes or traits involved in lateral transfer '. . . are often fundamental to cell physiology and 'lifestyle,' and are of the sort that microbiologists traditionally used to define such prokaryotic taxa, or employed in deducing phylogenetic relationships between them . . .' (Boucher *et al.* 2003). These 'phenotypes' may then be unreliable for use in prokaryotic systematics (e.g. see Figure 1.7).

As stated above, there have been significant advances made—in particular through the acquisition of DNA sequence data—in the ability to define species of prokaryotes. Yet, the occurrence of widespread lateral gene transfer still represents a significant barrier for systematic treatments, regardless of whether the phylogenetic and ecological, phylophenetic, or some other species concept is applied. For example, Doolittle and his colleagues have emphasized repeatedly the significance that must be placed on accounting for lateral gene transfer not only in prokaryotic origins, but in eukaryotic origins as well (Doolittle 1999; Boucher *et al.* 2003). Though a cautionary note for those who wish to construct prokaryotic phylogenies or, ultimately, the 'universal tree of life', the consideration/recognition of the role of genetic exchange in organismal origins reflects a wonderful illustration of the thesis of this book—that genetic exchange affects greatly evolutionary pattern and process. One of Doolittle's (1999) conclusions illustrates well the central role played by genetic exchange in the origins, and continued evolution, of the three domains of life. He states 'More challenging is evidence that most archaeal and bacterial genomes (and the inferred ancestral eukaryotic nuclear genome) contain genes from multiple sources . . . Molecular phylogeneticists . . .' may fail '. . . to find the 'true tree,' not because their methods are inadequate or because they have chosen the wrong genes, but because the history of life cannot properly be represented as a tree.' This conclusion then reflects

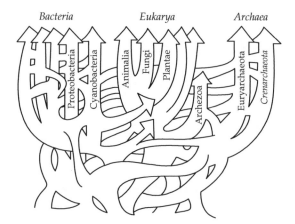

Figure 2.5 A representation of the history of biological diversification of all life illustrating the effect of reticulate and divergent evolution (from Doolittle 1999).

the reality of the web-of-life metaphor as the best representation of long-term evolutionary pattern and process (Figure 2.5; Arnold and Larson 2004).

2.3 Summary and conclusions

Over the past decade a paradigm shift has affected how species are defined in the context of evolution, conservation, and ecology (e.g. see Hey *et al.* 2003; Sites and Marshall 2004 for discussions of topics related to this conclusion). Many now realize that it is necessary to consider the possibility of ongoing genetic exchange as species arise and as they continue to evolve (see Levin 2000; Mallet 2005). As indicated by the topics in each of the chapters of the present book, genetic exchange must be considered as being of possible fundamental importance for organisms as varied as viruses and mammals. Notwithstanding the amount of data available to

support this conclusion, works still appear that presuppose a paradigm emphasizing a minor role for genetic exchange—works usually focused on 'higher organisms' (e.g. Coyne and Orr 2004). Though somewhat limited in their ability to decipher evolutionary pattern and process, such conceptual frameworks may be useful as null hypotheses against which to test various data sets, thus allowing much to be learned concerning the processes associated with genetic exchange.

As discussed in this chapter, some of the restrictive pigeon-holing of evolutionary phenomena—such that genetic exchange becomes unimportant—is due to the application of restrictive species concepts. However, this need not be the case. Instead, diversification that is coincident with genetic exchange among lineages can be incorporated into a consideration of evolutionary processes in general. Accepting this approach allows the recognition, rather than the rejection or ignorance of, the philosophical and scientific tension created by studying a dynamic process (Hey *et al.* 2003; Pigliucci 2003; Sites and Marshall 2004). This approach also allows the application of conceptually and mechanistically broadly based, rather than narrow, models for defining species. In this regard, any of the species concepts discussed in this chapter may be relaxed to incorporate reticulate evolution. However, models of the type encapsulated by the cohesion species concept allow the most straightforward opportunity for addressing all of the possible evolutionary processes. Using this category of concept, or relaxing other concepts, allows rigorous tests for the role of genetic exchange in the developing evolutionary trajectories (past and present) for organisms as different as bacteria and plants. Such an approach would seem to hold the most promise for deciphering the relative effects of all processes associated with evolutionary diversification.

CHAPTER 3

Testing the hypothesis

... chloroplast DNA in natural populations may be acquired from distantly related organisms, so that relationships shown by this molecule may not be representative of the genome as a whole ...

(Whittemore and Schaal 1991)

When gene copies are sampled from various species, the gene tree relating these copies might disagree with the species phylogeny. This discord can arise from horizontal transfer (including hybridization), lineage sorting, and gene duplication and extinction ... all of the gene trees are part of the species tree, which can be visualized like a fuzzy statistical distribution, a cloud of gene histories.

(Maddison 1997)

... there is growing evidence that lateral gene transfer has played an integral role in the evolution of bacterial genomes, and in the diversification and speciation of the enterics and other bacteria.

(Ochman et al. 2000)

Bifurcating phylogenies are frequently used to describe evolutionary history of groups of related species. However, simple bifurcating models may poorly represent the evolutionary history of species that have been exchanging genes.

(Machado and Hey 2003)

3.1 Genetic exchange as a testable hypothesis

In each of the following sections, I will examine different methodologies that have been used to test for the presence of genetic exchange. As each is discussed, it will become obvious that none are without weaknesses, mostly due to underlying assumptions of the likelihood of genetic exchange. For example, as discussed in Chapters 1 and 2, zoologists often assume that allele sharing between different taxa is due to incomplete lineage sorting while botanists are more likely to assume that the same pattern of allele sharing in plants is due to introgression. In addition, those working on 'lower' organisms—viruses, bacteria, and fungi—assume varying levels of lateral gene transfer, reassortment,

etc., depending upon the paradigm applied (e.g. Ge et al. 2005). Such seemingly idiosyncratic conclusions, drawn by different investigators studying the same organismic group, are widespread in the evolutionary literature. Several examples of alternate conclusions regarding the importance of lateral gene transfer, introgressive hybridization, and viral recombination were mentioned in Chapter 1, involving *Yersinia*, Louisiana irises, Darwin's finches, and influenza A. Since the present chapter reviews approaches for detecting evidence of genetic exchange, it is instructive to discuss in detail a species complex characterized by alternate conclusions concerning the evolutionary role of this process. The example I will use to illustrate how conclusions can vary from one study/investigator to the next involves the genus *Quercus*. This example will help communicate two points: (i) it will again illustrate the debate that surrounds the role of genetic exchange in evolution and (ii) it will emphasize the power of a comparative approach that incorporates all relevant information, drawn from as wide an array of different types of data (e.g. morphology and DNA sequences) as possible.

3.2 Controversy over genetic exchange: examples from oaks

3.2.1 North American oaks

Inferences of introgressive hybridization and hybrid speciation are replete for oaks in general. For example, Trelease (1917) reported over 120 hybrid derivatives from natural hybridization between various North American species. The data for this study came from his examinations of herbarium specimens and from reports of hybrids found in the scientific literature. In a reanalysis of hybridization among the same species examined by Trelease, Palmer (1948) arrived at the same conclusion, that introgression had been a major contributor to the evolution of these taxa. In the latter study more than 70 individual hybrid types (from different interspecific pairings) were identified. Grant also reflected such a conclusion when, in his classic *Plant Speciation*, he used white oak species as an example of a syngameon (Grant 1981, pp. 237–238).

Not only systematic analyses, but also numerous population studies have been undertaken to test for the effects of genetic exchange among North American *Quercus* species. For example, in the first issue of the journal *Evolution*, Stebbins *et al.* (1947) re-examined the hypothesis of Davis (1892) that *Quercus marilandica* and *Quercus ilicifolia* hybridized naturally on Staten Island, New York. Describing the population, Davis (1892) observed that there were hundreds of individuals easily identified as belonging to one or the other species. However, he also noted '... a number of trees whose place is not so evident, and it becomes a question when viewing them as to whether they exhibit more of the characters of *ilicifolia* or *nigra* [now *marilandica*]. They form when taken together a series leading from one species to the other ...' (Davis 1892). Stebbins *et al.*

(1947) tested this anecdotal report by collecting morphological data from allopatric and sympatric populations of *Q. marilandica* and *Q. ilicifolia*. These data were used for a *hybrid index* analysis. As Davis' insightful description suggested, the distribution of the species-specific characters made it clear that hybridization and introgression had impacted both of these species (Stebbins *et al.* 1947).

Like earlier, mainly taxonomic, descriptions (e.g. Trelease 1917; Palmer 1948), some population-level analyses had as their primary goal to test the hypothesis that certain taxa had arisen through natural hybridization. One such case involved the origin of the taxon known as *Quercus alvordiana*. Tucker (1952) examined herbarium specimens and sampled many natural populations of this species and its putative parents, *Quercus douglasii* and *Quercus turbinella*. Although he did not recognize *Q. alvordiana* as a separate species, Tucker (1952) did conclude that it was a geographically widespread hybrid derivative from crosses between *Q. douglasii* and *Q. turbinella*. In addition, he detected the effect of ongoing gene flow between the two progenitor species. Morphological analyses of populations from these species that were outside of, but adjacent to, the hybrid zones often detected 'modifications' (Figures 3.1–3.4). Tucker (1952) thus concluded, 'The fact that this modification is usually in the direction of the other species is taken as evidence of introgression.'

More recent tests for introgressive hybridization among North American *Quercus* species have utilized discrete, molecular markers. For example, Whittemore and Schaal (1991) assayed genetic variation for cpDNA and nuclear ribosomal DNA (rDNA) genes among eastern white oak species. They found that cpDNA variation, unlike that of rDNA, correlated well with geography rather than taxonomic identity of the samples. Thus species in sympatry shared cpDNA haplotypes. In contrast, the rDNA showed species-specific distributions (Whittemore and Schaal 1991). Furthermore, these species are well differentiated on the basis of morphology, allozymes, and ecological associations (Whittemore and Schaal 1991). The conclusion drawn from the cpDNA variation was that '... sympatric species of oak in the eastern United States do not represent fully isolated gene pools, but are actively exchanging genes'. For the present

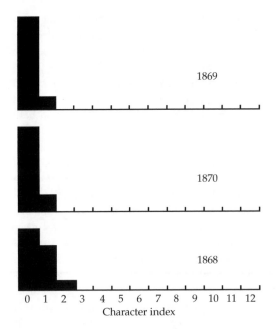

Figure 3.1 Hybrid index scores for allopatric populations of *Q. douglasii* (from Tucker 1952).

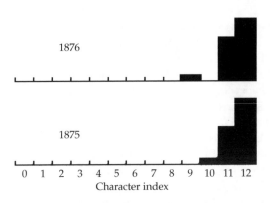

Figure 3.2 Hybrid index scores for allopatric populations of *Q. turbinella* (from Tucker 1952).

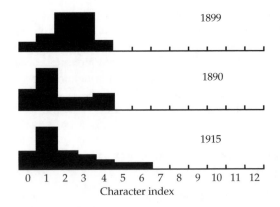

Figure 3.3 Hybrid index scores for populations of *Q. douglasii* adjacent to the putative hybrid species *Q. alvordiana* (from crosses between *Q. douglasii* and *Q. turbinella*; from Tucker 1952).

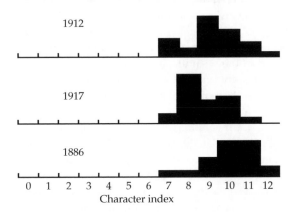

Figure 3.4 Hybrid index scores for populations of *Q. turbinella* adjacent to the putative hybrid species *Q. alvordiana* (from crosses between *Q. douglasii* and *Q. turbinella*; from Tucker 1952).

discussion of how contrasting conclusions can be drawn concerning the extent of genetic exchange, it is illustrative that based on genetic markers derived from nuclear loci (e.g. rDNA and allozymes), these same species could be used as an example of limited introgression. Indeed, this was the conclusion drawn from an analysis of nuclear DNA and morphological variation within and between the western white oak species, *Quercus grisea* and *Quercus*

gambelii (Howard *et al.* 1997). These workers detected the introgression of nuclear markers in a majority of individuals from some populations near sympatric regions. However, these authors also found that species-specific nuclear DNA markers and morphological characteristics were significantly correlated, in the same individuals, even in sympatry. Thus introgression could be seen as being of minor importance with regard to the molecular markers and morphological characteristics sampled (Howard *et al.* 1997). Yet, a separate analysis by these same authors (Williams *et al.* 2001) detected patterns consistent with increased fitness of certain hybrids in specific habitats. Furthermore,

Howard *et al.* (1997) did not negate the possibility of introgression involving many portions of the genome; rather they pointed out that the few portions of the genome that controlled species-specific morphological characteristics were not apparently being exchanged. Finally, Howard *et al.* (1997) cited the findings of Whittemore and Schaal (1991) and suggested that cytoplasmic elements might be expected to introgress at a much higher frequency than elements of the nuclear genome. If, as expected, the cpDNA shows greater levels of introgression in the western species individuals as well, then the nuclear plus cytoplasmic introgression would indicate extensive, rather than limited, genetic exchange.

Taken together, the studies from Davis (1892) to those of Whittemore and Schaal (1991), Howard *et al.* (1997) and Williams *et al.* (2001) reveal the need for a broad-based approach that takes into consideration all of the available data for robust tests of hypotheses concerning the extent and evolutionary effect of genetic exchange. A single data set might result in either support for or rejection of the hypothesis of significant amounts and evolutionary effects from genetic exchange. For example, focusing on only nuclear-based markers, or only population genetics without fitness estimates, could lead to the conclusion that introgression between North American *Quercus* species was limited in its effect. This would hardly seem reasonable with the weight of evidence now available. Examples of conflicting viewpoints concerning the role of genetic exchange among oak species are, however, not restricted to North American taxa. The literature on the evolution of European species also contains an example of contradictory conclusions.

3.2.2   European oaks

Hybridization among Old World species of *Quercus*, as with New World forms, has often been inferred (e.g. Ferris *et al.* 1993; Muir *et al.* 2000; Belahbib *et al.* 2001; Jiménez *et al.* 2004; Valbuena-Carabaña *et al.* 2005). One species pair in particular, *Quercus robur* and *Quercus petraea*, has been the focus of many studies involving a wide array of morphological and molecular characteristics (e.g. Grant 1981, p. 241; Ferris *et al.* 1993; Dumolin-Lapègue *et al.*

1997, 1998, 1999; Petit *et al.* 1997, 2003; Streiff *et al.* 1999; Muir *et al.* 2000, 2001; Muir and Schlötterer 2005). There seems to be no question that these two European oak species hybridize. Indeed, discussions of the evolutionary, population genetic, and even conservation biology implications are replete. Citing earlier surveys, Grant (1981, p. 241) reflected the already accepted fact that '. . . where it is native, *Q. robur* hybridizes with *Q. petraea* . . .' More recent analyses have focused on such topics as (i) the role of separate refugia and post-glacial expansion on patterns of introgression (Ferris *et al.* 1993; Dumolin-Lapègue *et al.* 1997; Petit *et al.* 1997), (ii) the effect of mating system on patterns of introgression (Bacilieri *et al.* 1996), (iii) the role of introgression as a dispersal mechanism (Petit *et al.* 2003), (iv) the extent of interspecific cytoplasmic gene flow as assayed for both mitochondrial and chloroplast DNA, (v) molecular mechanisms that may affect the expression of introgressed genes (Muir *et al.* 2001), and (vi) the effect of extensive introgression—resulting in the absence of species-specific genetic markers—on taxonomy and thus forest management (Muir *et al.* 2000).

Notwithstanding all of the above evidence indicating introgression between *Q. petraea* and *Q. robur*, Muir and Schlötterer (2005) concluded the following: 'Based on our results we propose that the low genetic differentiation among these species results from shared ancestry rather than high rates of gene flow'. For their study, Muir and Schlötterer (2005) examined genetic variation among microsatellite loci for several populations of each species. This analysis allowed a test for (i) species-specific markers and (ii) genetic differentiation of populations that were in geographic proximity compared with those more distant. From this analysis it was discovered that there were no microsatellite loci that were species-specific (although there were private alleles) and that populations that were closer geographically did not show significantly higher levels of interspecific gene flow (based upon indirect estimates) than those separated by greater distances.

Since this is only a single study, compared with many previous analyses, it is instructive for the present discussion to consider how these authors came to an alternative conclusion. To do this, these authors had to account for the results from a wide range of characters and from many studies, including their own (Muir *et al.* 2000, 2001). The data suggesting high levels of introgression between *Q. robur* and *Q. petraea* that Muir and Schlötterer (2005) had to contend with was based upon morphological and molecular data sets, the latter including both nuclear and cytoplasmic (i.e. mitochondrial DNA (mtDNA) and cpDNA) sequence information. They first suggested that direct estimates of hybrid formation drawn from seed progeny (Streiff *et al.* 1999) were overestimates because these samples did not reflect post-zygotic selection during establishment. This conclusion is almost certainly correct (see Cruzan and Arnold 1994; Cornman *et al.* 2004 for examples of such postzygotic selection). However, Streiff *et al.* (1999) detected 7% hybrid progeny. Such a large proportion is consistent with the presence of high levels of gene flow between these species. Second, Muir and Schlötterer (2005) concluded that although each of the many studies of phylogeographical variation for cytoplasmic genomes (including cpDNA and mtDNA) detected high rates of introgression, '... gene flow is difficult to quantify in the chloroplast genome as the latter represents a single locus.' In contrast, these workers did not account satisfactorily for the potentially high level of homoplasy contained within their own (microsatellite) marker data set. In summary, these authors dismiss findings from morphological surveys and molecular markers (including their own previous work) in favor of a single data set that is likely affected by homoplasy. This seems to be a non-parsimonious approach. More importantly, to dismiss numerous data sets and thus reject a comparative methodology seems antithetical to a robust examination of evolutionary process and pattern.

3.3   Methods to test for genetic exchange

As with the oak studies, the patterns of variation in each of the species systems discussed within the following sections could be accounted for partially by (i) incomplete lineage sorting (i.e. sharing of ancestral polymorphisms), (ii) genetic exchange, or (iii) a combination of both processes. These are likely only partial causes of the patterns detected

because we must also assume that selection and drift may play ongoing roles in the structuring of genetic variation. To understand the potential effect of genetic exchange on evolutionary lineages, I will discuss various examples of the following methodologies: (i) trans-generational sampling of populations associated with hybrid zones, (ii) tests for phylogenetic discordance caused by introgressive hybridization and horizontal gene transfer, (iii) the combined use of gene genealogies and models of speciation, (iv) the assessment of intragenomic divergence, and (v) the application of nested clade analysis.

3.3.1   Concordance across data sets: hybrid zone analyses

Robust tests for the presence of ongoing genetic exchange are made possible through examinations of genetic or phenotypic variation and habitat×genotype associations within hybrid zones, especially if made across generations. In particular, such data collected from populations within and outside of putative hybrid zones allow tests for reoccurring admixtures of alleles/characteristics normally found in one or other of the parental forms. The utility of such studies, for testing hypotheses concerning the occurrence and evolutionary effect from genetic exchange, has already been illustrated in Chapter 1. It is thus clear that without multi-data set/multiyear (i.e. generational) sampling, the conclusions drawn concerning the evolutionary role of genetic exchange for Darwin's finches, Louisiana irises, and influenza A would not have been possible. Yet, each of these examples was used for other topics and not for illustrating the power of such research designs in testing for genetic exchange. In the following sections I will use two complexes of insect species—the Australian grasshopper genus *Caledia* and the North American cricket genus *Allonemobius*—to illustrate the utility of multiple data sets for detecting genetic exchange and for testing for the effects of exchange on the lineages involved.

Caledia captiva

The Australian grasshopper genus *Caledia* is a remarkable model system for studying evolutionary phenomena. Shaw and his colleagues have addressed such diverse topics as population genetics, chromosomal evolution, speciation, adaptation, and the role of climate change in organismal distributions—all examined and illustrated by a genus that ostensibly contains only two species, *Caledia captiva* and *Caledia species nova 1* (Shaw 1976). The presence of two recognized species (but see below for the discussion of a sibling species) within this genus belies the remarkable genetic, ecological, behavioral, and developmental diversity housed within various subspecific taxa (e.g. Shaw 1976; Shaw *et al.* 1976, 1980; Shaw and Wilkinson 1978; Daly *et al.* 1981; Arnold *et al.* 1986; Kohlmann *et al.* 1988; Groeters and Shaw 1992).

Some of the earliest studies of *Caledia* discovered cytologically distinct forms that occurred in populations stretching continuously from southern Victoria, up the east coast of Australia to northern Queensland, along the north coast to the Northern Territory and across the Torres Strait to Papua New Guinea (Shaw 1976; Shaw *et al.* 1976). Subsequent analyses of chromosome, allozyme, highly repeated DNA, morphological, and experimental hybridization data further differentiated this genus into eight recognizable taxa (Shaw 1976; Shaw *et al.* 1976, 1980, 1988; Moran and Shaw 1977; Daly *et al.* 1981; Arnold and Shaw 1985; Arnold 1986; Arnold *et al.* 1986, 1987a, b; Marchant *et al.* 1988). These included the southeast Australian, Moreton, southern Torresian, northern Torresian, Northern Territory Torresian, Papuan Torresian, Daintree, and *C. species nova 1* forms. Further assignment of populations to evolutionary groupings was made possible by employing the numerous available data sets. For example, although a north–south cline in centromere position for each chromosome (Figure 3.5) resulted in completely acrocentric linkage groups in Victoria, and metacentric chromosome complements in the extreme northern populations (Shaw *et al.* 1988), the southeast Australian and Moreton taxa were recognized as belonging to the same subspecies. Thus similarities in chromosome banding patterns (Shaw *et al.* 1988), allozyme variation (Daly *et al.* 1981), mtDNA (Marchant and Shaw 1993; Shaw *et al.* 1993), and highly repeated DNA sequences (Arnold and Shaw 1985), along with reproductive compatibilities

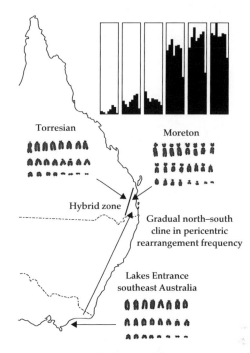

Figure 3.5   The distribution of chromosome variation within the Torresian, Moreton, and southeast Australian races. The location of the contemporary hybrid zone between Torresian and Moreton races is noted. The changeover in the frequency from the Moreton to the Torresian chromosomal form is indicated by the histogram at the top. This changeover occurs over a distance of only 200 m (from Shaw and Coates 1983).

detected through experimental hybridization (Shaw *et al.* 1980), indicated that the populations located along this cline belonged to a single evolutionary clade. The same type of data also indicated that all of the Torresian forms should be placed into a second subspecies, but that the Daintree form was a sibling species of *C. captiva* and that *C. species nova 1* represented a third, morphologically differentiated species (Arnold *et al.* 1987b). The richness of the *Caledia* species complex for tests of evolutionary pattern and process is thus seen clearly. However, these organisms are likely best known for their utilization in detailed, multi-year analyses elucidating hybridization, introgression, and reproductive isolation within and between taxa.

As with the complexity of chromosomal rearrangement and highly repeated DNA sequence differences (reflected in vastly different chromosome banding patterns), levels of reproductive isolation

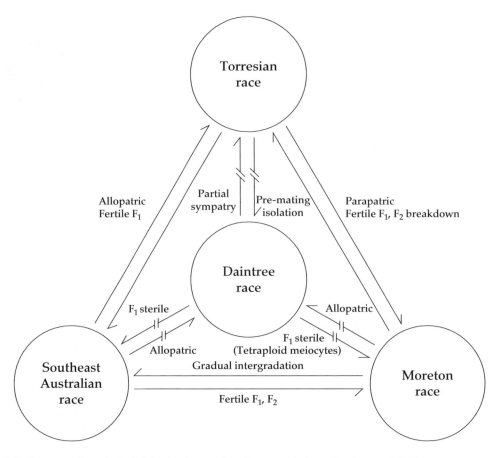

Figure 3.6   The pattern of reproductive isolation found among the various *C. captiva* forms (from Shaw *et al.* 1980).

also vary greatly depending upon the taxa examined (Figure 3.6; Shaw *et al.* 1980, 1982; Shaw and Coates 1983). For example, the Daintree sibling species produces few F_1 progeny when paired with any of the other *C. captiva* taxa and these hybrid progeny are sterile (Figure 3.6; Shaw *et al.* 1980). In contrast, the viability and fertility of F_1 individuals produced by crossing Moreton/southeast Australian and Torresian individuals is not significantly reduced compared to control (within-subspecies) crosses (Shaw *et al.* 1980). However, a significant reduction in the viability of the reciprocal BC_1 generations and the F_2 generation is detected (Figure 3.6; Shaw *et al.* 1980, 1993). Indeed, crosses between Torresian individuals and the various chromosomal races of the southeast Australian/Moreton result in BC_1 viabilities ranging from approx. 40 to 70% and F_2 viabilities

of approx. 0–45% (Shaw *et al.* 1993). It is apparent that barriers to genetic exchange among the subspecies and species are quite strong (Figure 3.6). In the context of the present discussion, we must therefore ask the question of whether multiple data sets support or reject a hypothesis of genetic exchange between the various taxa. In particular, does multigenerational sampling from inside and outside contemporary hybrid zones detect patterns of genetic variation suggesting introgressive hybridization? Furthermore, if the hypothesis of genetic exchange is supported, does the pattern of genetic variation suggest processes that affect the extent of introgression for different genomic (cytoplasmic and nuclear) elements?

Hybrid zone analyses within the *C. captiva* species complex have focused on the parapatrically distributed Moreton and Torresian subspecies. These subspecies were found to be diagnostically different for chromosome rearrangements, chromosome banding, cytological distribution of highly repetitive DNA families, DNA sequence variation for highly repetitive DNA families, rDNA sequences, allozymes, and mtDNA (Shaw 1976; Shaw *et al.* 1976; Daly *et al.* 1981; Arnold and Shaw 1985; Arnold *et al.* 1986, 1987a, b; Marchant *et al.* 1988; Marchant and Shaw 1993). The first data collected and used to test for genetic exchange involved investigations of the chromosomal-rearrangement differences between these two subspecies near and within the region of parapatry (Moran and Shaw 1977; Moran 1979). Populations of the two subspecies, near the area of parapatry, differ cytologically in that Moreton individuals have a preponderance of submetacentric or metacentric chromosomes while Torresian individuals possess acrocentric or telocentric chromosomes (Figure 3.5). By sampling from one side of the zone of parapatry to the other, Moran and Shaw (1977) detected (i) individuals polymorphic for the 'diagnostic' chromosome rearrangements and (ii) the presence of Torresian-specific chromosome variants on the Moreton side of the zone of parapatry, but no Moreton-specific variants on the Torresian side of the zone. These findings led to the conclusion that introgressive hybridization was indeed occurring and that it was asymmetric; that is, only from Torresian into Moreton (Moran and Shaw 1977).

A second analysis of the diagnostic rearrangement differences was made during a subsequent generation. This study involved the assaying of chromosomal variation for both the original transect across the zone of parapatry assayed by Moran and Shaw and also for a second transect (Shaw *et al.* 1979). In addition, allozyme variation was surveyed across the hybrid zone. In the case of the chromosomes, asymmetric introgression from Torresian into Moreton populations was once again supported, this time reflected by two transects. Like the chromosome markers, the allozyme variation also suggested introgression between the subspecies. However, unlike the cytological markers, introgression was symmetrical with the allelic variants typical for each subspecies found in populations on both sides of the zone of overlap (Shaw

et al. 1979; Moran *et al.* 1980). The asymmetry of the chromosome introgression was hypothesized to reflect the selective advantage for the Torresian chromosomes (Shaw *et al.* 1979, 1993). It was concluded that the advantage was environmentally mediated. In this case the ecological trigger was a drought that led to environment × genotype interactions that favored introgression from the xeric-adapted Torresian individuals into the mesic-adapted Moreton grasshoppers (Kohlmann *et al.* 1988). Consistent with this conclusion was the observation that the asymmetric introgression was reversed in generations of grasshoppers that experienced more rainfall (Shaw *et al.* 1985, 1993). This reversal suggested that greater rainfall favored components of the Moreton (mesic-adapted) genome and thus caused their introgression into populations on the Torresian side of the zone of overlap (Shaw *et al.* 1993). As with the diagnostic cytological (chromosomal-rearrangement) characters, molecular markers distinguished an area of increased polymorphism in the region of parapatry. This added evidence was also consistent with a hybridization hypothesis. These findings, in concert with all the other data sets, were thus conclusive in regard to the location of a zone of secondary contact.

Given the support for the hypothesis of genetic exchange between the Moreton and Torresian subspecies it is then possible to ask whether the various markers demonstrate similar patterns of introgression. Like the chromosomal markers, an asymmetry was also detected when the Moreton and Torresian sides of the hybrid zone were compared for both nuclear and cytoplasmic markers (i.e. rDNA and mtDNA; Arnold *et al.* 1987a; Marchant *et al.* 1988). When these same individuals/populations were surveyed for the diagnostic allozymes the same asymmetrical pattern was found for these markers as well (Shaw *et al.* 1990). It is important to note that earlier studies resolved allozyme variation, suggesting symmetrical introgression of allozyme markers (e.g. Moran *et al.* 1980). The finding of asymmetry in the later studies likely reflects the more intensive sampling regime of these analyses. Regardless, the asymmetry detected with the rDNA, allozyme, and mtDNA markers was consistent in that introgression largely involved the transfer of Moreton markers into Torresian populations near the zone (Shaw *et al.* 1990, 1993). However, the extent of introgression (i.e. the distance of introgressed markers from the current hybrid zone) of molecular markers compared to that of chromosome structural differences was significantly different (Arnold *et al.* 1987a; Marchant *et al.* 1988). Thus near the zone of contact the chromosome rearrangement and C-band differences changed from 100% Moreton to 100% Torresian within 1 km of the hybrid zone. In contrast, the Moreton molecular markers were found up to 450 km from the present-day hybrid zone (Shaw *et al.* 1990).

The most parsimonious explanation for the coincidental occurrence of Moreton allozyme, rDNA, and mtDNA markers up to 450 km north of the present-day zone of overlap is that the zone has moved, leaving the Moreton markers in its wake (Marchant *et al.* 1988). Given the well-established pattern of environmental change that has caused a switch from mesic to xeric habitats north of the present-day hybrid zone, it would be expected that the xeric-adapted Torresian form would have displaced Moreton populations. However, the frequency of the introgressed allozyme, rDNA, and mtDNA markers is also of interest. As expected for the trailing behind of neutral markers as a hybrid zone moves, the Moreton allozymes are not in high frequency in the more northerly populations (Marchant *et al.* 1988). In contrast, the Moreton mtDNA and rDNA markers are in the majority in these populations—indeed fixed in some of these chromosomally Torresian populations (Shaw *et al.* 1990). Various scenarios can be proposed to explain this pattern. However, comparison with other systems and experimental analyses have suggested that (i) the introgressing Moreton mtDNA may have been selected for in the Torresian nuclear background (Arnold 1997) and (ii) the Moreton rDNA acted as the template for gene-conversion events, leading to its preponderance in the introgressed Torresian populations (Arnold *et al.* 1988).

The trans-generational surveys of the *Caledia* species complex reflect a robust method for testing genetic exchange. In addition, such multi-year analyses allowed numerous inferences concerning the pattern and underlying causes of introgression between the hybridizing taxa. Similarly, studies of another model animal system—the cricket genus *Allonemobius*—illustrate well the strength of such research programs for discerning evolutionary pattern and process.

Allonemobius

Like the *Caledia* research program, the work of Howard and his colleagues on the cricket genus *Allonemobius* has encompassed evolutionary ecology and evolutionary biology. In the course of analyses over the past quarter of a century, these workers have addressed topics as diverse as habitat choice, systematic relationships, hybrid zone structure, gamete competition, reinforcement of premating barriers to reproduction, and the effect of endosymbionts on speciation (e.g. Howard and Harrison 1984a, b; Howard 1986; Howard and Waring 1991; Howard *et al.* 1993, 1998; Britch *et al.* 2001; Marshall *et al.* 2002; Marshall 2004). Another similarity with the *C. captiva* work is that the research on *Allonemobius* by Howard *et al.* is an elegant model of how to detect the presence and effects of genetic exchange using data across generations. Indeed, results from their long-term analyses have transformed this genus, and particularly *Allonemobius fasciatus* and *Allonemobius socius*, into another clear exemplar of evolutionary diversification coincident with ongoing genetic exchange. However, the unexpected finding that the taxon known as *A. fasciatus* contained two cryptic species necessarily preceded initial evidence for interspecific hybridization within *Allonemobius*. In this case allozymic differentiation was defined, suggesting concordant and strong discontinuity at several allozyme loci between northern and southern populations of *A. fasciatus* (Howard 1982, 1983). From these data it was concluded that the northern and southern forms should be considered different species, *A. fasciatus* and *A. socius*, respectively (Howard 1982, 1983). In addition to electrophoretic variation, these largely cryptic species could also be differentiated based upon morphometrics and analyses of calling or mating songs (Howard and Furth 1986). Like the cryptic Moreton and Torresian forms of *C. captiva*, the parapatrically distributed *A. fasciatus* and *A. socius* were also candidates for investigating genetic exchange via introgression.

The initial evidence consistent with genetic exchange between the two *Allonemobius* species, as with the original detection of *A. socius*, came from allozyme analyses (Howard 1986). The first indication that natural hybrids were being formed between *A. fasciatus* and *A. socius* came from hybrid index scores (the character-index methodology of Sage and Selander 1979 as adapted by Howard 1986) for populations collected along transects (i) on the eastern seaboard of the USA and (ii) inland across the Allegheny mountains (Figure 3.7). Along each transect, populations were detected in which crickets occurred that demonstrated admixtures of alleles characteristic for *A. fasciatus* and *A. socius* (Howard 1986). In fact the frequency of putative hybrid genotypes within the zone of overlap in the mountain transect ranged from 20 to 33%. Furthermore, genetic variation in one population separated by approx. 100 km from the nearest mixed population indicated the effect of introgressive hybridization from *A. fasciatus* into *A. socius*; 3% of the genotypes found in this population could be assigned to hybrid classes (Howard 1986).

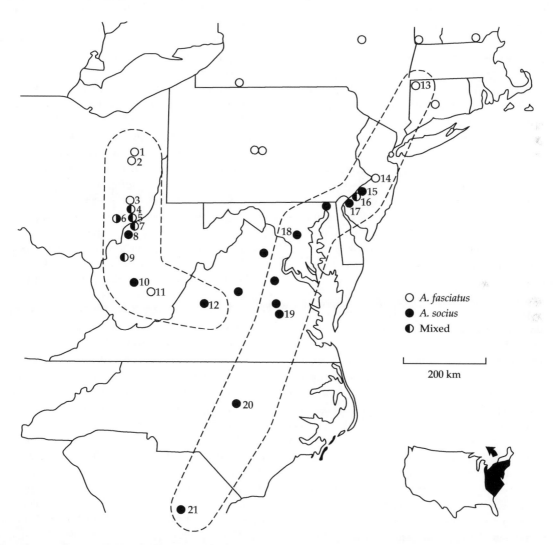

Figure 3.7 Map showing the distributions of the *Allonemobius* populations in the mountain (1–12) and east coast (13–21) zones of overlap. The unnumbered circles are additional samples intended to reflect the range of the two species (from Howard 1986).

It is likely that different numbers of generations produced the genetic variation sampled along the two *Allonemobius* transects. Thus the population samples reflected indirect estimates of the contributions of multiple generations to the population genetic structure. Howard (1986) also carried out indirect trans-generational sampling when he detected significant deviations from Hardy–Weinberg expectations consistent with a genetic-exchange hypothesis. In this case, the deviations resulted from the deficit of 'hybrid' genotypes. This is the expected outcome if two species hybridize to produce a certain proportion of recombinant genotypes possessing lower fitness. Indeed, this is the pattern observed in numerous plant and animal hybrid zones (Barton and Hewitt 1985; Arnold 1997). However, it is important to note that the parental and hybrid classes do not always show increased fitness (as measured by relative survivorship) of *A. fasciatus* and *A. socius* genotypes relative to hybrid genotypes (Howard *et al.* 1993).

In addition to the results from the allozyme analyses, studies of other types of genetic markers and the species-specific calling song parameters also detected variation indicative of genetic exchange between these two taxa. First, the definition of RAPD marker variation among populations near and within an area of overlap in Illinois detected banding phenotypes similar to those found in experimental hybrids (Chu *et al.* 1995). These same individuals were also assayed for species-specific allozyme markers, confirming the hybrid RAPD genotypes. However, Chu *et al.* (1995) argued that these results, along with earlier findings, suggested only limited genetic exchange between *A. fasciatus* and *A. socius*. Along with the molecular marker data, calling-song variation in natural populations outside and within the zones of overlap between these two species also reflects apparent genetic admixtures (Mousseau and Howard 1998; Roff *et al.* 1999). In this case, some populations that are 'hybrid' on the basis of genetic markers also possess 'intermediate' calling songs (e.g. Mousseau and Howard 1998). Further support for the hybridization hypothesis came from the observation that the proportion of admixture in the hybrid genotypes was strongly correlated with the calling-song phenotype. Thus, hybrid

individuals (not just populations) possessed intermediate songs (Mousseau and Howard 1998).

Each of the above studies reflects direct (in the case of survivorship of the parental and hybrid classes) or indirect (in the case of population genetic samples) assays across generations that can be used to test for genetic exchange between *A. fasciatus* and *A. socius*. Each data set supports the hypothesis of introgressive hybridization (albeit argued to be limited in extent, e.g. see Howard 1986). Yet the findings from Howard and his colleagues are applicable to not only the detection of genetic exchange, but also to answering many questions concerning the processes and effects from such exchange. For example, many of the studies and reviews from this group have focused on tests for the presence of reproductive character displacement caused by reinforcement (e.g. Howard 1986, 1993; Benedix and Howard 1991; Howard and Waring 1991; Howard and Gregory 1993; Howard *et al.* 1993; Cain *et al.* 1999; Britch *et al.* 2001).

In regard to the role of reinforcement in the evolution of reproductive isolation between *A. fasciatus* and *A. socius*, Howard's group has found evidence from various studies that is consistent with its impact on this species complex. First, patterns of genetic variation and levels of reproductive isolation have been found to correlate well with the expectation that a wider area of overlap between these two species is accompanied by a greater level of reproductive isolation (Howard 1986). This is expected if the width of the overlap reflects a greater opportunity for hybridization, leading to stronger selection for premating isolating barriers (Howard and Waring 1991). Second, the detection in *A. fasciatus* males of a calling-song 'displacement'—a change, in sympatry, in song characteristics found in allopatric populations—suggested character displacement for this trait thought to be important for mate choice by females (Benedix and Howard 1991). Such a displacement was thought to reflect selection for *A. fasciatus* males that had maximum calling-song differences from *A. socius* in areas of sympatry, allowing females to choose the correct (i.e. conspecific) mate and thus increase their fitness (Benedix and Howard 1991). Third, Howard (1982, 1986) and his colleagues (e.g. Harrison 1986) had defined the concept of *mosaic hybrid zones*—zones that resemble a patchwork quilt with alternating populations containing one or other of the parental species and others containing parentals and hybrids (Figure 3.7). In mosaic zones, relative to other types of hybrid zone, reinforcing selection is expected to occur under a greater array of conditions (Cain *et al.* 1999).

Several data sets have thus suggested a role for reinforcement in the evolution of the two *Allonemobius* species. However, other findings have indicated that this process may have been of limited effect. These findings include the following: (i) though, in earlier studies, correlations between zone width and degree of reproductive isolation seemed consistent with reinforcement, later analyses did not support this conclusion (Howard and Waring 1991); (ii) male calling-song displacement was apparently not due to differential female preference because females from neither species showed a preference for the conspecific male songs (Doherty and Howard 1996); and (iii) the fact that populations of the two species, along with mixed populations within the overlap zone, are found in a mosaic has not seemed to result in a strengthening of reproductive isolation, at least during a decade and half of observations (Britch *et al.* 2001).

Notwithstanding the remaining questions concerning the overall importance of reinforcement in the evolutionary history of *A. fasciatus* and *A. socius*, the research program built up around years of study indicates the power of multi-year/multi-generational analyses of genetic exchange. For example, by studying hybrid zones across generations and between different geographical locales, Howard constructed the basis for what is known as the mosaic hybrid zone concept (Howard 1982, 1986; Harrison 1986, 1990). Inherent in this concept of hybrid zones is the hypothesis that environment-dependent selection affects greatly the genetic structure and habitat associations of various populations. This model has become the touchstone for analyses of genetic exchange via introgressive hybridization (e.g. see Arnold 1997 for a review) and has resulted in an appreciation of the key role of environmentally mediated selection.

Of equal importance to reframing our concept of hybrid zones was Howard and colleagues' discovery that gamete competition formed a strong barrier to genetic exchange between the *Allonemobius*

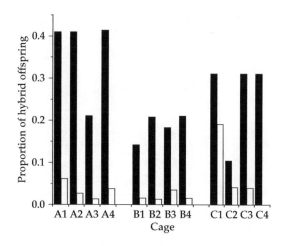

Figure 3.8   Predicted (filled bars) and observed (open bars) proportions of F₁ hybrid offspring between *A. fasciatus* and *A. socius*. The A cages contained equal numbers of the two species. The B cages contained 80% *A. socius* individuals. The C cages contained 80% *A. fasciatus* individuals (from Howard *et al.* 1998).

taxa (Howard and Gregory 1993; Howard *et al.* 1998, 2002). This conclusion would not have been possible without the multi-generational data that demonstrated limits to genetic exchange. Indeed, their work reflects a general model for deciphering processes associated with post-fertilization barriers for introgression. Specifically, their studies demonstrated that the pre-zygotic process of conspecific sperm precedence (i.e. gamete competition) resulted in few hybrid offspring formed when females mated with both conspecific and heterospecific males (Figure 3.8; Howard and Gregory 1993; Howard *et al.* 1998, 2002). I would argue that their findings, and the findings from a diverse array of other taxa (e.g. Carney *et al.* 1994; Chang 2004; Geyer and Palumbi 2005), have yielded a renaissance in our understanding of the importance of prezygotic barriers in affecting the patterns of genetic exchange.

Recently, the long-term studies of the *Allonemobius* hybrid zones have likewise resulted in the detection of hybrid-zone evolution consistent with a causal role for global environmental changes. Britch *et al.* (2001), using the cumulative 14 years-worth of data for the east coast transect and the mountain transect described above detected differential hybrid-zone responses (Figure 3.9). In particular, the mountain transect showed an

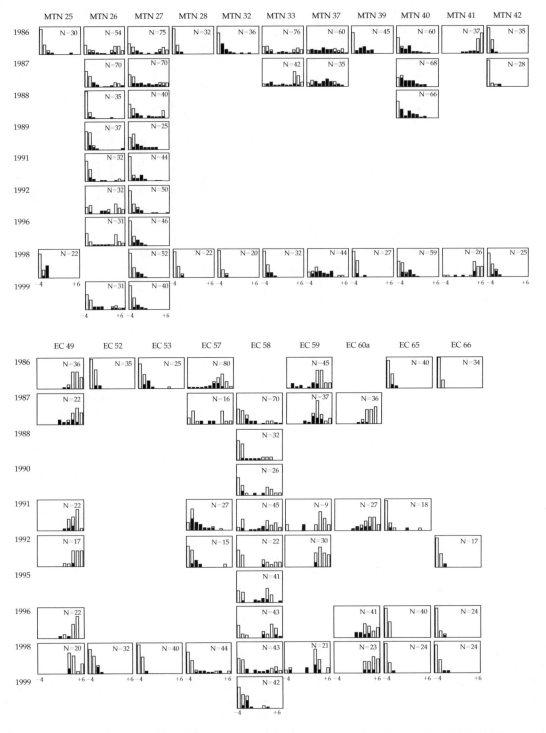

Figure 3.9 Hybrid index profiles for mountain (MTN) and east coast (EC) populations over 14 years. The *x*-axis indicates hybrid index score, with *A. socius* individuals reflected in the negative scores at the left of each histogram, *A. fasciatus* reflected in the positive scores at the right, and the proportion of hybrid individuals indicated by the filled portion of the bars (from Britch *et al.* 2001).

increase in the frequency of *A. socius* genotypes in many of the populations. Britch *et al.* (2001) inferred that 'These increases suggest a northward movement of this southern cricket and may reflect response to warming of the climate in the Appalachian Mountains.' In contrast, *A. socius* did not increase in frequency along the east coast transect (Figure 3.9). In answer to the question of why there were not similar increases in frequencies along both transects, Britch *et al.* (2001) suggested that '. . . the Atlantic Ocean buffers temperature and moisture fluctuations brought about by climate warming . . .' The transects reflect different patterns of hybrid-zone evolution, yet both are determined environmentally.

It is important to note that although I have focused on two insect systems to highlight the power of trans-generational studies in tests for the occurrence and evolutionary consequences of genetic exchange, as stated in the introduction to this section I could have instead chosen influenza A, Darwin's finches or Louisiana irises (see Chapter 1)—or a number of other examples from widely divergent species complexes (Arnold 1997). Thus, wherever multi-generational observations have been implemented, important findings in regard to genetic exchange have been made. In a similar way, I will use a small set of species complexes in the next section to illustrate how discordance in phylogenies constructed from different data sets can be used to test for the evolutionary effects of genetic exchange.

3.3.2 Phylogenetic discordance

Various processes can cause non-concordance (i.e. different branching patterns and taxa associations) between phylogenies derived from different character sets. However, the retention of ancestral polymorphism and reticulate events (caused by introgressive hybridization and lateral gene transfers) are viewed as being two of the main causal factors. In the following examples—including putative cases of both horizontal transfer and introgressive hybridization within plants, animals, and an apicomplexan parasite—different approaches were used to test whether retention of ancestral polymorphisms or genetic exchange best accounted for the resolution of discordant phylogenetic

hypotheses. For some species complexes, genetic exchange involved different domains of life, while in other instances the exchange involved congeners. In all cases, the footprints of genetic exchange were detected with some form of comparative approach.

Senecio
The plant genus *Senecio* has proven to be another excellent system for illustrating the descriptive and predictive value of the web-of-life metaphor. In particular, the species contained within this genus form the basis for understanding the processes of hybrid speciation, the semi-permeable nature of species boundaries, barriers to reproduction, adaptive trait introgression and the relative effects of incomplete lineage sorting and genetic exchange on phylogenetic hypotheses (Abbott 1992; Abbott *et al.* 1992, 2003; Comes and Abbott 1999, 2001; Lowe and Abbott 2000, 2004; Abbott and Forbes 2002; Abbott and Lowe 2004; Chapman *et al.* 2005). This system is well known for the discovery of two newly formed (i.e. approximately 100–200 YBP) polyploid, hybrid taxa: (i) the tetraploid *Senecio eboracensis* formed from introgressive hybridization between the British Isles-native, tetraploid *Senecio vulgaris* and the introduced diploid species *Senecio squalidus* (Abbott *et al.* 1992; Lowe and Abbott 2004; James and Abbott 2005) and (ii) the hexaploid *Senecio cambrensis* also formed from hybridization between *S. vulgaris* and *S. squalidus* (Ashton and Abbott 1992). Work by Abbott and his colleagues has confirmed both the mode of formation of these species as well as the degree of reproductive isolation between parents and hybrid derivatives and the processes that contribute to this reproductive isolation (Abbott 1992; Lowe and Abbott 2000, 2004; Abbott and Lowe 2004). In addition, these workers have demonstrated that the introgression of the loci causing ray florets in the hybrid species *S. eboracensis* is likely adaptive, although there are fitness tradeoffs due to the cost of outcrossing (rather than the high frequency of self-fertilization seen in the parental species) by the hybrid genotypes (Abbott *et al.* 2003).

In the context of the current topic—using phylogenetic discordance to discern the evolutionary effects of genetic exchange compared with other processes such as incomplete lineage sorting—the

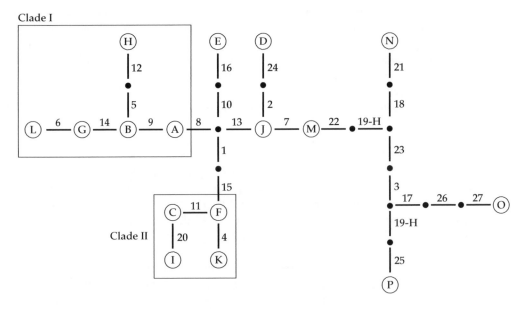

Figure 3.10 Gene tree (maximum parsimony) derived from cpDNA haplotypes found in *Senecio* populations. Each branch reflects a single mutational step, with numbers along the branches merely indicating the specific polymorphism involved. The black dots indicate unobserved, intermediate haplotypes. Clades I and II have associated bootstrap values of 64 and 89%, respectively (from Comes and Abbott 2001).

work of Abbott and colleagues on the genus *Senecio* is also illustrative. In particular, inferences concerning species relationships based on nuclear (internal transcribed rDNA [ITS rDNA] sequences and RAPD loci), cpDNA, and morphological characters were considered in a geographic context (Figures 3.10 and 3.11; Comes and Abbott 2001). Through this analysis it was possible to test for the role of genetic exchange and retention of ancestral polymorphisms among species of Mediterranean *Senecio* (section *Senecio*). However, these authors reflected the limitations of such studies when they stated, 'Obviously, these alternative processes [i.e. reticulations and incomplete lineage sorting] are not necessarily mutually exclusive and are notoriously difficult to disentangle.'

The illustrative nature of Comes and Abbott's (2001) study is thus not solely in what they were able to conclude with some measure of confidence, but also in what they were not able to discern. Indeed, these authors made the common discovery that they were limited by the distribution of genetic variation, both within and between the taxa sampled. This limitation made it possible only to infer how incomplete lineage sorting or reticulation may

have affected the evolutionary diversification for a portion of the species studied (Comes and Abbott 2001). However, these authors did conclude that incomplete lineage sorting and reticulations were both likely causal factors in the evolution of section *Senecio*. For example, they concluded that the phylogenetically discordant (relative to relationships derived from morphology) cpDNA variation found in populations of the widespread diploid species, *Senecio gallicus*, was best accounted for by the retention of ancestral polymorphisms, rather than introgressive hybridization. The findings supporting this conclusion were that (i) although *S. gallicus* is a distinct species based upon breeding experiments and phenotype, it shares the three cpDNA haplotypes with two other species (Figure 3.11), (ii) unlike most other species within its cpDNA subclade, *S. gallicus* is genetically unified, as reflected by allozyme variation, and (iii) there is no evidence for present-day sympatry or hybridization with the other diploid species or the tetraploid *S. vulgaris* that also harbor the cpDNA variation found in *S. gallicus* (Figure 3.11; Comes and Abbott 2001). In regard to *S. gallicus*, Comes and Abbott (2001) considered the caveat that

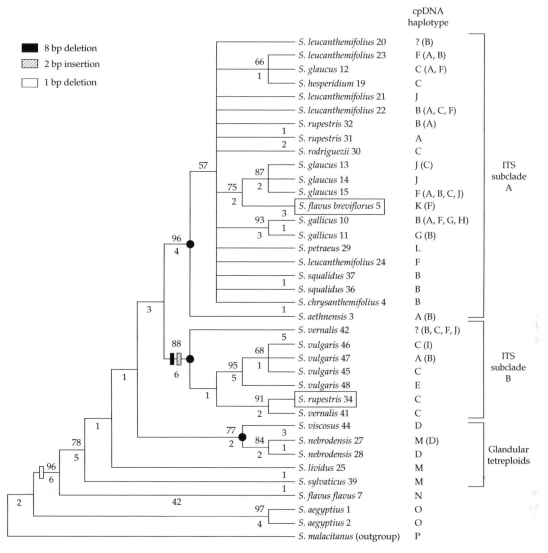

Figure 3.11 Parsimony tree derived for 37 accessions and 18 species of *Senecio* based on nuclear ITS rDNA sequences. Three informative insertion/deletion events are listed in the key and their placement is indicated on the phylogeny. Numbers located above and below the branches indicate the bootstrap values and the number of nucleotide changes, respectively. Boxed species indicate two examples of discordance between ITS rDNA, cpDNA, and species taxonomy (from Comes and Abbott 2001).

ancient reticulation, resulting from Pleistocene climatic perturbations, brought this species into contact with the other diploid species located across the Strait of Gibraltar (Comes and Abbott 2001). Notwithstanding this cautionary note, these workers suggested a stronger inference that incomplete lineage sorting was causal in the pattern of cpDNA variation.

Comes and Abbott (2001) did not only infer effects from incomplete lineage sorting in producing the phylogenetic discordance among *Senecio* species. They also concluded that both past and contemporary hybridization was reflected in two species pairs (Figures 3.10 and 3.11). The evidence for past reticulate evolution was resolved for the species *Senecio flavus* and *Senecio glaucus*. In this case, populations of *S. flavus* from Israel have recently been found to not be diploid, but rather tetraploid. Though morphologically highly similar to *S. flavus* subsp. *flavus*, this taxon possesses ITS rDNA and cpDNA sequences that are '*glaucus*-like' (Comes and Abbott 2001). This suggests that either *S. glaucus* was the maternal (cpDNA) parent in the cross forming the new (allo-)polyploid species, or that introgressive hybridization occurred between the polyploid form and *S. glaucus* subsequent to the polyploidy event (Comes and Abbott 2001). In either case, genetic exchange was apparently involved in the evolution of these species. The second case for which reticulate evolution has been inferred involved what Comes and Abbott (2001) referred to as 'contemporary hybridization and introgression'. Specifically, they found that *Senecio rupestris* and *Senecio vernalis*, on the basis of allozyme variation, morphology, and life-history differences were (i) highly divergent from one another and (ii) sister taxa of species belonging to separate cpDNA clades. Surprisingly, Comes and Abbott (2001) discovered cpDNA and ITS rDNA variation in one sample of *S. rupestris* that was indistinguishable from some *S. vernalis* populations (Figures 3.10 and 3.11). Considering all of the data, Comes and Abbott (2001) concluded that, like the *flavus/glaucus* example, one species (in this case *S. vernalis*) had been the donor for both the ITS rDNA and cpDNA sequences.

In general the *Senecio* species complex reflects numerous outcomes associated with the web of life. Comes and Abbott (2001) demonstrated, however, that reticulate events are likely accompanied by other processes such as incomplete lineage sorting. The application of tests for phylogenetic discordance thus allow the detection of patterns that are consistent with one or other of these evolutionary phenomena. In the next section I will consider studies of phylogenetic discordance that have also allowed the detection of genetic exchange in a parasitic eukaryote, in this case exchange involving the process of lateral gene transfer.

Cryptosporidium

The eukaryotic phylum Apicomplexa contains several human parasitic pathogens including the genera *Plasmodium, Toxoplasma,* and *Cryptosporidium.*

Unlike other members of this phylum, *Cryptosporidium parvum* lacks the relictual plastid (and its genome) known as the apicoplast (Zhu *et al.* 2000b; Abrahamsen *et al.* 2004). Furthermore, this organism contains a degenerate mitochondrion lacking a genome (Abrahamsen *et al.* 2004). Although *C. parvum* is usually only fatal to immunocompromised individuals, the lack of an effective treatment and its resistance to normal chlorine disinfection in water have resulted in the concern that this organism could be used as a water-borne bioterrorism agent (see the US Centers for Disease Control Bioterrorism Agents/Diseases list; www.bt.cdc.gov/agent/agentlist.asp).

Of particular interest for the topic under consideration is the observation that depending on which gene sequences are examined, different phylogenetic relationships between *C. parvum* and other apicomplexan species are resolved (Carreno *et al.* 1999; Zhu *et al.* 2000a; Leander *et al.* 2003). First, the use of various combinations of sequences derived from small-subunit rRNA, large-subunit rRNA, protein-encoding and β-tubulin genes have indicated that *C. parvum* is (i) a sister species to the gregarine (e.g. parasites of annelids and insects) complex (Carreno *et al.* 1999; Leander *et al.* 2003) or (ii) basal to the entire apicomplexan clade (Zhu *et al.* 2000a). Most importantly, however, is the finding that numerous genes in the *C. parvum* genome (Abrahamsen *et al.* 2004) place this species alternately with cyanobacteria, eubacteria, or plant/algae species (Striepen *et al.* 2002; Huang *et al.* 2004). For example, the gene encoding leucine aminopeptidase associates *C. parvum* with cyanobacteria, the glucose-6-phosphate isomerase gene sequences place it with algal/plant taxa, and its tryptophan synthetase β subunit gene indicates its relationship with eubacteria (Figure 3.12; Huang *et al.* 2004). Indeed, based upon phylogenetic discordance, approx. 30 genes were identified as having entered the *C. parvum* genome through horizontal gene transfer (Huang *et al.* 2004). The donors of gene sequences for this eukaryotic parasite's genome thus include '. . . α-, β-, and ε-proteobacteria, cyanobacteria, algae/plants and possibly Archaea' (Huang *et al.* 2004). Furthermore, many of these transfers have likely led to adaptive evolution, either due directly to the original

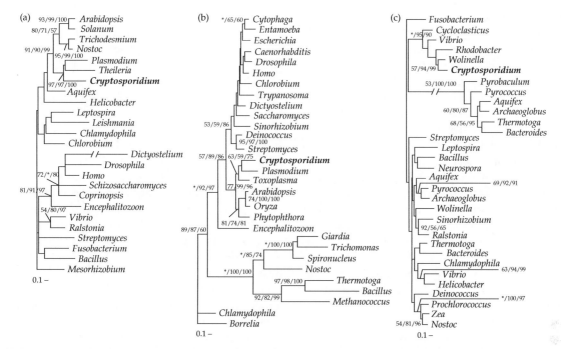

Figure 3.12 Phylogenetic placement of *C. parvum* based on (a) leucine aminopeptidase, (b) glucose-6-phosphate isomerase, and (c) tryptophan synthetase β subunit. These genes indicate horizontal transfers from (a) cyanobacterial, (b) algal, and (c) eubacterial sources. Numerical designations above the branches indicate the puzzle frequency (using TREE-PUZZLE) and bootstrap values for maximum parsimony and neighbor-joining methods, respectively. Asterisks are placed on branches for which bootstrap values are less than 50% (from Huang *et al.* 2004).

transfer, or through gene duplications and change of gene function subsequent to the transfer event (e.g. Striepen *et al.* 2002, 2004; Madern *et al.* 2004).

The resolution of genetic exchange afforded by comparisons of completely sequenced genomes using phylogenetic discordance as the litmus test is illustrated well by the studies of *C. parvum* (Figure 3.12). However, as indicated by the *Senecio* results, phylogenetic discordance as a methodology for detecting such exchanges is not limited to those species for which complete genome sequences are available. This is also demonstrated clearly by an example involving the African electric fish belonging to the superfamily Mormyridae.

Mormyridae

The African fish families Mormyridae and Gymnarchidae (superfamily Mormyroidea) encompass the largest taxonomic assemblage of freshwater electric fish species (Taverne 1972, as cited by Lavoué *et al.* 2003). The mormyrid species are noted for having species-specific electrical discharges

(electric organ discharges, or EODs) that appear to aid in the recognition of conspecifics (Hopkins and Bass 1981). Differentiation in EOD type has been found between morphologically cryptic and co-occurring forms suggesting a role for sympatric divergence in this group of fish (Arnegard *et al.* 2005). The enormous diversity of the mormyrid group is reflected in its being the most abundant fish species complex in some African riverine systems (Lavoué *et al.* 2000). Indeed, the mormyrids are an example of explosive radiation resulting in species flocks with hundreds of member species (Sullivan *et al.* 2002). These species flocks appear analogous to those of the African rift lake cichlid complexes in that there are numerous morphologically defined species that show low levels of interspecific genetic divergence (e.g. compare Hey *et al.* 2004 with Sullivan *et al.* 2004).

The similarity of at least some of the genera and species of mormyrids is reflected in phylogenetic discordance between (i) accepted taxonomic placement (based upon morphological characteristics) and phylogenies derived from molecular markers and (ii) phylogenies built from different molecular markers (e.g. Lavoué *et al.* 2000, 2003; Sullivan *et al.* 2000). However, the explanations for these discordances, as with those for any species clade, are likely complex. For example, a study of mtDNA cytochrome *b* sequences for 27 species and 15 genera of mormyrids (Lavoué *et al.* 2000) detected polyphyletic relationships for members of the genera *Marcusenius, Pollimyrus,* and *Brienomyrus.* Lavoué *et al.* (2000) concluded that taxonomic inaccuracies were likely the cause of the polyphyly defined for the latter two genera. However, they argued for the impact of introgression between sympatric *Marcusenius, Brienomyrus,* and *Paramormyrops* species to explain the similarity of *Marcusenius* mtDNA sequences with members of *Brienomyrus* and *Paramormyrops* (Lavoué *et al.* 2000). Yet this conclusion was not supported by a subsequent analysis of sequences derived from the nuclear RAG2 locus (Sullivan *et al.* 2000). This latter study also placed the same species of *Marcusenius* with *Brienomyrus* and *Paramormyrops,* suggesting that the taxonomic (i.e. morphological) placement was the incorrect inference (Sullivan *et al.* 2000).

Although taxonomic treatments may be a source of experimental error in tests for phylogenetic discordance among mymorid clades, incomplete lineage sorting and introgression have also apparently contributed to present-day population-genetic variation and thus to many of the discordant hypotheses. For example, both processes were implicated in the initial discovery and phylogenetic description of the 'Gabon-clade *Brienomyrus' species flock* (Sullivan *et al.* 2002). In this analysis, 38 operational taxonomic units (OTUs) were defined on the basis of morphology and EOD characteristics were examined phylogenetically using mtDNA cytochrome *b* sequence data. Although the monophyly of the Gabon-clade *Brienomyrus* species flock was supported by these data, monophyly of samples representing many of the OTUs was not detected (Figure 3.13; Sullivan *et al.* 2002). From these results, Sullivan *et al.* (2002) first observed that 'Incomplete mitochondrial lineage sorting is suggested by the presence of divergent haplotypes within single populations . . .' Indeed, these authors argued that species relationships within some clades of the Gabon-clade phylogeny (e.g. clade G, Figure 3.13) were hopelessly confused due to the effect of incomplete lineage sorting (Sullivan *et al.* 2002).

In addition to incomplete lineage sorting, Sullivan *et al.* (2002) recognized as well the likely role played by introgression among many of the OTUs. Thus they found several instances in which the haplotypes of different, but sympatric, OTUs were more similar (or indeed identical) than they were to allopatric samples of their own OTU (e.g. clade H, TEN; clade I, BP6/BN2 and SN7/SN2; clade J, BP1/SN3; Figure 3.13). In contrast to their conclusions concerning the large impact of incomplete lineage sorting on clade G (see above), Sullivan *et al.* (2002) stated, 'Strangely, throughout much of clade G, geographical proximity seems to be a more consistent predictor of clade membership than is OTU identity' (Figure 3.13). This conundrum was addressed in a subsequent study using nuclear amplified fragment length polymorphism (AFLP) markers (Sullivan *et al.* 2004). Unlike the mtDNA haplotype data, the phylogenetic trees reconstructed using the AFLP markers generally supported the OTUs defined by EOD and morphology. Sullivan *et al.* (2004) concluded from this that the AFLP phylogeny reflected more accurately—than the cytochrome *b* phylogeny—OTU/species relationships. Given this result, and the observation of shared mtDNA haplotypes between morphologically divergent, sympatric OTUs (Figure 3.13), Sullivan *et al.* (2004) concluded that mtDNA introgression had been a persistent process in the evolution of this species flock, but that this introgression had not caused the loss of species-specific morphological and EOD characteristics or the phylogenetic signal provided by some nuclear loci. Thus, as with any example of horizontal transfer or introgressive hybridization, these processes affected differentially, separate genomes and portions of the same genome of the participating species. Furthermore, certain species-specific characteristics remain detectable; otherwise the organisms receiving the donated genomic material would not be recognized as divergent from the donor species. In the next section I consider results suggestive of frequent horizontal transfer involving plant species. Specifically, there is now evidence of bacterium-to-plant nuclear genome exchange as well as extensive mitochondrial gene transfer between widely divergent angiosperm species. In

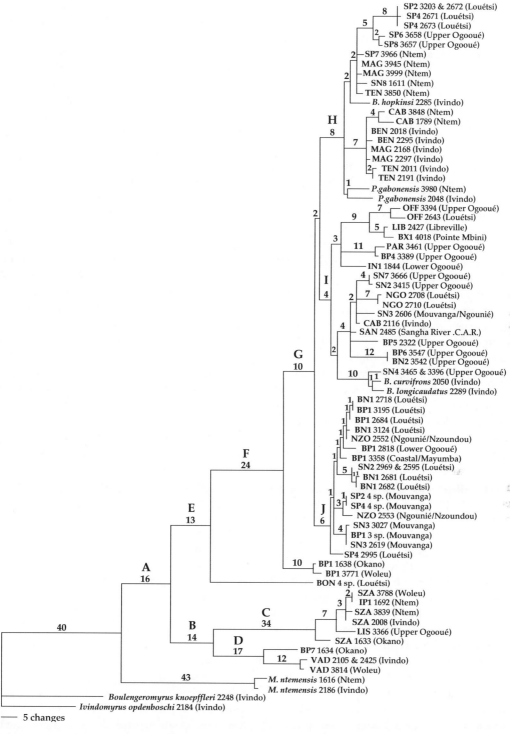

Figure 3.13   A strict consensus tree constructed using the complete cytochrome *b* gene sequences for Gabon-clade *Brienomyrus* electric fish and outgroup species (from Sullivan *et al.* 2002).

each of these cases, like those already discussed, genetic exchange, though widespread and significant, has apparently affected only a portion of the genomes of the participating taxa.

Horizontal transfer and plants

Frequent genetic exchange among plant species is almost as widely accepted as for prokaryotes (Anderson 1949; Stebbins 1959; Grant 1981; Arnold 1997; Rieseberg 1997). However, the form of genetic exchange assumed as the major mode for gene transfer has been introgressive hybridization and not horizontal gene transfer. Thus although the *intracellular* transfer of organellar genes into plant (and, more generally, eukaryotic) chromosomes has been described as a 'deluge' (Timmis *et al.* 2004), most avenues for *interspecific* lateral gene transfer are still considered closed in the case of higher plants. For example, the frequency of horizontal gene transfer involving the chloroplast genome appears to be extremely low (e.g. Goremykin *et al.* 2003; Bergthorsson *et al.* 2004; Martin 2005). Furthermore, there have been only two examples that I have found involving horizontal transfer of individual genes into plant nuclear genomes, both involving the movement of bacterial genes into the plant genomes (Intrieri and Buiatti 2001; Zardoya *et al.* 2002).

In contrast to the above, studies of some mobile elements as well as the mitochondrial genomes of some plant species (e.g. Won and Renner 2003) have detected horizontal transfers, in some cases revealing remarkable frequencies of exchange and incredible diversity in the donor species. In regard to transposable element transfer, Feschotte and Wessler (2002) detected phylogenetic discordance suggestive of lateral transfer of *mariner*-like class 2 transposable elements. For example, *mariner*-like sequences from the dicotyledon *Arabidopsis* were placed phylogenetically with sequences from the monocotyledon Louisiana iris species, rather than with other dicotyledon species. Though consistent with a horizontal gene transfer model, these authors cautioned that it would be necessary to undertake a much broader taxonomic sampling for a rigorous test of whether these elements had indeed been involved in lateral transfer (Feschotte and Wessler 2002). In contrast, with regard to the

mobile group I introns, Cho *et al.* (1998) concluded that these elements had been transferred more than 1000 times from a non-plant donor into the mtDNA *cox1* genes of angiosperms. These authors were able to make such an assertion due to their sampling of 278 genera (281 species) of angiosperms. The estimate of over 1000 separate lateral transfers of the group I introns was an extrapolation from Cho *et al.*'s (1998) detection of 32 separate transfers (approx. 11%) among the taxa sampled. In contrast to Feschotte and Wessler (2002), these authors inferred a widespread, and important, role for lateral transfer of mobile elements among angiosperms.

As stated at the start of this section, it is possible that, unlike either the nuclear or chloroplast genomes, lateral transfer involving plant mtDNA—whether or not it involves mobile elements—is frequent. Indeed, numerous instances of the horizontal transfer of mtDNA sequences among plant species have now been recorded. One type of transfer (analogous to that found in some animal groups, e.g. see Houck *et al.* 1991 and Kidwell 1993) detected by tests for phylogenetic discordance has involved gene transfer between host and parasitic plant taxa. Davis and Wurdack (2004) and Mower *et al.* (2004) described three separate instances of this type of transfer. In the first study, the transfer involved the apparent acquisition of the host-plant (*Tetrastigma*) mtDNA *nad1B-C* sequences by the parasitic plants *Rafflesia* and *Sapria* (Figure 3.14; Davis and Wurdack 2004). Mower *et al.* (2004) also detected plant–plant transfers involving parasitic and non-parasitic species. Both cases involved hosts from the genus *Plantago* and were discovered through analyses of the mtDNA gene *atp1*. The two examples involved different parasitic lineages and, unlike *Rafflesia* and *Sapria*, the exchange was from the parasite to the host (Mower *et al.* 2004).

Several additional examples of phylogenetic discordance involving mtDNA have been recorded that indicate horizontal transfer (i) from angiosperms into *Gnetum*, (ii) from parasitic angiosperms into ferns, (iii) from monocotyledons into eudicotyledons, (iv) from eudicotyledons into the earliest angiosperm lineage (represented by *Amborella*), and (v) between eudicotyledon species (Bergthorsson *et al.* 2003; Won and Renner 2003;

(a) *matR, PHYC, 18S*   (b) *nad1B-C*

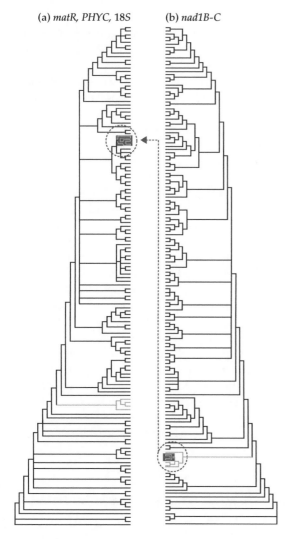

Figure 3.14 Phylogenetic trees demonstrating conflicting placement of the parasitic plant group Rafflesiaceae. In (a) Rafflesiaceae (indicated by red branches) is placed in Malpighiales (based upon combined mitochondrial *matR* and nuclear *PHYC* and ribosomal genes), but in (b) is placed near its host genus *Tetrastigma* in the basal eudicotyledon family Vitaceae (indicated by yellow branches). The dashed red line indicates the hypothesized mtDNA horizontal-transfer event from host to parasite that resulted in the alignment of Rafflesiaceae near *Tetrastigma* (from Davis and Wurdack 2004). See also Plate 2.

Davis *et al.* 2005). Phylogenetic discordances thus indicated a diverse array of taxonomic groups impacted by lateral gene exchange and led Bergthorsson *et al.* (2003) to conclude that horizontal gene transfer 'occurs at an appreciable frequency for

plant mitochondrial genes in general.' However, no other species involved in mtDNA horizontal transfer studied thus far approaches the frequency of exchange detected for the basal angiosperm species, *Amborella trichopoda* (Bergthorsson *et al.* 2004). Specifically, one or more copies of 20 of 31 known mitochondrial protein genes have been acquired from other land plant species. Seven of these gene transfers apparently involved *Amborella* acting as the recipient of genes donated by moss species (Bergthorsson *et al.* 2004). At least three of these gene transfers came from separate moss lineages. The remainder of the genes received by *Amborella* were inferred to be of angiosperm (mostly eudicotyledon) origin (Bergthorsson *et al.* 2004).

The four examples used in this section illustrate how tests for phylogenetic discordance can be used to detect genetic exchange—exchange through either introgressive hybridization or horizontal gene transfer. In each case, a comparative approach was taken by the investigators to detect genetic signatures specific for exchange rather than, for example, incomplete lineage sorting. Thus in Mediterranean *Senecio* and the Gabon-clade *Brienomyrus* species flock, geographical data were taken into consideration when choosing between the various potential causes of discordance. In addition, it was possible to mine extensively or completely sequenced genomes and thus compare numerous gene trees to detect the lateral gene transfers associated with *Cryptosporidium* and plant nuclear and mtDNA. I am certain that phylogenetic discordance will be utilized even more extensively as data from genome sequencing (partial and complete) becomes available for additional species. In the next section I review an example from *Drosophila* to highlight the utility of detailed genetic information, in combination with the application of (i) gene genealogical approaches and (ii) models that assume isolation, or isolation and migration during the divergence of lineages to test for genetic exchange (in this case via introgressive hybridization).

3.3.3 Gene genealogies and models of speciation

Dobzhansky and his colleagues developed *Drosophila pseudoobscura* and allied species as a premier model system for evolutionary studies (e.g. see Dobzhansky 1970). For example, the sibling species *D. pseudoobscura* and *Drosophila persimilis* are now considered an excellent example of the process of speciation ('biological speciation') via reinforcement (Noor 1995, 1999; Kelly and Noor 1996; Noor *et al.* 2000; Servedio and Noor 2003; Ortiz-Barrientos and Noor 2005). However, with regard to the topic of this book, it is interesting that this species complex and in particular *D. pseudoobscura* and *D. persimilis* have been viewed as examples of taxa that are reproductively well isolated (e.g. Mayr 1946) and not examples of taxa exhibiting evidence of significant genetic exchange since divergence from a common ancestor. Indeed, Dobzhansky (1973)—having identified only one F_1 individual from many thousands collected in areas of sympatry for the two species—gave a value of 1 in 10 000 for interspecific matings. Powell (1983), in a study of the same species, also detected extremely low frequencies of natural F_1 hybrid formation, but concluded that high levels of genetic exchange had occurred, as reflected by cytoplasmic (i.e. mtDNA) introgression in areas of sympatry. However, Powell (1991) later suggested that the sharing of the mtDNA haplotypes in sympatry might instead reflect incomplete lineage sorting rather than introgression. This latter conclusion enforced the idea of limited, or no, gene exchange between these species (Dobzhansky 1973) and, as an aside, illustrated the difficulty (already discussed in this chapter) in separating the effects of genetic exchange from retention of ancestral polymorphisms (Powell 1991).

Given the above conclusions it is surprising that the *D. pseudoobscura/D. persimilis* complex is now considered an excellent example of divergence in the face of gene flow. Specifically, Hey and his colleagues have documented repeatedly the signature of introgressive hybridization between these two species (Wang and Hey 1996; R.-L. Wang *et al.* 1997; Machado *et al.* 2002; Machado and Hey 2003; Hey and Nielsen 2004). The data collected came from numerous nuclear and mitochondrial genes and supported Key's (1968) and Harrison's (1986) concept of the *semi-permeable genome*. An initial study by Hey *et al.* (Wang and Hey 1996) examined the molecular population genetics of the period (*per*) locus—in *Drosophila* this locus is known to affect both circadian rhythm and male courtship song

(Konopka and Benzer 1971; Kyriacou and Hall 1980). In their analysis of *per* sequences from *D. pseudoobscura* and *D. persimilis* (as well as *Drosphila miranda* and *Drosphila pseudoobscura bogotana*), Wang and Hey (1996) detected a putative 'hybrid' sequence in *D. persimilis* (*persimilis*-40; Figure 3.15). The effect of this sequence on the gene genealogy that included all the *per* sequences was to make *D. persimilis* paraphyletic. Furthermore, consistent with an introgression, rather than an incomplete lineage-sorting hypothesis, the calculated time of the introgression event was well after the initial divergence (*c*.1 million YBP) of *D. pseudoobscura* and *D. persimilis* from a common ancestor, but was also estimated to be ancient (i.e. this introgression event occurred a few to several hundreds of thousands of years before the present; Wang and Hey 1996).

In a subsequent analysis of the molecular population genetics of *D. pseudoobscura* and *D. persimilis* (R.-L. Wang *et al.* 1997), the *per* locus data from Wang and Hey (1996) were augmented with sequence information from *Adh* (much of these data derived from studies by Schaeffer and Miller, e.g. see Schaeffer and Miller 1993) and the heat-shock protein gene *Hsp82*. In this second study, R.-L. Wang *et al.* (1997) used the methodology of Wakeley and Hey (1997) to fit general allopatric

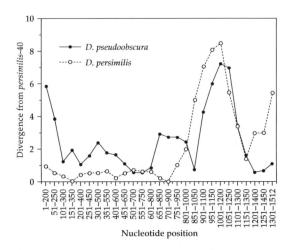

Figure 3.15 Plot of the average pairwise DNA sequence divergence found in comparisons of the putative 'hybrid' period locus gene (*persimilis*-40) with *D. pseudoobscura* and *D. persimilis* (excluding *persimilis*-40) period locus sequences (from Wang and Hey 1996).

models of speciation to the data from the three gene loci. The assumptions of this model are that (i) two populations (or species) diverged from an ancestral form at a single time point and (ii) there has been no gene flow since divergence (Wakeley and Hey 1997). The major finding from R.-L. Wang *et al*'s. (1997) study was that all of the isolation models of divergence (i.e. divergence in allopatry) were incompatible with at least some of the population genetic data. It thus appears that divergence of *D. pseudoobscura* and *D. persimilis* was accompanied by gene flow between the two lineages. However, the three loci differed with regard to the *amount* of genetic exchange inferred; the *Adh* gene trees indicated large amounts of recent gene exchange (Figure 3.16); the *per* locus indicated limited, ancient introgression; data for *Hsp82* were consistent with no genetic exchange since divergence of the two lineages (R.-L. Wang *et al*. 1997). Thus isolation divergence models are insufficient to explain the entirety of the molecular population-genetic data for *D. pseudoobscura* and *D. persimilis*. However, a model that can simultaneously assume

divergence and gene flow (the isolation with migration or IM model) fits well the multilocus data for these taxa (and for other species as well: e.g. cichlids, Hey *et al*. 2004; Won *et al*. 2005; chimpanzees, Won and Hey 2004). In particular, an extension of the IM model proposed by Nielsen and Wakeley (2001) is consistent with the hypothesis that the two *Drosophila* species have been involved in genetic exchange (Hey and Nielsen 2004).

It is important to note that Machado *et al*. (2002) and Machado and Hey (2003) also detected various levels of introgression between *D. pseudoobscura* and *D. persimilis* by applying estimates of linkage disequilibrium and a phylogenetic approach, respectively. In particular, these studies demonstrated that genes closely linked to genomic regions containing hybrid sterility loci were not exchanged, while those not associated with such loci demonstrated variation suggestive of introgression. This suggests one cause of the semi-permeable nature of the *D. pseudoobscura* and *D. persimilis* genomes. Thus, as with the other methodologies applied to test for genetic exchange, the work by Hey and colleagues—on a species complex considered to be a paradigm for descriptions of reproductive isolating barriers—indicates everything from no introgression to extensive introgression. One of the approaches used by Wang and Hey (1996) in their analysis of the *per* locus was, instead of searching for sequence similarity between divergent species, to look for islands of sequences that showed high intraspecific *divergence* (Figure 3.15). In the next section, I will review how this methodology has also been applied to prokaryotic systems to detect genetic exchange through lateral gene transfer.

3.3.4 Intragenomic divergence

As discussed throughout this book, the emphasis of tests for genetic exchange is normally placed on detecting similarities between divergent evolutionary lineages. For example, when complete genomes are sequenced they are searched for similarities with other genomes. What can be considered the other side of the methodological coin are searches that examine within-gene-/-genome sequences in an effort to find intraspecific divergences that are much higher than expected by chance. To illustrate

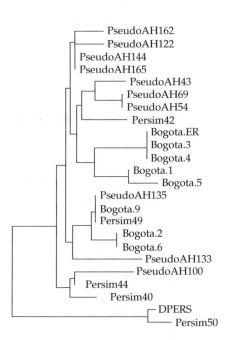

Figure 3.16 Gene genealogy for *D. pseudoobscura pseudoobscura*, *D. pseudoobscura bogotana*, and *D. persimilis* based upon approximately 200 bp of sequence from the *Adh* gene (from R.-L. Wang *et al*. 1997).

this approach, I will discuss four studies involving various pathogenic bacterial species. Each of these studies tested for islands of intraspecific sequence divergence that would indicate recombinational (i.e. genetic exchange) events between divergent organisms.

The first analysis involved pathogenic (enteropathogenic and enterohemorrhagic) *E. coli* strains and the gene *eae* that encodes the outer membrane protein intimin. McGraw *et al.* (1999) examined the evolutionary genetics of this putative virulence factor by sequencing the *eae* genes from three enteropathogenic *E. coli* and four enterohemorrhagic *E. coli* strains. In terms of genetic exchange, it is important to first note that this gene is part of the LEE pathogenicity island, a chromosomal region apparently transferred laterally between different bacterial strains (see McGraw *et al.* 1999 for a discussion). It is also significant that McGraw *et al.* (1999) argued that large regions within the LEE island demonstrated sequence signatures indicative of genetic transfer. The focus of the study by McGraw *et al.* (1999)—the *eae* genes—also demonstrated the effects of lateral transfer/recombination between *E. coli* strains and between these strains and '. . . sources outside of the *E. coli* population.' Thus genetic transfer, particularly within the 3' end of the various intimin genes leads to mosaic genes that show high interstrain divergence in the 3' portions of the *eae* gene, while retaining relatively low levels of divergence in other portions of the gene (McGraw *et al.* 1999).

The argument that intragenic mosaicism is likely caused by genetic exchange through recombination was also made by Feil *et al.* (2000) in a study of *Neisseria meningitidis* and *Streptococcus pneumonidae*. Indeed, these workers gave a clear description of the pattern and cause of mosaic gene structure: 'For example, most of a gene may be identical in sequence for two isolates of a species, whereas a 500-bp region in the middle may differ at 5% of nucleotide sites. Significant mosaic structure is indicative of recombinational exchanges . . .' (Feil *et al.* 2000). By examining multilocus sequence typing data sets, Feil *et al.* (2000) were able to estimate the number of alleles that have arisen through genetic exchange compared with point mutations. The mosaic nature of the two genomes was reflected

in the ratios of 10:1 and 4:1 for *S. pneumonidae* and *N. meningitidis*, respectively. As with the single-locus analysis of the *E. coli* strains, this latter study reflected the widespread effects of genetic exchange in producing numerous mosaic alleles in *S. pneumonidae* and *N. meningitidis* (Feil *et al.* 2000).

The last two studies I wish to discuss utilized a common approach to test for mosaicism in prokaryotic genes (Hughes *et al.* 2002; Hughes and Friedman 2004). In both cases the proportion of synonymous substitutions per site (p_S) and the proportion of nonsynonymous substitutions per site (p_N) were used as one means to test for mosaicism. In a comparison of two completely sequenced *Mycobacterium tuberculosis* genomes, Hughes *et al.* (2002) placed genes into two categories (high or low p_S) using a probabilistic (Bayesian) model. The fact that a small proportion of gene pairs from the two genomes showed high levels of divergence was seen as evidence for the possible role of several different processes, including horizontal gene transfer. Indeed, Hughes *et al.* (2002) concluded that a majority of the examples of highly divergent gene regions was due to genetic transfer (with the remainder explained by 'differential deletion' of genes). In a similar study, Hughes and Friedman (2004) compared sequence divergence in 5' intergenic regions with genes linked to these nongenic sequences. In the context of this section—testing for genetic exchange by examining patterns of intragenomic sequence divergence—two findings are illustrative. First, the species *Streptococcus agalactiae* possessed numerous genes and 5' spacers that were divergent from surrounding regions, indicative of their being received as a unit from a distantly related genome. In contrast, *Chlamydophila pneumoniae* also possessed numerous spacers and gene units that have been transferred from divergent genomes, but few of these spacer/gene combinations appeared to have been donated by the same distant genome (Hughes and Friedman 2004). These two divergent patterns, however, reflect the pervasive effects of genetic exchange on prokaryotic evolution and also indicate the utility of tests of intragenomic/intraspecific sequence divergence for detecting such exchange.

In the next section I return to the discussion of tests for introgressive hybridization, in this case through the use of gene genealogical approaches linked to geographic data, and termed nested clade analysis. As will be seen, such an approach provides a test for the affect of genetic exchange on the population genetic structure and evolutionary trajectories of clades of related organisms.

3.3.5   Nested clade analysis

Although the *nested clade analysis* methodology was formalized by Alan Templeton and colleagues in the mid-1990s (e.g. Templeton 1993; Templeton *et al.* 1995), the theoretical and informational underpinnings for this approach can be traced further back to (i) Templeton's earlier research on factors affecting population-genetic structure, methodologies for constructing phylogenetic inferences, tests for associations between phenotypic and genotypic variation, and considerations of species concepts (Templeton 1981, 1983, 1989; Templeton *et al.* 1987, 1992), (ii) the foundational work of Avise *et al.* (1979a, b) that formed the basis for the field of phylogeography, and (iii) so-called vicariance biogeography (e.g. Rosen 1978). In regard to the latter concept, Rosen (1978) concluded that 'Geographic coincidence of animal and plant distributions to form recognizable patterns suggests that the separate components of the patterns are historically connected with each other and geographic history.' He then went on to propose that a comparison of biological (i.e. species) phylogenies to phylogenies of geologic events for the region under investigation would test the effects of past geologic perturbations on evolutionary diversification (Rosen 1978). Similarly, nested clade analysis correlates genetic divergence, as reflected by phylogenetic branching, with geographic distribution to determine the roles of dispersal, extinction, vicariance, and reticulation in the evolution of species and species complexes (Templeton *et al.* 1995). The steps followed in nested clade analysis are: (i) conversion of a cladogram into a nested design with identical genotypes placed together in 0-step clades and genotypes differing by numbers of mutations into 1-step and 2-step clades (Templeton *et al.* 1987; Templeton 1993); (ii) the calculation of the geographical locations of clade types (i.e. 1-step etc.; Templeton *et al.* 1987; Templeton 1993); and (iii) the testing for associations between clades and

geographical locations using permutation analysis in a nested design (i.e. 'clade types within a nested category vs. geographical location'; Templeton *et al.* 1995) or via a more elaborate test performed between clades and geographical locations, which includes data on geographical distances or positions among the samples collected (Templeton *et al.* 1995).

Numerous analyses have utilized nested clade analysis to test for events and processes that have affected the geographic and evolutionary trajectory of species and species complexes. Some analyses have emphasized the utility of this methodology for disentangling past (and ongoing) events affecting the species under study (e.g. Templeton 1993, 1998, 2001, 2002, 2004a, b; Templeton *et al.* 1995; Hutchison and Templeton 1999; Gómez *et al.* 2000; Bernatchez 2001; Nielsen *et al.* 2001; Byrne *et al.* 2002; Pfenninger and Posada 2002). Other studies have discovered either restricted power in nested clade analysis for detecting evolutionary processes or the unreasonableness of some of the evolutionary interpretations generated by this methodology (e.g. Turner *et al.* 2000; Irwin 2002; Knowles and Maddison 2002; Paulo *et al.* 2002; Masta *et al.* 2003).

The attraction of nested clade analyses is that they allow inferences concerning a wide range of factors (both recent and ancient) that can affect the evolutionary trajectory of lineages. One of the more controversial groups that has been examined with this method is the clade that contains our own species. In particular, nested clade analysis has been applied to test for (i) the most likely geographic point of origin of *Homo sapiens*, (ii) patterns of dispersal (and gene exchange) away from this point of origin, and (iii) population fragmentation (Templeton 1993, 2002, 2005). From these studies a number of conclusions have been drawn (Figure 3.17). First, the lineage leading to *Homo* arose in Africa. Second, multiple migration events from the point of origin occurred over the evolutionary history of this lineage. Third, migration events also occurred from areas previously invaded by populations from African sources back into Africa. Fourth, migration events were not marked by a replacement of previous occupants of the regions invaded, but rather involved migration and introgression between the native populations and the migrants.

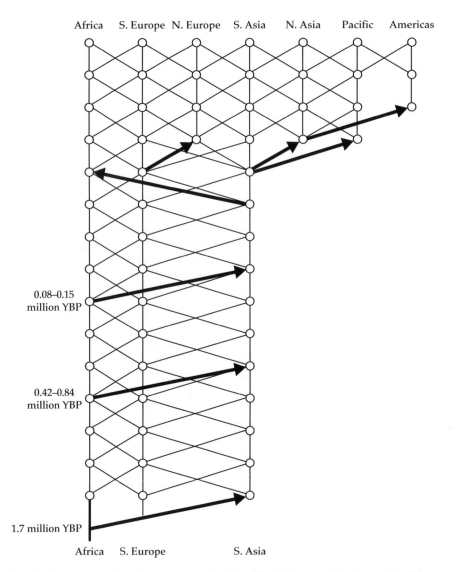

Figure 3.17   A model reflecting expansion/genetic exchange events involving different *Homo* populations/species. The thick arrows indicate major expansion events. Vertical and diagonal lines reflect genetic descent and genetic exchange, respectively. The original expansion from Africa 1.7 million YBP involved *Homo erectus*, as supported by fossil evidence (from Templeton 2002).

The implications of these conclusions are obviously of primary importance for an understanding of the evolution of our species. Thus if genetic exchange among various archaic populations and more modern variants occurred, the human population is the product of genetic exchange, just as is influenza A or Darwin's finches, etc. (see Chapter 9 for a detailed discussion of this topic). Also, we see that the *Homo* lineage was not a geographically static clade, but rather went through expansions and contractions, once again just like any other species. The implications of these findings also relate to the applicability of nested clade analyses. In this case, nested clade analysis detected patterns suggestive of processes that to some degree affect all species—migration, fragmentation, etc. (Figure 3.17). However, for the present discussion, we are particularly interested in the finding of substantial levels of genetic exchange among the various lineages, through time (Figure 3.17; Templeton 2002, 2005). Thus, as with all other methodologies, the use of nested clade analysis can provide a window for examining evolutionary processes including reticulate events between divergent lineages.

3.4   Summary and conclusions

The purpose of this chapter was to illustrate a number of methodologies for testing the hypothesis of genetic exchange. Examples were reviewed for which different workers had come to alternate conclusions concerning the impact of such exchange (i.e. species of oaks). However, even when there were questions concerning the role played by genetic exchange, a comparative approach gave the most insight into the evolutionary patterns and processes associated with introgressive hybridization and lateral gene exchange. Comparative approaches of one type or another can be seen in each of the methods explained in section 3.3. Whether trans-generational hybrid zone analyses of animal taxa, gene genealogical or phylogeographic-based analyses of plants, or tests for phylogenetic discordance involving the primitive eukaryote, *Cryptosporidium*, multiple data sets were utilized to derive tests for various evolutionary processes, including genetic exchange.

The studies reviewed above again illustrate (i) the broad phylogenetic distribution of lateral gene transfer and introgressive hybridization and (ii) the many ways in which strong inferences can be derived. Thus the web-of-life metaphor is consistent with the above examples. In addition, it is clear (and should be heartening for many evolutionary biologists) that it is not necessary to limit oneself to organisms with sequenced genomes to study the evolutionary processes and effects associated with genetic exchange. Neither are researchers limited by the need to choose organisms that are amenable to experimentation. This is obvious not only from studies of higher organisms that are difficult to keep in captivity, but is also true for microorganisms that are difficult to culture. With the widely varying methodologies that can be brought to bear to address the effect of genetic exchange on the evolutionary trajectory of organisms, there would seem to be a method of choice available for any organismic clade. If we wish to test rigorously the occurrence and effects of genetic exchange, we do not need to limit ourselves to those organisms currently exchanging genes (i.e. forming hybrid zones) or to microorganisms that can be cultured and sequenced rapidly. Any organismic group is amenable to one or more of the approaches outlined above.

CHAPTER 4

Barriers to gene flow

The coincident and sharply stepped clines at each of five enzyme loci, together with the strong linkage disequilibria between these loci, imply that there is a substantial barrier to gene flow ...

(Szymura and Barton 1986)

Strong conspecific sperm precedence would serve most effectively as a reproductive barrier when two species occur with equal abundance and females mate many times.

(Gregory and Howard 1994)

... we found that YBF30, the only fully sequenced example of HIV-1 group N ... is a recombinant of divergent viral lineages ... This mosaic genome structure of YBF30 implies previous co-infection and recombination of divergent SIVcpz strains in a *P. t. troglodytes* host.

(Gao *et al.* 1999)

Eretmodus is an algae scraper, whereas *Tanganicodus* is a specialized invertebrate picker ... the observed pattern is more likely to be a consequence of asymmetric introgression than of shared ancestral polymorphism.

(Rüber *et al.* 2001)

Empirical and theoretical research has revealed two important patterns in the evolution of reproductive isolation in animals: isolation typically arises as a result of disrupted epistatic interactions between multiple loci and these disruptions map disproportionately to the X chromosome.

(Payseur *et al.* 2004)

4.1 Barriers to exchange form a multi-stage process

The above quotations reflect the two major concepts that I wish to emphasize regarding barriers to genetic exchange: (i) reproductive isolation (or recombinational isolation in the case of viruses and bacteria) consists of not a single barrier but a series of processes each contributing only partially to whatever level of isolation is present, and (ii) genetic exchange can occur in spite of very stringent barriers. The idea that reproductive barriers are multistaged is not new. This was, for example, a major tenet of Dobzhansky's 1937 classic, *Genetics and the*

Origin of Species. Yet, one conclusion that can be drawn from the concept of reproductive isolation as a series of stages is that it is somewhat inaccurate to speak and write of 'speciation genes'.

It is an observation (rather than a hypothesis) that barriers to genetic exchange are often genetically controlled. It is also an observation that individual barriers are one of many processes, and yet organisms exchange genes in spite of these barriers (e.g. *Mimulus guttatus* and *Mimulus nasutus*; Kiang and Hamrick 1978, Diaz and Macnair 1999; Sweigart and Willis 2003; N.H. Martin and J.H. Willis, unpublished work). Because of this, it is likely to be more important to study the potentially creative aspects of genetic exchange, in concert with the genetic architecture of the phenotypes that are held together in spite of hybridization/lateral gene transfer/viral recombination, than merely how genetic exchange may be limited by certain barriers. However, reproductive isolation is still one factor that can affect evolutionary diversification, and it is most instructive that it be examined as a multi-stage process. In this regard, Szymura and Barton (1986) reflected on the genetic variation in hybrid zones between fire-bellied toads, variation that is likely affected greatly by the differential ecological settings favored by the two parental species, *Bombina bombina* and *Bombina variegata*. Likewise, the formation of recombinant HIV viruses is only possible when the ecological barrier of existing in two different host individuals is breached, resulting in the cocirculation of different viral lineages (Gao *et al.* 1999). Similarly, both behavior and ecological sorting contribute to the prezygotic barrier between the two Lake Tanganyika cichlid species from the genera *Eretmodus* and *Tanganicodus* (Rüber *et al.* 2001). Prezygotic (or in the case of viruses and prokaryotes, the stage of pre-recombination) isolation is also emphasized in the quote of Gregory and Howard (1994), but in this case it is post-insemination, gamete (i.e. sperm) precedence that is seen as the major barrier to crosses between the cricket species *A. fasciatus* and *A. socius*. Finally, the 'disrupted epistatic iteractions' mentioned by Payseur *et al.* (2004)—as tested for in a hybrid zone between *Mus musculus* and *Mus domesticus*—result in post-zygotic inviability.

In addition to considering barriers to genetic exchange as multi-staged, such barriers should also be recognized as varying episodically in intensity. For example, in an elegant analysis of reproductive isolation between the plant species, *Mimulus cardinalis* and *Mimulus lewisii*, Ramsey *et al.* (2003) stated, 'In aggregate, the studied reproductive barriers prevent, *on average* [my emphasis], 99.87% of gene flow, with most reproductive isolation occurring prior to hybrid formation.' The key, I believe, is the phrase 'on average'. Thus, as I emphasized in Chapter 2, rare events can be extremely important. If evolutionary biologists did not believe this, they would not point to the vanishingly infrequent mutations (or recombination events) that result in

increased fitness as the foundation for evolutionary change. Thus, average effects should be considered a null hypothesis, and not necessarily predictive of evolutionary pattern or process. It is likely that, as with many other similar plant examples, *M. cardinalis* and *M. lewisii* go through rare, episodic bouts of gene exchange through introgressive hybridization, regardless of what the average strength of barriers might be at any given time. The future detection of the genetic footprints from such gene-exchange events between these two species should thus be expected.

In the following sections I will illustrate why reproductive isolation can be defined as multi-staged. However, though I will consider many factors, this list is obviously not exhaustive regarding the types of processes that cause varying levels of reproductive isolation. Furthermore, some stages do not appear to have easily definable counterparts in all taxonomic groups. For example, the category of 'behavioral barriers' as I have defined it is absent from plant taxa (see below). I will first exemplify the multi-stage character of reproductive barriers by once again turning my attention to work on species of Louisiana irises. I will then emphasize the similar effects from the various processes on widely divergent organismic groups. In keeping with the thesis of this book, I consider organisms that not only exemplify the multi-stage nature of reproductive isolation, but also, in spite of these barriers, demonstrate genetic exchange.

4.2 Genetic exchange: Louisiana irises and reproductive barriers

4.2.1 Ecological setting as a barrier

Microhabitat associations are the first topic I wish to consider in the multi-stage process of reproductive isolation. The Louisiana irises demonstrate distributions among habitats that contribute to limitations to interspecific gene flow. For example, while *I. fulva* and *I. brevicaulis* occur in shaded, bayou, hardwood forest, and swamp habitats, *I. hexagona* is found in open, freshwater marshes (Viosca 1935; Bennett and Grace 1990; Cruzan and Arnold 1993; Johnston *et al.* 2001). Likely because of its occurrence near coastal environments, *I. hexagona* demonstrates a higher tolerance of salinity stress, relative to at least *I. fulva* (Arnold and Bennett 1993; Van Zandt and Mopper 2002, 2004; Van Zandt *et al.* 2003). Furthermore, *I. fulva* and *I. brevicaulis* sort into different habitats defined by their ability to grow in more or less soil moisture, respectively (Cruzan and Arnold 1993; Johnston *et al.* 2001; Martin *et al.* 2006). Although the habitat differences present limitations for genetic exchange it is also apparent that the mosaic nature of microhabitat distributions allows the different species to grow in close proximity and thus hybridize (Viosca 1935; Arnold *et al.* 1990a; Cruzan and Arnold 1993; Johnston *et al.* 2001).

Another component of the ecological setting of Louisiana irises involves the response of pollen vectors to different floral syndromes. Viosca (1935) suggested that hummingbirds preferred the pollination syndrome possessed by *I. fulva* (i.e. flowers of this species are solid red with protruding anthers and highly reflexed sepals, and lack nectar guides and a strong scent—characteristics normally associated with hummingbird pollination). Bumblebees were thought to be the major pollinators for *I. brevicaulis* and *I. hexagona*, as reflected by their floral characteristics (i.e. blue flowers marked with prominent white and yellow nectar guides, stiff upright sepals, and strong scent—characteristics normally associated with insect pollinators; Viosca 1935). In contrast, the nectar rewards (a major benefit gained by the pollinators) present in each of the three species do not possess characteristics that clearly place them into classical nectar-pollen vector categories (Burke *et al.* 2000b; Emms and Arnold 2000; Wesselingh and Arnold 2000). For example, although *I. fulva* possesses lower nectar sugar concentrations than *I. brevicaulis*, as expected for a hummingbird- versus a bumblebee-pollinated floral syndrome, the nectar concentration of *I. brevicaulis* (i.e. a bumblebee-syndrome-type floral pattern) falls well within concentrations found in numerous supposedly hummingbird-pollinated species (Burke *et al.* 2000b; Wesselingh and Arnold 2000). Furthermore, no significant differences were found in the nectar concentrations of *I. fulva* and the other bumblebee-type floral syndrome species, *I. hexagona* (Emms and Arnold 2000).

To determine the effect of different floral syndromes on pollinator preferences, our group

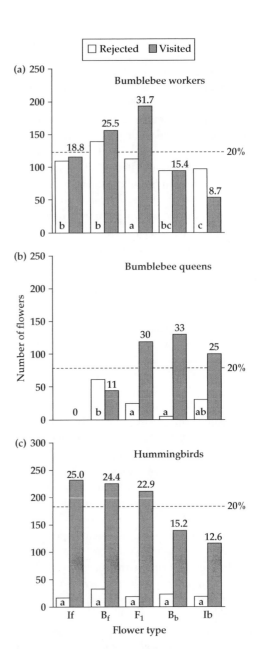

Figure 4.1 Distribution of approaches to each of the flower/genotypic classes for (a) bumblebee workers, (b) bumblebee queens, and (c) hummingbirds. The approaches are divided into flowers rejected (white bars) and flowers visited (shaded bars). The percentage of the total number of flowers visited is given above the shaded bars. The 20% line indicates the expected number of flowers visited when visitation rates would be equal for all flower types. At the bottom of the white bars, letters indicate significant differences in the fraction of flowers rejected between flower types. If, *I. fulva*; Ib, *I. brevicaulis*; B$_f$, first backcross generation toward *I. fulva*; B$_b$, first backcross generation toward *I. brevicaulis* (from Wesselingh and Arnold 2000).

designed two experiments involving (i) *I. fulva, I. hexagona*, and experimental F_1 hybrids (Emms and Arnold 2000) and (ii) *I. fulva, I. brevicaulis*, and experimental F_1 and BC_1 hybrids (Wesselingh and Arnold 2000). These studies allowed a test for pollinator preference as a pre-pollination barrier to genetic exchange. The findings from both studies indicated that different pollinator classes had preferences for certain floral syndromes, and that this preference would lead to some assortative mating. However, the pollinators, particularly bumblebees, were also a bridge for the initiation of interspecific hybridization (e.g. Figure 4.1; Wesselingh and Arnold 2000).

4.2.2 Gamete competition as a barrier

In concert with the pollinator behavior studies, our group also designed several analyses to test for the role of gamete competition in limiting the frequency of F_1 production. Only a single natural hybrid, out of hundreds of natural hybrid individuals analyzed thus far, has been found to possess a genotype consistent with that predicted for an F_1 individual (Johnston *et al.* 2001). Our research has identified the process of gamete competition as a major factor contributing to such low frequencies of F_1 formation. This in turn acts as a barrier to introgression through backcrossing.

Darwin (1859, p. 98), citing results from the German physician turned botanist Karl Friedrich Gärtner, reflected the outcome of gamete competition when he stated, '. . . if you bring on the same brush a plant's own pollen and pollen from another species, the former will have such a prepotent effect that it will invariably and completely destroy . . . any influence from the foreign pollen'. In the case of the Louisiana irises, and indeed for many plant species, such a mixture can arise when pollen vectors visit more than one floral syndrome (i.e. species or hybrid class). If gamete competition affects F_1 production in the Louisiana irises, we would expect that conspecific pollen on a stigma would father a disproportionate number of the resulting offspring, relative to the interspecific gametes present on the same stigma. In general, this is the observed pattern. Arnold *et al.* (1993) and Carney *et al.* (1994) demonstrated a significant reduction in F_1 offspring produced in either *I. fulva* or *I. hexagona* fruits when any

proportion of conspecific pollen was present simultaneously with heterospecific pollen (Figure 4.2; Carney *et al.* 1994). Similarly, Emms *et al.* (1996) discovered a significantly greater number of conspecific, over F_1, progeny when they applied 50:50 mixtures of *I. fulva* and *I. brevicaulis* pollen onto stigmas of both species. Additional experiments demonstrated that these results could be explained largely by a race between the conspecific and heterospecific pollen tubes, with the conspecific tubes growing faster and thereby winning the race to the ovules significantly more often (Carney *et al.* 1996; Emms *et al.* 1996). For example, Carney *et al.* (1996) found significant increases in the frequency of F_1 progeny formed when heterospecific pollen tubes were given a head start. Specifically, when *I. fulva* and *I. hexagona* pollen were applied to the alternate species' stigmas 1 and 24 h, respectively, before conspecific pollen was applied, there was a significant increase in F_1 seed formation (Carney *et al.* 1996).

Results from the above studies indicated that gamete competition restricted F_1 formation among Louisiana irises. Yet, the above analyses also discovered underlying complexities in the pattern of hybrid formation. Specifically, an asymmetry was discovered in which (i) *I. fulva* was always a more restrictive maternal parent (i.e. significantly fewer F_1 progeny were formed in *I. fulva* fruits for any given

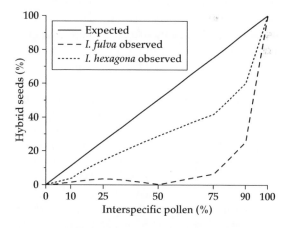

Figure 4.2 Percentage of hybrid seeds across pollen ratio treatments for *Iris fulva* and *I. hexagona* parents. The solid line indicates the expected percentage of hybrid seeds assuming random fertilization by pollen grains present on stigmas. The observed values were significantly lower (at the $P = 0.005$ level) than the expected proportions, except for 0 and 100% interspecific pollen (from Carney *et al.* 1994).

pollen mixture, relative to either *I. brevicaulis* or *I. hexagona* and a much longer head start for heterospecific pollen applied to *I. fulva* stigmas was required to ensure an increase in F_1 formation—Carney *et al.* 1994, 1996; Emms *et al.* 1996) and (ii) *I. fulva* was also a more successful paternal parent in terms of fathering F_1 seeds in either *I. brevicaulis* or *I. hexagona*. This fertilization pattern predicts an asymmetry in hybrid zones, with introgression from *I. fulva* into either of the other two species being more likely. Asymmetrical introgression discovered in a natural hybrid zone between *I. fulva* and *I. brevicaulis* is consistent with this prediction; a higher frequency of genetic exchange proceeded from the former into the latter species (Cruzan and Arnold 1994).

4.2.3 Hybrid viability as a barrier

Louisiana irises illustrate well cases of genotype-specific hybrid inviability. As an example, several experimental and field analyses have detected hybrid inviability in crosses between *I. fulva* and *I. brevicaulis*. The first indication of genotype-specific inviability among *I. fulva* × *I. brevicaulis* advanced-generation hybrids came from a study of genetic variation in a natural hybrid zone (Cruzan and Arnold 1994). In this analysis, Cruzan and Arnold (1994) examined the genotypic classes present in established, flowering plants and in the seed cohorts produced by these plants. These authors detected significant differences in the disequilibrium estimates for the adult plants and seeds, with the adults demonstrating significantly higher levels of disequilibrium. This pattern suggested that selection had acted against certain recombinant genotypes, between the seed and adult life-history stages, thus leading to more conspecific allele associations in the adults than in the seed population. To test whether such selection could be due partly to inviability during the life-history stage at which the progeny were collected, Cruzan and Arnold (1994) compared the viability of the seeds to their genotypes. Consistent with inviability as one form of selection, hybrid genotypic classes most dissimilar to the parental genotypes (i.e. *I. fulva* or *I. brevicaulis*) had the highest levels of inviability (Figure 4.3). Thus viability selection was implicated as a major isolating factor for these taxa.

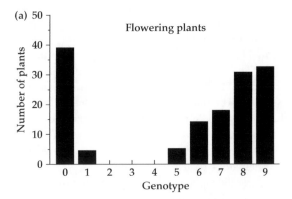

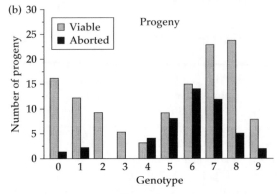

Figure 4.3 Distribution of genetic markers among (a) adult, flowering plants and (b) seeds—both viable and aborted—collected from the individuals in (a). 0 and 9 are the genotypic scores for *I. fulva* and *I. brevicaulis* plants, respectively. Scores of 1–8 are various hybrid classes (from Cruzan and Arnold 1994).

The above conclusion was supported by a series of experimental crosses, including one study that examined reciprocal F_2 populations. Burke *et al.* (1998b) detected a significant deficit of intermediate hybrid genotypes, but an excess of *I. fulva*-like and *I. brevicaulis*-like hybrid genotypes, in the *I. fulva* × *I. brevicaulis* F_2 progeny. The under- and over-representation of certain genotypes was apparently due to both maladaptive nuclear–nuclear and cytonuclear interactions, leading to post-mating barriers, including post-zygotic inviability (Burke *et al.* 1998b). Indeed, one hybrid genotype was absent from both F_2 cohorts, indicative of a recombinant lethal genotype.

A more recent set of analyses also used experimental crosses to test for viability affects on genetic exchange, but with added resolution afforded by linkage mapping. First, Bouck *et al.* (2005) tested for segregation distortion as an indicator of post-zygotic (i.e. viability) selection on reciprocal, first-generation backcross hybrids. These workers detected regions that reflected significantly lower than expected rates of introgression, i.e. portions of the genomes for which selection acted against introgression. Second, Martin *et al.* (2005), examining long-term survivorship in these same BC_1 hybrid populations, also detected the effect of interactions between introgressed and non-introgressed regions that lowered viability under greenhouse conditions. Third, a study of survivorship, once again for the same BC_1 hybrid genotypes, but this time in a natural setting (i.e. the Atchafalaya basin in southern Louisiana) also detected genotype-specific viability, with some hybrid genotypes demonstrating significantly lower viability under natural conditions. (Martin *et al.* 2006)

Each of the studies described above indicates that inviability plays a significant role as a barrier to genetic exchange between *I. fulva* and *I. brevicaulis*. In addition, reduced viability of certain hybrid genotypes is also likely a factor impeding introgression between these two species and a third species, *I. hexagona* (Arnold and Bennett 1993). In the sequence of reproductive isolating barriers, viability apparently acts as a sieve for the introgression of genetic material among Louisiana irises. Yet, as described in Chapter 1 these species were used as the paradigm for the process of introgressive hybridization (Anderson 1949). Recent analyses have confirmed the evolutionarily significant role played by genetic exchange among these species, including the derivation of a stabilized hybrid species (Arnold 1993). In contrast to the restriction imposed by the lower viability of some hybrids—and indeed imposed by a whole set of impediments to intertaxonomic hybridization between Louisiana irises—genetic exchange has apparently played a significant role in structuring the population genetics, ecological amplitude, and taxonomic diversity of this plant species complex.

4.2.4   Hybrid fertility as a barrier

Though more limited, the data on hybrid fertility in Louisiana irises reflect the expectation that this stage can also act as a barrier for gene flow among the various species (Riley 1938, 1939; Anderson 1949; Randolph *et al.* 1967; Bouck 2004). For example, Anderson (1949, p 3) reported pollen fertilities for *I. fulva, I. hexagona*, and naturally occurring hybrids. Pollen fertilities for individuals of the two parental forms ranged from 89 to 99% (mean, 95.7%), while the hybrid plants possessed fertilities of 52–98% (mean, 85%).

Randolph *et al.* (1967) collected pollen fertility data from numerous natural populations, including (i) individuals from allopatric populations of *I. fulva, I. brevicaulis*, and *I. hexagona*, (ii) individuals of *I. fulva, I. brevicaulis*, and *I. hexagona* from sympatric populations, and (iii) hybrid plants from natural hybrid zones. These investigators found that the allopatric, sympatric, and hybrid samples had pollen fertilities of 45–94% (mean, 93.5%), 15–95% (mean, 93%), and 15–95% (mean, 89%), respectively. Similarly, two *I. fulva* × *I. brevicaulis* BC_1 mapping populations also demonstrated a range of pollen fertilities (Bouck 2004). However, there was an asymmetry in the reduction in pollen fertility with the backcross population toward *I. brevicaulis* demonstrating a much lower (i.e. mean, 65%) value than the backcross population toward *I. fulva* (mean, 91%).

The Louisiana irises illustrate well the sequential and multi-staged nature of reproductive isolation. Yet, they are one of countless such examples. In the following sections I will again highlight the multiple stages that act as reproductive barriers using a variety of microorganisms, plants, and animals.

4.3   Genetic exchange: five stages of reproductive isolation

4.3.1   Ecological setting

As for the Louisiana irises, the first step in the sequence of barriers considered is that of ecological adaptations. I realize that it may seem unusual to consider the requirement for cocirculation of different variants of human immunodeficiency virus (HIV) as an ecological barrier. However, I believe that it is conceptually similar to the earliest steps of hybrid zone formation when two types of animal or plant first enter into a parapatric or sympatric association. The fire-bellied toads *B. bombina* and *B. variegata*, and the plant species *Ipomopsis aggregata* and *Ipomopsis tenuituba* reflect additional examples of the breaching of ecological barriers, thus leading to genetic exchange. In both of these cases genetic exchange is apparent, in spite of strong ecological diversification.

Bombina

Studies of natural hybridization between the fire-bellied toad species *B. bombina* and *B. variegata* have been developed by Szymura, Barton, and their colleagues into a classic model for discerning factors that affect hybrid-zone evolution and the development of barriers to reproduction (e.g. Szymura 1976, 1993; Arntzen 1978; Szymura *et al.* 1985, 2000; Szymura and Barton 1986, 1991; Gollmann *et al.* 1988; Sanderson *et al.* 1992; Nürnberger *et al.* 1995, 2005; MacCallum *et al.* 1998; Vines *et al.* 2003). One of the earliest conclusions drawn from these studies was that the pattern of clinal variation occurring across some *Bombina* hybrid zones (Figure 4.4)

indicated the action of strong selection against hybrids in the face of continued immigration of parental forms into the area of overlap; that is, the interactions between the *Bombina* species resulted in a *tension zone*. Consistent with this conclusion— of hybrid zones that demonstrate uniform selection against hybrids balanced by continual immigration by parental genotypes into the zones, thus keeping the zones from going extinct—are geographic regions that show coincident and concordant clinal variation among different nuclear, cytoplasmic, morphological, and behavioral characters (Szymura 1993). Also consistent are hybrid zones from different geographic regions that demonstrate similar clinal structure for the same set of genetic and morphological markers (Szymura and Barton 1991; Szymura 1993). In addition, significant linkage disequilibria often exist between genetic markers and morphological characters in hybrid

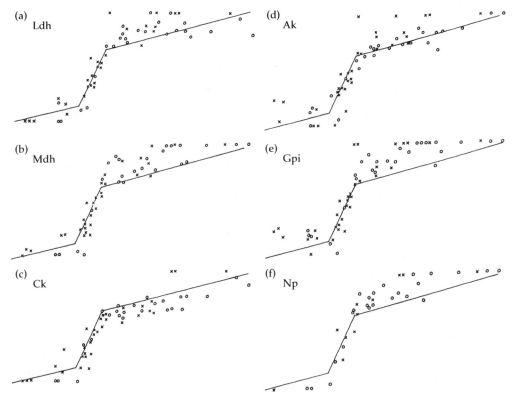

Figure 4.4 Frequencies of the *B. variegata* alleles at six allozyme loci across two hybrid zones. Crosses and open circles indicate data points for the Przemysl and Cracow transects, respectively. The solid lines in each graph reflect the best fit of the data in terms of the transition from mainly *B. variegata* to mainly *B. bombina* alleles (from Szymura and Barton 1991).

populations (Szymura 1993). Finally, studies of hybrid and parental development have detected lower fitness for some hybrid genotypes (e.g. Nürnberger *et al*. 1995).

In regard to the topic of this section—the influence of ecology on genetic exchange—the above observations suggest that the *Bombina* hybrid zones are indeed tension zones and thus not affected greatly by ecological factors. However, other findings seem to belie this conclusion. For example, Szymura (1993) concluded that not every hybrid transect demonstrated the expected pattern (for a tension zone) of simple clinal changeover. In fact, many demonstrated genetic structures other than smooth clinality (Szymura 1993) that were attributed to differences between *B. bombina* and *B. variegata* in their habitat selection and mating preferences (Szymura 1993). Arnold (1997) argued that consideration of all the data available at that time led to a conclusion that the areas of overlap for *Bombina* were better defined as mosaic hybrid zones. Subsequent experimentation supported this conclusion (Figure 4.5). For example, MacCallum *et al*. (1998) found that the pattern of species-specific allelic variation at five allozyme loci across a hybrid zone in Croatia was mosaic. Furthermore, the mosaic genetic variation correlated well with a mosaic habitat structure within the zone. Specifically, *B. bombina*-like hybrids and *B. variegata*-like hybrids showed a preference for more permanent (i.e. ponds) and less permanent (i.e. puddles) breeding habitats, respectively (MacCallum *et al*. 1998).

Environmentally mediated selection affecting the habitats chosen by different parental and hybrid genotypes was also supported by data collected by Nürnberger *et al*. (2005) in a study that inferred mating patterns in the same hybrid zone studied previously by MacCallum *et al*. (1998). These authors reported observations consistent with (i) the pattern of habitat choice detected by MacCallum *et al*. (1998), (ii) no assortative mating within the hybrid zone, (iii) a lower contribution than expected by more recombined genotypes to subsequent generations, and (iv) 'natural selection against hybrids' (Nürnberger *et al*. 2005).

Even with the obviously strong barrier to reproduction afforded by ecologically mediated selection, this barrier breaks down frequently, resulting

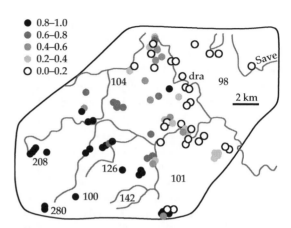

Figure 4.5   Location of *Bombina* populations in a Croatian hybrid zone with the frequencies of *B. variegata* DNA marker alleles (microsatellite and simple sequence conformational polymorphism [SSCP] loci) indicated. Elevations (e.g. "104") in meters also indicated on map. A mosaic pattern of genetic variation, rather than a simple clinal transition, is present (from Vines *et al*. 2003).

in numerous hybrid zones (Szymura 1993). The failure of the microhabitat associations/preferences of *B. bombina* and *B. variegata* as a reproductive barrier is likely due to the mosaic structure of natural settings in which ponds and puddles are intermixed (MacCallum *et al*. 1998) in areas of overlap. Furthermore, though some level of selection against certain hybrid genotypes has been detected, this has not resulted in a lack of genetic exchange (Vines *et al*. 2003).

As already illustrated in Chapters 1–3, the observation of a mosaic genome due to genetic exchange is easily seen for many microorganisms. However, as I will argue in the following section, one can also think in terms of ecological barriers for microorganisms—in this case the necessity of co-infection of the same host. To highlight this, I will use the evolution of HIV lineages.

HIV

The acquired immune deficiency syndrome (AIDS) pandemic is causing catastrophic damage to human populations, both in terms of human fatalities (approx. 8000/day worldwide) as well as the social and economic fabric of countries and entire regions (see the World Health Organization website, www. who.int/en/). Understanding the evolution of HIV, the causal agent of AIDS, is thus of chief concern both for reducing human suffering and for remediation of social systems. Of particular interest for the general topic of evolutionary diversification via genetic exchange and for the specific topic of ecological barriers that limit such exchange, it is now understood that HIV evolution often involves recombination between divergent viral lineages. In fact, it has been suggested that HIV is freed from constraints that limit the virulence of other viruses, due largely to its highly recombinant nature (Hutchinson 2001). Yet, genetic exchange between divergent lineages, leading to some of the most prevalent and devastating forms of HIV, reflects the passage across multiple ecological barriers.

Crossing the first ecological barrier involved zoonotic (i.e. cross-species) infections from other primates into humans (Hutchinson 2001). Such zoonotic infections do not, however, occur at random. Specifically, simian immunodeficiency viruses (SIVs; the ancestral lineages of HIV) are apparently most likely to be transmitted among closely related primates (Moya *et al.* 2004). Given the phylogenetic distance between humans and some of the primate sources of SIV/HIV, the zoonotic event(s) must be considered a major barrier for the initial transfer of SIV/HIV and thus the subsequent genetic exchange among HIV lineages. In particular, it has been concluded that HIV-1 (the major variant found in the human population) derived from a SIV from chimpanzees (specifically *Pan troglodytes troglodytes*; Gao *et al.* 1999), while HIV-2 (mainly found in West Africa and India) was transferred from sooty mangabeys (Hutchinson 2001).

The opportunity for the genetic exchanges that are the basis for so much of the evolutionary diversification of HIV resulted from a significant host/ecological shift from one primate species to another. However, the host shifts necessary for the genetic exchange among SIV/HIV types must also have involved transfers *within* primate lineages and *within* the same individual. First, for viral recombination between divergent types to occur, the types must be cocirculating in the same individual. Such cocirculation means by definition that either the divergent lineages arose independently in the same individual, or there was a dispersal event by one or both into the same habitat (i.e. an individual organism). The coinfection of an individual host is

therefore similar to the dispersal by different species of animals or plants across ecological barriers into a geographic zone of overlap. Second, in the case of the diploid HIVs (containing two RNA molecules per *virion*), hybrid RNA molecules can only arise when (i) virions from two or more divergent lineages infect the same cell and (ii) RNA molecules from divergent virions are packaged into a single virion (Hu and Temin 1990; Hutchinson 2001; Chin *et al.* 2005).

The failure of ecological barriers to isolate the various HIV forms is itself a multi-phase process. First, the crossing of the between species/within species/within individual ecological barriers to form hybrid SIV/HIV lineages predates the transfer into *H. sapiens*. Thus, viral exchange—likely occurring within chimpanzees—resulted in a mosaic, ancestral HIV-1 (Figure 4.6; Gao *et al.* 1999; Paraskevis *et al.* 2003). Second, subsequent to the chimpanzee→human transfer, recombination was rife, resulting in hybrid HIV-1 lineages accounting for 10–40% of the infections in Africa and 10–30% of the infections in Asia. Furthermore, of the three phylogenetically distinct subclades (M, N, and O), group M is the predominant type and it alone

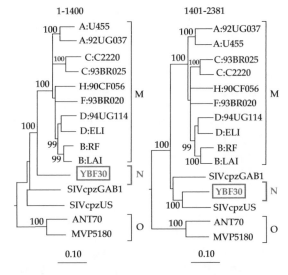

Figure 4.6 Phylogenetic discordance for the viral form (YBF30) that is ancestral to the human HIV-1 virus. The different phylogenies are based on different portions of the YBF30 genome and indicate its recombinant origin and thus its discordant placement relative to other HIV and SIV forms (from Gao *et al.* 1999).

contains 16 circulating recombinant forms (Chin et al. 2005). Finally, within individual humans, intertype genetic exchanges are extremely frequent, leading to the production of new genotypes at frequencies that equal or exceed mutation rates (Shriner et al. 2004).

The ecological barriers to genetic exchange among HIV lineages have not prevented such exchange. In fact, recombination between divergent HIVs provides the basis for a large proportion of their diversity and evolution. It is important to note that these contrasting observations—strong barriers to genetic exchange and the circumvention of the barriers—may be due to positive selection favoring the recombinant HIV genotypes (Bonhoeffer et al. 2004). Indeed Moya et al. (2004) hypothesized that 'Frequent recombination [between HIV variants from divergent lineages] seems advantageous because it can create high fitness genotypes more rapidly than mutation alone.'

Ipomopsis

Diane Campbell and her colleagues have, with the plant genus *Ipomopsis*, provided a wonderful model of how to study evolutionary processes. Many of their earlier experiments were designed to test for environmentally mediated natural selection (e.g. Campbell 1997). Of significance to the topic of genetic exchange in spite of ecological barriers, a number of Campbell et al.'s earlier studies also examined the role of pollinators, particularly hummingbirds, as a key component of the ecology that can affect (i) floral trait evolution (Campbell et al. 1991, 1994; Campbell 1996) and (ii) gene flow within and among populations of *I. aggregata* (e.g. Campbell and Waser 1989; Campbell 1991, 1998; Campbell and Dooley 1992).

In addition to their work on within-species selection, Campbell and her colleagues have also examined the effect pollinator-mediated selection has on reproductive isolation, hybrid fitness, and hybrid-zone evolution (e.g. Meléndez-Ackerman 1997; Meléndez-Ackerman et al. 1997; Alarcón and Campbell 2000; Wolf et al. 2001; Campbell et al. 2002a, 2002b, 2003, 2005; Aldridge 2005). In particular, the species *I. aggregata* and *I. tenuituba* have been developed as a model system for deciphering processes that affect natural genetic exchange. For example, it was discovered that pollinators, specifically hummingbirds, demonstrated *floral preference* for the *I. aggregata* floral type compared to the *I. tenuituba* form (Figure 4.7; Campbell et al. 1997, 2002b; Meléndez-Ackerman et al. 1997; Campbell 2003). This strong preference may be due to a suite of floral characteristics—for example, red floral color and shorter/wider corolla tubes—demonstrated by *I. aggregata* that acts as an advertisement to the hummingbirds of flowers containing a moderate volume of nectar (mean, 1.8 μl/day; Meléndez-Ackerman et al. 1997). In contrast, *I. tenuituba* possesses floral traits—light-colored flowers, longer/narrower corolla tubes, and low volumes of nectar (mean, 0.25 μl/day)—that are more attractive to insect pollinators such as hawkmoths (Meléndez-Ackerman et al. 1997). Regardless as to the exact cues that are causing the differential pollinator preferences for either *I. aggregata* or *I. tenuituba*, this component of the ecological setting for *Ipomopsis* forms a barrier for interspecific gene flow.

In spite of the pollinator preferences that drive a reproductive wedge between *I. aggregata* and *I. tenuituba*, these species form numerous hybrid zones in areas of sympatry (Aldridge 2005). As will be discussed in more detail in Chapter 5, once the ecological barrier presented by pollinator preference is overcome, some hybrid genotypes produced by

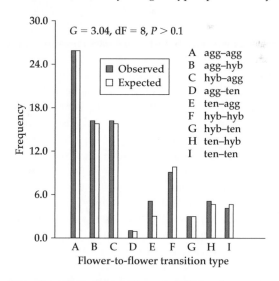

Figure 4.7 Observed/expected frequencies of flower–flower transitions by hummingbirds between *I. aggregata* (*agg*), *I. tenuituba* (*ten*), and natural hybrids (*hyb*) (from Meléndez-Ackerman et al. 1997).

crosses between these taxa demonstrate high fitness relative to their parents (Campbell and Waser 2001; Campbell 2003). Furthermore, once F_1 hybrids are formed, the original barrier to gene flow becomes a highly effective bridge for genetic exchange because pollinators transition between the F_1 hybrids and the parental types as frequently as they do between conspecific plants (Figure 4.8; Campbell *et al.* 2002b). This results in the transfer of pollen from hybrids to the stigmas of other plants at frequencies equaling those found for the transfer within *I. aggregata* and *I. tenuituba* (Campbell *et al.* 1998). Overall then, the patterns of pollinator behavior provide a significant, but incomplete, barrier for pollen flow between *I. aggregata* and *I. tenuituba*. However, these rare pollen-transfer events are followed subsequently by a favoring, by pollinators and through other forms of natural selection, of the hybrids leading to the evolution of numerous, genotypically rich hybrid zones (Wu and Campbell 2005).

4.3.2   Behavioral characteristics

Behavioral differences as impediments to genetic exchange have been a classical focus of studies by evolutionary zoologists (e.g. see Dobzhansky 1937; Mayr 1942; Coyne and Orr 2004). Yet, even the genus *Drosophila* and the order Aves, used for so many decades as paradigms for the allopatric development of reproductive barriers and thus 'biological speciation', are now seen as exemplars of evolutionary diversification in the face of gene flow (e.g. Grant and Grant 1992; Noor 1995). In this section I will discuss two examples from the zoological literature that highlight the observations made for groups such as *Drosophila* and birds. Furthermore, I will consider what can be argued to be the 'next' barrier to reproduction after ecological settings/adaptations, that of behavioral differences. To illustrate the presence of significant behavioral differences, I will discuss findings for the genus *Papio* (i.e. baboons) and the family Cichlidae (i.e. cichlid fish).

Papio

Jolly (2001) has argued that the Papionina – the African subtribe of primates that includes

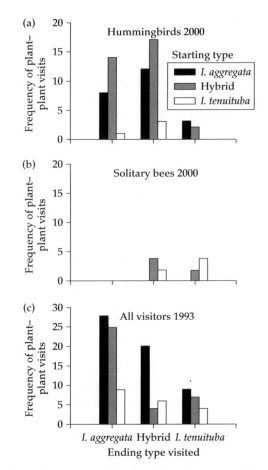

Figure 4.8 Frequency distribution of plant–plant flight transitions between *I. aggregata*, *I. tenuituba*, and natural (1993) and experimental (2000; F_1 and F_2 generation) hybrids for bees and hummingbirds (from Campbell *et al.* 2002b).

baboons, mangabeys, and mandrills – is a model system for understanding the evolution of *H. sapiens*. More specifically, Jolly (2001) contended that '. . . the papionins are a fruitful source of analogies for hominins [the tribe containing *H. sapiens*] because they are phylogenetically close enough to share many basic attributes by homology, yet far enough that homoplastic modifications of these features are easily recognized as such.' Significantly, Jolly and his colleagues consider hybridization among species of papionins to be a significant aspect of their evolutionary history (e.g. Jolly *et al.* 1997; Szmulewicz *et al.* 1999; Jolly 2001; Dirks *et al.* 2002; Newman *et al.* 2004; Nystrom *et al.* 2004). In Chapter 9 I will discuss evidence for the role of introgressive hybridization in the evolution of the clade that includes *Homo*, *Pan*, and *Gorilla*. In this regard then, Jolly's (2001) placement of papionins as a paradigm for understanding human evolution extends also to the role of genetic exchange.

The papionin taxa, like many primates, have well-defined behavioral characteristics that should act as a significant stage in reproductive isolation. An example of such diagnostic differences is illustrated by the taxa known commonly as hamadryas (*Papio hamadryas hamadryas*) and olive (*Papio hamadryas anubis*) baboons. (Note: these taxa are considered separate species by some workers, *P. hamadryas* and *P. anubis*, respectively; Alberts and Altmann 2001) Szmulewicz *et al.* (1999) have summarized the behavioral differences between the hamadryas and olive taxa. *P. h. anubis* males exist in natal, multimale–multifemale assemblages (groups), their males emigrate from these groups at maturity, their females demonstrate philopatry and thus remain in their natal group with their matrilineal family members, and during estrus, *P. h. anubis* females may mate with several males. Most male *P. h. hamadryas* display a behavior that is very rare among papionins, by remaining and breeding in their natal group. *P. h. hamadryas* males develop and maintain a one-male unit (i.e. a polygynous harem); groups of males plus their one-male units form clans that routinely travel and forage together; a band of *P. h. hamadryas* is made up of a set of these clans (the bands appear analogous to the olive baboon groups); several bands can share a sleeping site and form an ephemeral troop; female *P. h. hamadryas* are often recruited before reproductive age into a one-male unit that is within the females clan or band; the female recruitment often takes place through a young male; and there are no distinctive matrilineal groups at any of the *P. h. hamadryas* societal levels.

Given the differences between these two *Papio* taxa, it is apparent that the initiation of hybridization might be severely limited. Although this is likely, there are numerous studies that indicate introgressive hybridization between not only the olive and hamadryas forms, but between other species/subspecies combinations and even between different genera of papionins. In regard

to intergeneric hybridization and introgression, it has been established that natural hybrids occur between *P. hamadryas* and *Theropithecus gelada* (Jolly *et al.* 1997). Indeed, Harris and Disotell (1998) argued that introgression among the ancestral lineages leading to different papionin genera (such as *Papio* and *Theropithecus*) may be the explanation for the phylogenetic discordance detected in comparisons of evolutionary hypotheses resulting from different molecular markers. Similarly, the observation that olive (*P. h. anubis*) and yellow (i.e. *Papio hamadryas cynocephalus* or *Papio cynocephalus*) baboon mtDNA haplotypes are interspersed in a single clade (Figure 4.9; Newman *et al.* 2004) was consistent with introgression of this maternal marker as a consequence of their contact and hybridization in East Africa (Alberts and Altmann 2001). Animals from the hybrid zone between the olive/yellow taxa showed a continuous distribution from the yellow to olive morphologies (Alberts & Altmann 2001). Furthermore, the frequency of hybridization and introgression is increasing in this zone, likely due to increased immigration by *P. h. anubis* and hybrid individuals fleeing from habitat destruction (Alberts & Altmann 2001).

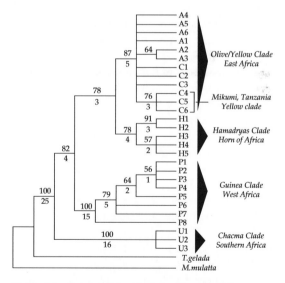

Figure 4.9 Phylogenetic hypothesis, based upon mtDNA sequences, illustrating the introgressed lineages within the olive/yellow baboon clades. Numbers above branches are bootstrap scores, and those below are numbers of synapomorphies supporting each major clade (from Newman *et al.* 2004).

Returning to the olive and hamadryas forms, there are various data sets indicating a significant role for introgression in their evolutionary trajectories as well—this in spite of the obvious behavioral barriers to intertaxa reproduction (e.g. Hapke *et al.* 2001; Dirks *et al.* 2002; Bergman and Beehner 2004; Nystrom *et al.* 2004). The consequences of genetic exchange can be seen clearly in a zone of overlap and hybridization in the Awash National Park in Ethiopia. Analyses of morphology, blood proteins, *Alu* repeat elements, and mtDNA markers diagnostic for the two taxa have uncovered evidence for bi-directional, but asymmetric, introgression, with gene flow proceeding largely from hamadryas into olive baboons (Shotake *et al.* 1977; Szmulewicz *et al.* 1999; Jolly 2001; Newman *et al.* 2004). This asymmetry is likely due, at least partially, to the behavioral attributes that act to separate hamadryas and olive gene pools. In this regard, Szmulewicz *et al.* (1999) suggested that the asymmetrical gene flow might be a consequence of olive baboon males attempting (unsuccessfully) to mate only with in-estrus, hamadryas females, females that are closely guarded by hamadryas males.

As with any limitation for genetic exchange, it is possible for behavioral barriers to fail to isolate completely. This is seen in the papionin taxa, but also is evident for other animal systems as well (Arnold 1997). In the next section I discuss a system known mainly as an evolutionary model for understanding adaptive radiations—the species that belong to the family Cichlidae. More recently, however, this marvelously diverse group of organisms has begun to be appreciated as an exemplar for testing the evolution of species complexes as they undergo genetic exchange.

Cichlidae
It is an understatement that members of the family Cichlidae demonstrate an astounding level of evolutionary and ecological diversification. Possibly they are best known for the apparent rapidity, and recency, of their radiation (Meyer *et al.* 1990). This explosive, adaptive radiation has resulted in widely varying trophic specializations reflected in morphological characteristics associated with different feeding strategies (Albertson *et al.* 1999, 2003; Rüber *et al.* 1999; Albertson and Kocher 2005; Albertson *et al.*

2005). Although the most widely recognized examples of cichlid adaptive radiations come from the African rift lakes (Johnson *et al.* 1996), New World forms from this family also demonstrate similar patterns and levels of diversification and adaptation (Barluenga and Meyer 2004). Both Old and New World clades have been used to test different models of speciation—including allopatric divergence (Pereyra *et al.* 2004), diversification in refugia formed by fluctuations in lake water levels (Johnson *et al.* 1996; Sturmbauer *et al.* 2001; Baric *et al.* 2003; Verheyen *et al.* 2003, but see Stager *et al.* 2004 and Abila *et al.* 2004; Verheyen *et al.* 2004) and sympatric divergence accompanied by gene flow (Barluenga and Meyer 2004; Hey *et al.* 2004; Seehausen 2004; Won *et al.* 2005; Barluenga *et al.* 2006).

Hypotheses abound concerning the mechanisms that have resulted in the radiation of the African cichlid fishes, a radiation apparently occurring in the last 50 000 years and resulting in thousands of named species (Kornfield and Smith 2000). Yet there are numerous studies to suggest that (i) evolutionary radiations are correlated often with male color variation and (ii) that these color differences limit hybridization between species as a consequence of female choice. Numerous lines of evidence support the strength of this behavioral barrier. First, experimentation has demonstrated female preference that is affected by male coloration (Knight *et al.* 1998; Seehausen and van Alphen 1998; Turner *et al.* 2001; Haesler and Seehausen 2005). Second, the masking of male coloration, by either the experimental manipulation of light wavelength or the occurrence of turbidity that restricts light in natural settings, results in interspecific matings (Figure 4.10; Seehausen *et al.* 1997; Seehausen and van Alphen 1998; Streelman *et al.* 2004). Third, parallel radiations of cichlid species have occurred, leading to i) numerous, closely related and often sympatric species pairs that differ in male coloration and ii) the same color patterns arising repeatedly in unrelated species pairs (Allender *et al.* 2003). Fourth, genes affecting spectral sensitivity demonstrate fixed differences and parallel evolutionary change in African cichlids, indicative of strong selection on genes that affect color recognition (Terai *et al.* 2002; Sugawara *et al.* 2005). Fifth, long-wavelength-sensitive pigment varies among species and is correlated with differences in the long-wavelength-sensitive opsin genes and large shifts in male body color (Carleton *et al.* 2005). Sixth, genes that may affect pigmentation show much greater mRNA diversity in lineages that have undergone high levels of speciation (Terai *et al.* 2003). It is thus apparent that mate selection, controlled by female choice of male coloration, likely forms a strong behavioral barrier to genetic exchange among the cichlid fauna, and indeed may be the primary factor limiting interspecific matings (Kornfield and Smith 2000).

Given the strength of reproductive isolation due to mate choice, it is possible that sympatric divergence has been of primary importance in the Cichlidae. However, if such divergence involved some level of genetic exchange in the form of introgressive hybridization then phylogenetic treatments should show paraphyly rather than

monophyly. In contrast to this prediction, some studies of certain species flocks have resolved phylogenetic trees that demonstrate monophyletic origins (e.g. Meyer *et al.* 1990; Mayer *et al.* 1998; Takahashi *et al.* 1998, 2001). However, as discussed in the previous chapter, genetic exchange may result in phylogenetic discordance between phylogenies constructed using different data sets. This is the case for the Cichlidae (e.g. Figure 4.11; Salzburger *et al.* 2002; Smith and Kornfield 2002). The cichlid clades thus reflect the results from incomplete reproductive isolation, in this case associated with mate preference. In the next section I turn my attention to a subsequent stage involving the interaction of gametes from individuals from the same evolutionary lineage and those of individuals from different evolutionary lineages (e.g. subspecies, species, etc.). In particular, I consider the effects of such interactions on barriers to introgression between species of *Drosophila*.

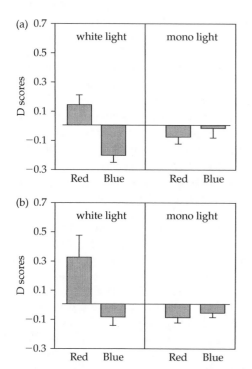

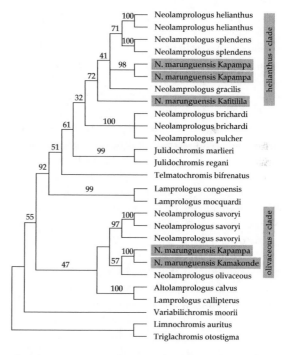

Figure 4.10   Relative scores for female mate choice experiments for two *Haplochromis* species (indicated as Red and Blue). The values were calculated by taking the proportion of encounters where females approached (a) or followed (b) a red male (after he displayed) minus the proportion of encounters where females responded to blue males (from Seehausen and van Alpen 1998).

Figure 4.11   Phylogenetic tree, based on two mtDNA regions, for species of Lake Tanganyika cichlids. The numbers above the branches are bootstrap values. Two divergent mtDNA lineages were detected in *Neolamprologus marunguensis* resulting in its placement into two separate clades (from Salzburger *et al.* 2002).

4.3.3   Gamete competition

Gamete competition can occur when gametes from two or more males are vying for the same set of female gametes. However, in the context of this book, gamete competition has been defined by 'the finding that contaxon gametes are superior to heterotaxon gametes in fathering progeny, when both gamete types are present'. Such competition can occur in animals, regardless of whether fertilization is external or internal, and can affect plant crosses regardless of the mechanism of pollen transfer. The minimum requirement for the occurrence of gamete competition is the presence of multiple, divergent, male gamete types. Below I consider an example for which numerous data sets have been collected. This example involves the vinegar fly species *Drosophila simulans* and *Drosophila mauritiana*. For these species, gamete competition is but one reproductive isolating mechanism in a complex sequence. However, experimental results indicate that this barrier is quite strong.

Drosophila

Various forms of gamete competition have been documented for numerous plant and animal complexes (e.g. Smith 1968, 1970; Gregory and Howard 1994; Robinson *et al.* 1994; Rieseberg *et al.* 1995; Levitan 2002; Geyer and Palumbi 2005). For species of *Drosophila*, gamete competition is evident even in crosses involving genotypes from the same evolutionary lineage. Clark and colleagues have identified some of the processes—for example (i) the effect of female × male interactions and (ii) the effect of male genotype on the rank order of competitive ability—associated with gamete interactions when females are multiply mated with males from the same or different *Drosophila melanogaster* lines (Clark *et al.* 1999, 2000). However, *Drosophila* taxa also offer excellent illustrations of the role of competition between gametes that originate from different evolutionary lineages in limiting introgressive hybridization (e.g. Price *et al.* 2000; Chang 2004). For example, Price *et al.* (2000) found that *D. simulans* females, mated sequentially to *D. simulans* and *D. mauritiana* males, produced a preponderance of conspecific progeny, regardless of the order of the matings (whether the conspecific

male was the first or second mate; Price *et al.* 2000). Further, the analyses of Price *et al.* (2000) identified two processes that contributed to the conspecific sperm success. These processes were heterospecific sperm incapacitation and heterospecific sperm displacement from storage organs, both affected by the seminal fluid of the conspecific male. These authors concluded that the overwhelming competitive advantage of conspecific gametes reflected an evolutionary by-product from selection on gamete effectiveness within species (Price *et al.* 2000).

In spite of the barrier to genetic exchange presented by gamete competition, introgression involving *D. simulans* and *D. mauritiana*—along with a third species, *Drosophila sechellia*—is well-supported. Several phylogenetic analyses have defined relationships among *D. simulans*, *D. sechellia*, and *D. mauritiana* (Solignac and Monnerot 1986; Aubert and Solignac 1990). These three species are homosequential (they possess identical polytene chromosome banding patterns; Lemeunier and Ashburner 1984), are morphologically very similar, and hybridize readily under experimental conditions (David *et al.* 1974), leading to their definition as the most closely related taxa within the *D. melanogaster* subgroup (Solignac and Monnerot 1986). Furthermore, discordances have been detected between mtDNA haplotypes and the taxonomy of the populations. Specifically, the mtDNA haplotypes present in some of the *D. mauritiana* populations were identical, or more similar, to those found in some *D. simulans* samples than to other *D. mauritiana* haplotypes (Figure 4.12). Solignac and Monnerot (1986) proposed several hypotheses to explain these discordances, but concluded that the most likely explanation was introgression. In addition, they argued that the introgression of mtDNA was asymmetric, moving from *D. simulans* into *D. mauritiana*. This latter hypothesis was supported by results from experimental matings in which greatly reduced success in crosses involving female *D. mauritiana* and male *D. simulans* (David *et al.* 1974; Robertson 1983) were encountered, but with no limitations in the production of F_1 progeny in the reciprocal cross (Aubert and Solignac 1990). This latter cross did, however, result in sterile F_1 males, but fertile females (David *et al.* 1974; Robertson 1983). Finally, Aubert and Solignac (1990) demonstrated

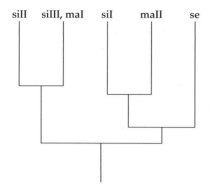

siII   siIII, maI   siI   maII   se

Figure 4.12 Phylogeny for *Drosophila simulans* (*si*), *D. mauritiana* (*ma*), and *D. sechellia* (*se*) based on mtDNA haplotypes (from Solignac and Monnerot 1986).

that mtDNA introgression from *D. simulans* and *D. mauritiana* could occur rapidly, suggesting that introgression was promoted by a selective advantage possessed by some hybrid genotypes (Aubert and Solignac 1990). Although gamete competition results in substantial limitations for introgression between *D. simulans* and *D. sechellia*, as with other such barriers, it is permeable to genetic exchange. In the next section I begin the discussion of post-zygotic barriers by examining the stage marked by hybrid viability.

4.3.4 Hybrid viability

In *Genetics and the Origin of Species*, Dobzhansky placed the mechanism of hybrid inviability into the category of reproductive isolating mechanisms that '. . . engender such disturbances in the development that no hybrids reach reproductive stage' (Dobzhansky 1937, p 231). Though understandable as a general description, I have already emphasized that all hybrids are not created equal (i.e. they are not always unfit relative to their parents). Instead, certain hybrid genotypes are expected to develop normally and demonstrate relatively high fitness (Arnold and Hodges 1995; Arnold 1997). Yet the presence of some frequency of hybrid inviability is a universal observation when matings occur between genetically divergent lineages. This breakdown in the zygotic development of some hybrid genotypes, but not in others, is one of the factors leading to the observed differential introgression

resulting in mosaic/recombinant genomes. The mosquito genus *Anopheles* provides an excellent example of how differential hybrid viability may affect organismic evolution.

Anopheles

As the vectors of malaria, species of *Anopheles* are of major interest in terms of genetic variation and evolutionary differentiation. Of particular importance is an understanding of the number of differentiated forms present in nature. This knowledge is needed to design control methods, particularly those that depend upon gene flow between introduced, genetically modified mosquitoes and naturally occurring individuals (Cohuet *et al.* 2005). Furthermore, whether there is a single, or many, species that must be controlled has obvious implications for the types of control measure—for example, the types of pesticide—that can be used effectively.

Two of the species of most interest as malaria vectors belong to the African assemblage known as the *Anopheles gambiae* species complex. This complex consists of seven recognized species that are morphologically indistinguishable (della Torre *et al.* 1997; Stump *et al.* 2005). Two of these species, *An. gambiae* and *Anopheles arabiensis*, are the two most important malarial vectors in sub-Saharan Africa (White 1971). It is now well established that *Anopheles* species complexes in general demonstrate the capacity for some frequency of intertaxonomic gene flow (e.g. Walton *et al.* 2000, 2001; Stump *et al.* 2005). This is particularly true for the *An. gambiae* complex, including *An. gambiae* and *An. arabiensis* (e.g. Lanzaro *et al.* 1998; Onyabe and Conn 2001; Coluzzi *et al.* 2002; Gentile *et al.* 2002; Tripet *et al.* 2005). It is also known that post-zygotic reproductive barriers likely contribute to limitations for genetic exchange both within and between *An. gambiae* and *An. arabiensis* lineages (White 1971; Slotman *et al.* 2005; Turner *et al.* 2005).

Significant inviability has been detected in the first backcross generation (as well as in the generation created by crossing individuals from within the same, reciprocal BC_1 population) between *An. gambiae* and *An. arabiensis*, as reflected by lower-than-expected frequencies of certain genotypes (Slotman *et al.* 2004). For example, in the BC_1 generation Slotman *et al.* (2004) found that the *An. gambiae* X chromosome (X_G) occurred at a frequency of approx. 10% (the expected frequency was 50%) in the backcross toward *An. arabiensis*. Thus the X_G chromosome causes inviability when placed onto a largely *An. arabiensis* genetic background. By examining autosomal loci, Slotman *et al.* (2004) were able to demonstrate that this large-scale inviability was due to an incompatibility between X_G chromosomal loci and one or more loci on each of the autosomes. Results from an examination of the progeny produced from crosses between the BC$_1$ individuals were consistent with the hypothesis that viability was significantly reduced (sometimes leading to the complete inviability of certain genotypes) by interactions between the X_G chromosome and autosomal loci (Slotman *et al.* 2004). It is also possible that interactions between the *An. gambiae* X chromosome and the *An. arabiensis* Y chromosome result in some level of inviability in the backcross populations (Slotman *et al.* 2004). Regardless, it is evident that gene flow between these two species is restricted by inviability in the backcross generations.

In spite of the genetic discontinuity, both within and between *An. gambiae* and *An. arabiensis*, caused by inviability (and infertility) of some hybrid genotypes, introgressive hybridization has impacted greatly the population genetics of these two species. This conclusion has been reached on the basis of data indicating the sharing of chromosomal inversions, mtDNA, autosomal loci, and X chromosomal loci in areas of sympatry (e.g. Figure 4.13; Besansky *et al.* 1997, 2003; Lanzaro *et al.* 1998; Gentile *et al.* 2002; Donnelly *et al.* 2004; Stump *et al.* 2005). Past and contemporary introgression results in a mosaicism displayed in the genomes of these species, similar to that found for influenza A, fire-bellied toads, Louisiana irises, etc. Genetic exchange has thus structured the genomic constitution of these (and other) *Anopheles* species. The patterns of introgression also reflect the role that selection plays in constraining and facilitating genetic exchange. I discussed above the selective sieve that does not allow the expected frequency of introgression for the X_G chromosome (Slotman *et al.* 2004). Furthermore, Turner *et al.* (2005) recently localized three genomic islands that confer some measure of reproductive isolation and thus provide a barrier for introgression within *An. gambiae* as well. However, selection also promotes the introgression of certain autosomal markers, as indicated by natural population variation and by experimental hybrids (della Torre *et al.* 1997). Indeed, it has been argued that selective introgression from *An. arabiensis* (a xeric-adapted species) into *An. gambiae* (a mesic-adapted species) may have led to the ecological expansion of *An. gambiae*, thus allowing it to become the predominant vector of malaria in sub-Saharan Africa (Besansky *et al.* 2003).

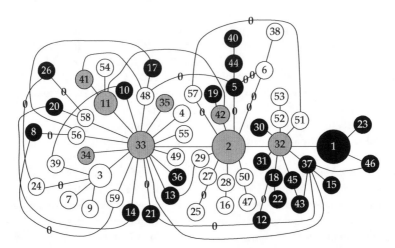

Figure 4.13 Parsimony network constructed using mtDNA haplotypes detected in *Anopheles* populations. Numbers in the circles indicate the haplotype designation, and the size of the circles is roughly proportional to the haplotype's frequency. Open and black circles indicate haplotypes unique to *An. gambiae* and *An. arabiensis*, respectively. Shaded circles reflect haplotypes shared between the species (from Besansky *et al.* 1997).

4.3.5 Hybrid fertility

One of the most widely estimated components of fitness is fertility. This is especially true for hybrid genotypes. The reason for the frequent use of this factor to estimate hybrid fitness is likely its relative ease of measurement. For plants and animals numerous methodologies have been applied to define the fertility of hybrid genotypes (e.g. Arnold and Jackson 1978). Often such methods allow inferences not only of the relative fitness (in terms of fertility) of hybrids and their parents, but also facilitate hypotheses concerning causality; for example, the presence of chromosomal structural heterozygosity leading to unbalanced meiotic products and eventually to nonviable gametes (Heiser 1947). However, genic components, as opposed to structural differences, can also lead to lower fertility in early- and later-generation hybrids. Once again considering Louisiana irises, we have detected male fertilities among experimental *I. fulva* × *I. brevicaulis* BC_1 hybrids that range from 0 to 100% (Bouck 2004). This effect on male fertility is apparently not due to chromosomal structural differences in the parental species since F_1 hybrids demonstrate nearly 100% fertile pollen.

Regardless of the methodology used to assay fertility, it is again a truism that different hybrid genotypes can vary in terms of their estimated fitness. I will use two examples to illustrate the effect of reduced fitness (i.e. fertility) on the pattern and degree of genetic exchange. The first example involves the well-studied *Mus musculus/Mus domesticus* species pair. The second comes from the equally well-investigated annual sunflower species, *Helianthus annuus* and *Helianthus petiolaris*.

Mus

The genus *Mus* has (and continues to be) developed as a model system for studying a diversity of topics from gene function to behavior. This species complex has also been used to study fundamentally important evolutionary hypotheses (Sage *et al.* 1993). A particular emphasis, in terms of research programs, has been placed on the dual processes of reproductive isolation and natural hybridization. For example, a recent issue of the *Biological Journal of the Linnean Society* (published in March 2005) was devoted to the genus *Mus* as a model for evolutionary studies. Each of the more than 25 papers discussed topics related to evolutionary diversification, reproductive isolation, and introgressive hybridization. In regard to the latter two topics, they were often discussed simultaneously, and in the context of limitations to gene flow posed by various reproductive barriers. The barriers discussed included both pre- and post-zygotic reproductive factors (e.g. Bímová *et al.* 2005; Chatti *et al.* 2005; Raufaste *et al.* 2005).

In terms of male and female fertility estimates, data are available from crosses between chromosomal races (i.e. involving *Robertsonian rearrangements*; Robertson 1916), subspecies, and species (Storchová *et al.* 2004; Britton-Davidian *et al.* 2005; Chatti *et al.* 2005; Trachtulec *et al.* 2005). For the present discussion, I will limit my considerations to data concerning the sterility barrier (and introgressive hybridization) between *M. musculus* and *M. domesticus* (sometimes identified as *M. musculus musculus* and *M. musculus domesticus*, respectively). It is well established that infertility of F_1 and later-generation hybrids occurs when these taxa hybridize. For example, Britton-Davidian *et al.* (2005) reported several observations indicating lower fertility in *M. musculus* × *M. domesticus* hybrids. These observations included (i) a significant proportion of F_1 females that did not reproduce, (ii) sterility of F_1 males in some crosses, (iii) 11–17% sterile BC_1 males, and (iv) decreased testis weight (suggestive of lower fertility) in a natural hybrid zone. Consistent with this, Storchová *et al.* (2004) found that introgressing the *M. musculus* X chromosome onto a mainly *M. domesticus* genetic background caused male sterility.

From the above it is seen that male and female infertility play a major role in limiting gene flow—particularly for genomic regions associated with sterility 'loci'—between *M. musculus* and *M. domesticus*. Indeed, the pattern of introgression of X-chromosome-specific markers (and also those on the Y chromosome) across natural hybrid zones indicates the restrictions that such fertility selection may place on genetic exchange; such markers frequently show reduced levels of introgression compared to mtDNA and autosomal loci (Vanlerberghe *et al.* 1986, 1988; Tucker *et al.* 1992;

Dod *et al.* 1993; Payseur *et al.* 2004; Payseur and Nachman 2005). Notwithstanding the significant sterility produced in crosses between *M. musculus* and *M. domesticus*, introgression and the hybrid origin of at least one subspecies have resulted from crosses involving these taxa (Vanlerberghe *et al.* 1986, 1988; Yonekawa *et al.* 1988; Tucker *et al.* 1992; Boursot *et al.* 1993; Dod *et al.* 1993; Sage *et al.* 1993). However, as with any genetic exchange event(s) the results are complex. For example, a comparison of mtDNA variation across a hybrid zone in the Czech Republic with one in Bavaria, Germany, detected reversed, asymmetric patterns of introgression (Bozíková *et al.* 2005). For the Czech hybrid zone introgression was from *M. musculus* into *M. domesticus*. mtDNA introgression was in the opposite direction across the Bavarian zone (Bozíková *et al.* 2005).

Even loci on the X chromosome—known to carry genes that affect male fertility—demonstrate differential patterns of introgression. For example, Payseur *et al.* (2004) and Payseur and Nachman (2005) detected candidate loci for reproductive isolation on the X chromosome, as indicated by lower-than-expected frequencies of introgression. However, Payseur *et al.* (2004) also detected X-chromosome loci that introgressed between the species at a greater frequency than expected. In combination, these results indicate that, while some loci are affected by negative selection and thus do not introgress, other loci introgress across the species boundary as a result of positive selection (Figure 4.14; Payseur *et al.* 2004).

Helianthus

As with *M. musculus* and *M. domesticus*, hybridization between the annual sunflower species *H. annuus* and *H. petiolaris* also demonstrates complex patterns of genetic exchange driven by natural selection. Natural selection acts negatively reducing recombination between some portions of the genomes of these species, while simultaneously favoring admixtures in other genomic regions. This, in turn, has led to hybrid products that have formed the basis for new evolutionary lineages.

The work of Heiser, and most recently of Rieseberg and his colleagues, has resulted in a monumental collection of publications on the annual sunflowers (genus *Helianthus*) that span the topics

of systematics, chromosome evolution, ethnobotany, transgenics, and speciation (e.g. Heiser 1954, 1965, 1979; Heiser *et al.* 1969; Schilling and Heiser 1981; Burke and Rieseberg 2003; Snow *et al.* 2003; Burke *et al.* 2004). However, this species complex is probably best known for the findings of Heiser and Rieseberg (and others) regarding the roles of introgressive hybridization and hybrid speciation (e.g. Heiser 1947, 1951a, b; Stebbins and Daly 1961; Rieseberg *et al.* 1988, 1990a, b, 1993, 1994, 1996, 2003; Rieseberg 1991; Dorado *et al.* 1992; Ungerer *et al.* 1998; Kim and Rieseberg 1999; Carney *et al.* 2000; Welch and Rieseberg 2002; Lexer *et al.* 2003a; Gross *et al.* 2004; Ludwig *et al.* 2004; Rosenthal *et al.* 2005).

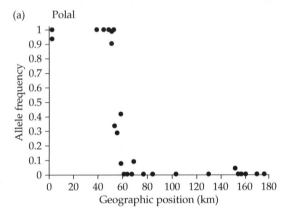

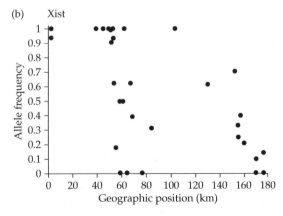

Figure 4.14 *M. domesticus* allele frequencies across a hybrid zone between this species and *M. musculus*. Pola1 (a) is a putative reproductive isolation locus and thus does not demonstrate introgression across the hybrid zone. In contrast, Xist (b) demonstrates significant polymorphism of the *M. domesticus* alleles in the *M. musculus* region (right side of the graph) consistent with adaptive trait introgression (from Payseur *et al.* 2004).

Regardless of the enormous amount of information concerning the widespread effects from genetic exchange, it is also well established that hybridization between various sunflower taxa is greatly restrained by fertility barriers. This is particularly true for the species *H. annuus* and *H. petiolaris*. For example, using experimental crosses, Heiser demonstrated that F_1 hybrids possessed pollen fertilities of 0–30% (mean , approx. 14%; Heiser 1947). In addition, crosses designed to form the F_2 and first backcross generations resulted in a maximum seed set of 1 and 2%, respectively (Heiser *et al.* 1969). This extremely low fertility is likely due to the numerous chromosomal rearrangement differences present between these species (Figure 4.15; Heiser 1947; Heiser *et al.* 1969). Indeed, the great limitations placed upon genetic exchange between *H. annuus* and *H. petiolaris* by such strong fertility selection is a partial cause of the repeated occurrence of similar hybrid genotypes from independent crossing experiments (Rieseberg *et al.* 1996). Furthermore, the observation that hybrid species tend to have the same genomic organization as found in the experimental hybrids suggests that the intense fertility selection against many hybrid genotypes has significantly affected natural genetic exchange as well (Rieseberg *et al.* 1996, 2003).

Contrary to expectations from the low fertility of early hybrid generations, natural hybridization between *H. annuus* and *H. petiolaris* has given rise to a plethora of hybrid zones and at least three hybrid species (*Helianthus anomalus*, *Helianthus deserticola*, and *Helianthus paradoxus*; Heiser *et al.* 1969; Rieseberg *et al.* 1990b, Rieseberg 1991). In addition, Rieseberg *et al.* (2003) have documented the direct role of hybridization in the formation of hybrid genotypes capable of invading novel habitats. Therefore, genetic exchange has not only played a major role in the genetic structuring of this species complex (Rieseberg 1991), but has also resulted in evolutionary and ecological novelty reflected in species with new adaptations (Lexer *et al.* 2003a; Rieseberg *et al.* 2003; Ludwig *et al.* 2004; Rosenthal *et al.* 2005). Again, it is apparent that a selective sieve—for *Helianthus* due to infertility—that allows the production of only a few hybrid genotypes does not necessarily destroy the opportunity for significant evolutionary effects from

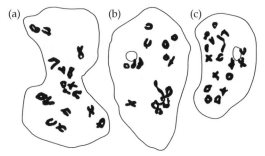

Figure 4.15 Drawings of metaphase I chromosomal configurations for (a) *H. annuus*, (c) *H. petiolaris*, and (b) the F_1 hybrid formed between these two species. Note that only bivalents occur in (a) and (c), but multivalents are present in (b), indicating chromosome-rearrangement differences between the two species (from Heiser 1947).

genetic exchange. Illustrating this conclusion, Rieseberg (1991) has typified the evolutionary history for annual sunflower species as reticulate rather than divergent.

4.4   Summary and conclusions

The rationale for this chapter, like reproductive isolation, has multiple parts. First, I have attempted to illustrate that reproductive isolation is almost always a sequential, multi-tiered process reflected in various factors that cause limitations for genetic exchange. Second, I have used a taxonomically broad array of examples to illustrate that reproductive isolation can be typified as a multi-staged process regardless of the organismic group. Third, I have emphasized the common finding that strong reproductive isolation does not necessarily remove the foundation for further evolutionary diversification, whether in terms of new lineages, novel gene combinations, or new adaptations. As I have stated before, this observation is one factor that undermines the dismissal of introgressive hybridization, hybrid speciation, or lateral gene transfer as evolutionarily unimportant based on the fact that there are a series of barriers to hybrid formation.

In conclusion, reproductive isolation may be best studied from the standpoint that barriers to genetic exchange are often permeable. If evolutionary diversification is indeed best described by a web-of-life metaphor, such an approach should offer the best opportunity to describe evolutionary process and pattern.

CHAPTER 5

Hybrid fitness

... hybrid genotypes are at least as successful as parental genotypes in a variety of frequently encountered natural aquatic habitat types ...

(Parris 2001a)

It appears that hybrids benefit from both the introgression of Fremont traits for larger seed size and narrowleaf traits, for increased asexual reproduction.

(Schweitzer *et al.* 2002)

... this ancient pathway has been extensively modified in *M. tuberculosis*, with HGT [horizontal gene transfer], duplication, and adaptive evolution contributing to these modifications. These results should help explain why *M. tuberculosis* is so accomplished at working with fatty acids.

(Kinsella *et al.* 2003)

The presence of tRNA, phage-related genes, and transposase genes near the extremities of this region suggest that it was acquired via horizontal transfer. This 'efflux island' may contribute to *Legionella*'s ability to flourish in plumbing systems and persist in the presence of toxic biocides.

(Chien *et al.* 2004)

Here, we demonstrate that regions of the Louisiana iris genome are likely to introgress across species boundaries, and that introgression results from fitness advantages (i.e. increased survivorship in highly-selective, flooded conditions) transferred concurrently with those chromosomal regions.

(Martin *et al.* 2006)

5.1 Components of hybrid fitness

In this chapter I will review data from studies of microorganisms (including bacteria and viruses), plants, and animals that help to illustrate that hybrid genotypes (produced by lateral gene transfer, viral recombination, or sexual reproduction) demonstrate a range of fitnesses depending on environmental setting. First, I must emphasize that genotype × environment interactions are not the only mechanism affecting hybrid fitness: some hybrids are maladapted regardless of environment (e.g. *Haldane's Rule or Law*; Haldane 1922,

1990; Tegelström and Gelter 1990; Zeng and Singh 1993; Palopoli and Wu 1994; Naisbit *et al.* 2002; Tao and Hartl 2003). For clear examples of such hybrid genotypes, I direct the reader to the previous chapter. However, notwithstanding the worldview of some authors (e.g. see Coyne and Orr 2004), hybrid genotypes are not different from other novel genotypes; the fact that individuals from divergent lineages produce them is irrelevant in terms of their potential for evolutionary importance. In other words, hybrids *are* created equal in that there is a probability that they will demonstrate lower, equivalent, or higher fitness relative to other genotypes, including their progenitors. Furthermore, even hybrids with lower fitness than their parents can act as the foundation for much evolutionary innovation (Arnold *et al.* 1999). One example of this is seen in annual sunflowers where highly infertile, early-generation hybrids apparently gave rise to numerous taxa (Rieseberg 1991).

Two classes of evidence that allow estimates of hybrid fitness will be emphasized in this chapter. The first involves those cases for which there are data suggesting the transfer of genetic factors that have contributed to novel adaptations. For example, the first section covers examples from microorganisms. Each of the species discussed in this section reflects exchanges that have at least partially facilitated the organism's invasion of novel environments through the acquisition of novel adaptations. This then leads to the inference of increased fitness of hybrid organisms (i.e. recombinant through lateral gene transfer or viral recombination) relative to the donor and recipient species. It is instructive to recall that we often speak and write of adaptive trait transfer as being between two taxa. This is not actually accurate. Rather, adaptive trait transfer produces introgressants that are by definition hybrid. Thus, in considering such transfers we should recognize that the highly fit products, some of which are the founders of new taxonomic lineages, are hybrid.

Adaptive genetic exchange can also be inferred through the examination of genetic markers across natural hybrid zones. These analyses compare the expected and observed frequencies of genetic transfer and thus test for a greater than expected frequency of introgressed markers. If markers are detected which introgress at a significantly greater frequency than predicted, they are recognized as candidates for adaptive trait loci. It is also important to note that whether the results of genetic exchange are referred to as adaptive trait transfer or hybrid speciation derives largely from the eye of the beholder (i.e. investigator). Thus both outcomes reflect the same set of processes—recombination between divergent lineages, environment-dependent and -independent selection leading to mosaic genomes—and qualitatively the same products; that is, new lineages that can be adapted to divergent habitats relative to their parents.

The second class of evidence allowing inferences concerning the relative fitness of hybrid genotypes comes from experimental analyses (e.g. Rhode and Cruzan 2005). Various hybrids, along with the parental genotypes from which they were formed, are placed into the same environmental setting (e.g. see Bolnick and Near 2005 for a review of hybrid embryo viability in crosses between centrarchid fish taxa). Fitness estimates are obtained and the hybrid and parental genotypes are given a ranking. The number and type of fitness components examined varies widely between studies. For example, Burke *et al.* (1998a) collected data for numerous components for the Louisiana iris species *I. fulva* and *I. hexagona* and for their F_1 hybrids. Furthermore, because these plants reproduce both asexually and sexually, fitness estimates reflecting both modes of reproduction were gathered (Burke *et al.* 1998a). In a second study, Burke *et al.* (1998b) inferred the fitness of F_2 hybrids formed in crosses of *I. fulva* and *I. brevicaulis* by examining the genotypic associations within the F_2 seedlings. In the latter study hybrid fitness was thus determined from the survivorship of only hybrid genotypes, rather than from data for both hybrids and parental individuals (as in Burke *et al.* 1998a).

In this chapter I have chosen an ad hoc division for the three sections—microorganisms, plants, and animals. It may seem odd that I refer to these groupings as ad hoc. I choose this phrase to again emphasize the point that, although the examples used can be grouped into these three taxonomic assemblages, with regard to the effect from genetic exchange, they could just as accurately be interspersed (e.g. see Chapters 1, 3 and 4). In addition, I would emphasize that almost any species complex affected by genetic exchange, including those discussed elsewhere in this book as well as those highlighted in separate reviews (see Arnold and Hodges 1995; Arnold 1997, 2004a, b; Mallet 2005) could be used to support the conclusion that hybrid fitness varies, and indeed is sometimes relatively high. The onus is now on those who hold to the paradigm that hybrids are (i) uniformly unfit and/or (ii) unimportant with regard to evolutionary innovation, to produce data rather than rhetoric in support of their framework.

Finally, it is important to recognize that hybrid genotypes demonstrating relatively high fitness—as with any 'fit' genotypes—are in constant danger of disruption through genetic recombination (i.e. viral recombination, lateral gene transfer, and sexual reproduction). In one sense then, the important question to ask is not whether a hybrid genotype is relatively fit, but whether a hybrid genotype facilitates (i) genetic exchange between previously isolated lineages or (ii) the formation of new evolutionary lineages. Yet, hybrid fitness is often used to predict future evolutionary contributions. It is thus important to consider the fitness consequences of genetic exchange.

5.2 Genetic exchange and fitness: microorganisms

5.2.1 *Mycobacterium tuberculosis*

Genetic exchange is only one of several processes that lead to the evolution and adaptation of microorganisms. For example, the evolutionary trajectory of bacterial species is affected greatly by gene duplication, deletion, inactivation, and ultimately erosion (Mira *et al.* 2001). Indeed, to explain the relatively small DNA content and lack of non-functional DNA in organisms that are regularly increasing DNA content through horizontal gene transfer and gene duplication (Zhaxybayeva and Gogarten 2004; Zhaxybayeva *et al.* 2004), Mira *et al.* (2001) concluded that there was a bias in which DNA deletions outnumbered additions. Similarly, Kurland (2005) argued, 'Genomes that continuously expand due to the uninhibited acquisition of horizontally transferred sequences are genomes that are earmarked for extinction.' Yet, as discussed in previous chapters, the transfer of genetic material is a major facilitator of evolutionary change for bacterial species (Ochman *et al.* 2000; Daubin and Ochman 2004). In addition, although there are barriers to lateral transfer among prokaryotes, as with barriers for all forms of genetic exchange, they tend to limit rather than prevent gene transfer (Bordenstein and Reznikoff 2005; Thomas and Nielsen 2005). Thus, horizontal transfer is indeed pervasive and acts as an agent for adaptive evolution in prokaryotes. This is seen clearly for the first of our examples, *Mycobacterium tuberculosis*.

It is estimated that in the Americas alone, in 2002, there were 370 000 new cases of tuberculosis and 53 000 deaths from the disease (World Health Organization, http://www.paho.org/english/hcp/hct/tub/tuberculosis.htm). The causative agent for tuberculosis, *M. tuberculosis*, has been alternatively thought to have (i) evolved from *Mycobacterium bovis* 15 000–20 000 YBP, possibly as a result of cattle domestication, or (ii) derived from an ancient human pathogen that also gave rise to *M. bovis* (Sreevatsan *et al.* 1997; Brosch *et al.* 2002; Hughes *et al.* 2002). Regardless of the exact evolutionary progression of *M. tuberculosis*, it demonstrates novel adaptations that are underlain by horizontal gene transfer. One of these adaptations involves the genetic architecture for fatty acid metabolism. Interestingly, this species possesses five times more enzymes (i.e. 250 compared with 50) that participate in fatty acid metabolism than the similarly sized *E. coli* genome (Cole *et al.* 1998). In particular, *M. tuberculosis* relies on mycolic acids (from elongated fatty acids) for the formation of a lipid layer in its cell wall that protects it while in the host environment (Barry *et al.* 1998).

Kinsella *et al.* (2003) suggested three mechanisms for the amplification of the number of genes involved in the fatty acid metabolism of *M. tuberculosis*. These included *de novo* gene origin, gene duplication, and/or lateral gene transfer. In regard to genetic exchange, lateral gene transfer from eukaryotic donors into *M. tuberculosis* has been postulated (Gamieldien *et al.* 2002). Furthermore, it has also been concluded that this species contains numerous eukaryotic—prokaryotic 'gene fusions' (i.e. genes that code for proteins that presumptively are made up of domains from the different kingdoms; Wolf *et al.* 2000). Consistent with the putative interkingdom transfers, Kinsella *et al.* (2003) detected lateral transfers involving *M. tuberculosis* and other bacterial species. Specifically, five protein families belonging to the fatty acid biosynthetic pathway placed the actinobacteria (the taxonomic group to which *M. tuberculosis* belongs) in a clade containing α-proteobacteria rather than, as expected, with other Gram-positive bacteria (Figure 5.1; Kinsella *et al.* 2003). It is important to note that copies of genes for the fatty acid pathway would have existed in this species' genome prior to the lateral transfers. The transfer of the genes from the α-proteobacteria reflects not

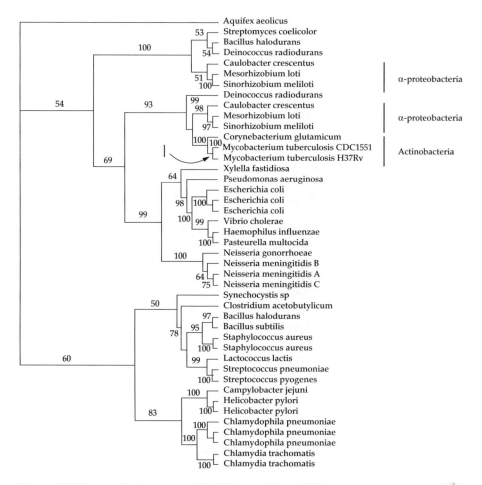

Figure 5.1 Phylogeny of bacterial, proteobacterial and actinobacterial taxa, including *M. tuberculosis*. The gene sequences used to construct the tree were from acetyl-CoA carboxylase carboxyl transferase β-subunit. The numbers on the branches indicate Bayesian clade probabilities and the arrow indicates the lineage analyzed for evidence of adaptive evolution (from Kinsella *et al.* 2003).

the addition of genes, but rather the addition and the replacement of native, actinobacterial genes (Kinsella *et al.* 2003). In this regard, Kinsella *et al.* (2003) also found that the pattern of nucleotide changes in some genes, along the phylogenetic branch leading to *M. tuberculosis* (Figure 5.1), suggested strong positive selection favoring some novel mutations. The acquisition of fatty acid biosynthesis genes from other prokaryotic lineages was apparently the basis for adaptations, indicative of increased fitness that facilitated *M. tuberculosis'* unique life history and thus its pathology. Similarly, in the next section I discuss adaptations in the pathogen *Legionella pneumophila*

that are likewise the result of horizontal gene transfer and natural selection.

5.2.2 *Legionella pneumophila*

Legionnaires disease has a false but enduring status as an exotic plague. In reality, this disease is a common form of severe pneumonia . . . (Fields *et al.* 2002)

With these words, Fields *et al.* (2002) began their review of 25 years of research on *L. pneumophila*, research spawned by the outbreak of severe pneumonia subsequent to a July 1976 American Legion convention in Philadelphia. Of the 182 cases linked to this convention, 29 were fatal (Fraser *et al.* 1977). Although isolates of this genus had been first collected in the 1940s (Fields *et al.* 2002), the taxonomic recognition of the genus *Legionella* was not to occur until 1979 (Brenner *et al.* 1979). Notwithstanding Fields *et al*'s. (2002) statement, *L. pneumophila*, like any other species of bacteria, virus, plant, or animal is indeed 'exotic' in that this species demonstrates unique adaptations to its environment. Furthermore, some of its uniqueness is directly attributable to genes brought in through horizontal transfer. Thus its ability to infect some of the participants of the Philadelphia meeting, and its lethality relative to other strains of the same species, was caused partially by genetic exchange (Chien *et al.* 2004). Some of the genomic elements transferred included (i) F-plasmid *tra/trb* genes (involved in the synthesis of pili—extracellular filaments which establish contact between donor and recipient cells during F-plasmid-mediated conjugation—leading to the transfer of the F-plasmid; Frost *et al.* 1994), (ii) several eukaryotic genes involved in recruitment of substrates (Nagai and Roy 2003) to the vacuole in which *L. pneumophila* resides within the host cell (Sturgill-Koszycki and Swanson 2000), and (iii) secretion system genes such as the *icm/dot* and *lvh/lvr* gene clusters (associated with conjugation and host cell death; Segal *et al.* 1998, 1999; Vogel *et al.* 1998).

Chien *et al.* (2004) have argued that among *L. pneumophila*'s most remarkable adaptations are its abilities to utilize so effectively the organelle-trafficking functions of a broad range of host cells (to create and maintain its vacuole), and to exist in extremely harsh environments, such as plumbing systems that are treated with biocides. As mentioned above, the first of these adaptations is underlain partially by horizontally transferred genes belonging to the *icm/dot* and *lvh/lvr* clusters. However, *L. pneumophila*'s ability to survive the harsh, human-made and -mediated aquatic environments associated with, for example, air-conditioning systems is also likely the result of genetic exchange. In particular, this species contains a 100-kb region that includes several genes for processing toxic compounds and heavy metals

(Chien *et al.* 2004). The adaptive significance of this region is seen in the fact that *L. pneumophila*'s natural hosts are protozoans that accumulate heavy metals from the environment (Chien *et al.* 2004). Mobile genetic elements are important components of numerous horizontal transfer events in prokaryotes (Batut *et al.* 2005; Frost *et al.* 2005). The role of lateral transfer in the acquisition of the 100-kb region in *L. pneumophila* was thus inferred from the observation that tRNA, phage-related, and transposase genes border it (Chien *et al.* 2004). It appears that the genetic architecture for at least two of the major adaptations that facilitate the unusual lifestyle and the lethality of some strains is foreign to the *L. pneumophila* genome (e.g. Brassinga *et al.* 2003; Chien *et al.* 2004). Once again, adaptive trait transfer is inferred and is indicative of an increase in fitness in the hybrid *L. pneumophila*, especially in its novel environmental setting. In the next example, that of Brazilian purpuric fever, we again see evidence of a transfer of the genetic machinery for an adaptation and a concomitant change in fitness leading to another highly effective pathogen.

5.2.3 Brazilian purpuric fever

Haemophilus influenzae is one of the most common microorganisms in the human upper respiratory tract (Kroll *et al.* 1998). It is also normally non-invasive, although it is capable of rare invasions leading most often to serious illnesses in children. For example, *H. influenzae* has been implicated as the causative agent of Brazilian purpuric fever. The designation of this disease reflects its original isolation from within a Brazilian population of children. An astounding 70% mortality rate occurred in the earliest outbreaks, with the average mortality now standing at 40% (Li *et al.* 2003). However, the description of the clinical features of this novel disease—meningococcal sepsis—suggested initially the involvement of a meningococcus species (Kroll *et al.* 1998). It was thus surprising that the pathological features of this disease were found instead to be the result of infection by *H. influenzae* biogroup *aegyptius*. Prior to the Brazilian purpuric fever outbreak, this organism had been known to cause nothing more serious than conjunctivitis ('pink eye'; Kroll *et al.* 1998). Brazilian purpuric fever *is* noted for producing a purulent conjunctivitis, indicating the connection with the organism's original phenotype. However, this initial stage is followed by high fever, vomiting, abdominal pain, a hemorrhagic skin rash, vascular collapse, and even to limb loss due to peripheral gangrene (Li *et al.* 2003).

The discovery of a previously benign organism that now caused a high death toll led to the hypothesis of trait transfer. Specifically, the novel *H. influenzae* was hypothesized to have acquired the genetic architecture necessary to produce the virulent meningococcal phenotype. Interestingly, each of the isolates of the virulent *H. influenzae* strain has been found to be nearly identical for numerous genotypic and phenotypic characteristics (Li *et al.* 2003). This observation has led to the multiple isolates being considered clones. Kroll *et al.* (1998) hypothesized that the evolution of this clonal lineage's phenotype was due to the transfer of virulence genes from *Neisseria meningitidis* into *H. influenzae*. Consistent with this hypothesis, these authors found that rRNA gene sequences (as well as other genes) from *H. influenzae* and *N. meningitidis* were highly divergent, while the *sodC* virulence genes isolated from the species demonstrated high sequence similarity (Figure 5.2). Subsequent to the initial analysis, Smoot *et al.* (2002) and Li *et al.* (2003) found evidence for the lateral transfer of additional genic regions between these two genera (and between *H. influenzae* and other bacterial species).

As is characteristic for any trait transfer, the acquisition of virulence genes that allowed the conjunctivitis-causing *H. influenzae* to demonstrate a new phenotype and thus invade a new niche within the human body likely reflects a concomitant increase in fitness, at least in that new habitat. Sadly, the derivation of pathogenic *H. influenzae* biogroup *aegyptius* illustrates the origin through genetic exchange of a '. . . clone with the phenotype of rapidly fatal invasive infection . . .' (Kroll *et al.* 1998).

5.2.4 *Entamoeba histolytica*

Like *M. tuberculosis*, *L. pneumophila*, and *H. influenzae*, the causative agent of amoebiasis, *Entamoeba histolytica*, is a microorganism adapted to the role of human pathogen due partially to lateral exchanges. Unlike the other three organisms, *E. histolytica* is a eukaryote; this species is a protist pathogen. After malaria, amoebiasis accounts for the most deaths from protist infections (Stanley 2003). *E. histolytica* utilizes bacteria in the lumen of the colon for food and can invade the intestinal wall, lyse the host epithelial cells, and spread to the liver, causing fatal abscesses (Huston 2004; Loftus *et al.* 2005). The name *E. histolytica* thus reflects the organism and the pathology of the disease; a tissue-lysing amoeba (Stanley 2003).

A common characteristic of many parasitic protozoans, and of *E. histolytica*, is that they are amitochondrial (e.g. León-Avila and Tovar 2004). It was once thought that such organisms represented the ancestral form of pre-mitochondrial eukaryotes (Cavalier-Smith 1991). However, with the discovery of organellar remnants of mitochondria (known as mitosomes) and mitochondrion-related genes in many amitochondrial parasitic protozoa (e.g. Clark and Roger 1995; Mai *et al.* 1999; Tovar *et al.* 2003; Williams *et al.* 2002; Riordan *et al.* 2003; Stanley 2003), it is now understood that these organisms are, instead, '. . . highly derived descendants of mitochondrion-containing ancestors . . .' (León-Avila and Tovar 2004).

Lateral gene transfer, as discussed throughout this book, is of primary importance in the evolution of prokaryotes. Less well established is its participation in the evolutionary histories of eukaryotic lineages (Lawrence and Hendrickson 2003). Yet evidence is accumulating that suggests an important role for this process in at least some eukaryotic clades (Huang *et al.* 2004). In the case of *E. histolytica*, genes that have been laterally transferred into this species may be performing the functions normally provided by mitochondria (e.g. van der Giezen *et al.* 2004). In their presentation of the *E. histolytica* genome, Loftus *et al.* (2005) argued for the central role of lateral gene transfer from prokaryotes in providing the genetic machinery necessary to replace mitochondrial functions; 58% of the 96 genes suggested to be of prokaryotic origin encoded metabolic enzymes (Figure 5.3). Furthermore, these authors concluded that a major effect of the horizontal transfers was to not only replace the metabolic functions normally provided by mitochondria, but to increase the range of

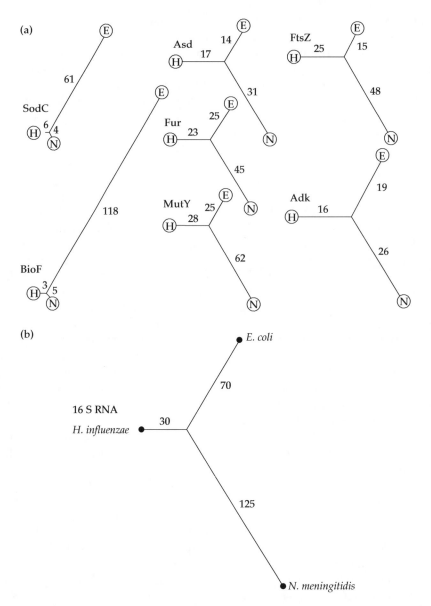

Figure 5.2 Gene trees derived from (a) protein and (b) 16 S RNA sequences of the bacterial species *H. influenzae* (H), *N. meningitidis* (N), and *E. coli* (E). Numbers along the branches indicate the number of mutational steps between terminal taxa (from Kroll *et al.* 1998).

substrates that *E. histolytica* was able to utilize (Loftus *et al.* 2005). For example, they discovered that glycosidases and sugar kinases had been transferred into *E. histolytica* with the likely result that this species could now utilize not only glucose, but also fructose and galactose (Loftus *et al.* 2005). This conclusion is reflected by Loftus *et al.* (2005) stating, 'It is clear that among the 96 genes, some result in significant enhancements to *E. histolytica* metabolism, thus contributing to its biology to a greater extent than indicated by the numbers alone.' In this species then, laterally transferred genes not only replaced those functions lost when mitochondria were lost, but also enlarged *E. histolytica*'s metabolic and thus adaptive capabilities (Loftus *et al.* 2005).

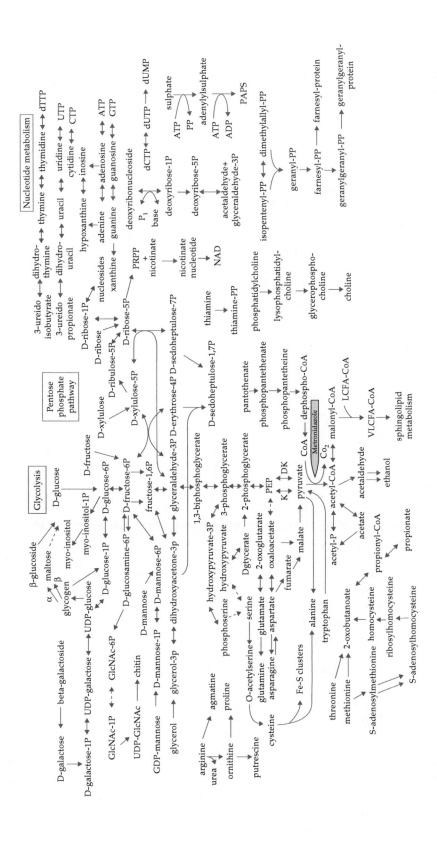

Figure 5.3 Predicted metabolic pathways for *E. histolytica*. Arrows indicate enzymatic reactions. Green arrows indicate the enzymes encoded by genes (96) transferred laterally into *E. histolytica* (from Loftus *et al.* 2005). See also Plate 3.

5.2.5   Bacterial viruses

Viruses of bacteria, otherwise known as bacterio-phages, are now understood to be genetically extremely diverse elements of the environment. A portion of this diversity is now known to be due to extensive horizontal genetic exchange (Hendrix *et al.* 2000; Filée *et al.* 2005; Silander *et al.* 2005). Such genetic exchanges include recombination between divergent bacteriophages, likely facilitated by co-infection of the same bacterial host (e.g. Waterbury and Valois 1993; Sullivan *et al.* 2003). However, hor-izontal transfers also occur between the viruses and their bacterial hosts (Hendrix *et al.* 2000; Filée *et al.* 2003). The exchanges in the direction of bacteria→bacteriophage have apparently resulted in the transfer of host genes that contribute to the propagation of the bacteriophages—including those associated with DNA replication, RNA tran-scription, and nucleic acid metabolism (Moreira 2000; Chen and Lu 2002; Casjens 2003; Filée *et al.* 2003; Miller *et al.* 2003b). There is a wide array of host genes that are not specifically involved in multiplication of the bacteriophages that are also found to reside within the genome of the viruses (Figueroa-Bossi and Bossi 1999; Rohwer *et al.* 2000; Miller *et al.* 2003a). It has been suggested that some of these may provide benefits to the bacteriophage by causing the host to function better prior to its lysis (Mann *et al.* 2003; Lindell *et al.* 2004). Horizontal transfer involving bacteriophage and host is not, however, uni-directional. There are much data to suggest that, like the transfer of genes from host to bacteriophage, transfers from bacterio-phages to their bacterial hosts have shaped not only the genomic structure of the host, but also their adaptive responses to the environment (Forterre 1999; Wagner and Waldor 2002; Filée *et al.* 2003; Pedulla *et al.* 2003; Lindell *et al.* 2004).

One example of bi-directional transfer between bacteriophages and their hosts—transfers that have affected the evolution of both participants—involves viruses belonging to the families Myoviridae and Podoviridae that infect marine cyanobacteria such as *Synechococcus* and *Prochlorococcus* (Mann *et al.* 2003; Lindell *et al.* 2004). These cyanobacterial genera are of fundamental importance for marine systems in that they account for up to 90% of the primary production of oligotrophic oceanic ecosystems (Liu *et al.* 1997). Lateral transfer between the bacterio-phages and their cyanobacterial hosts is indicated by the presence of host photosynthesis genes in the bac-teriophage genomes (Mann *et al.* 2003; Lindell *et al.* 2004; Millard *et al.* 2004). Lindell *et al.* (2004) suggested that the presence of these host photosyn-thesis genes was common among the associated bacteriophages. Although the host species and the bacteriophages that infect them cluster phylogeneti-cally based upon photosynthesis gene sequences (Figure 5.4; Lindell *et al.* 2004), the genomic organi-zation of the genes in the hosts and pathogens is very different. In the hosts the genes are dispersed throughout the genome, but they are clustered in the bacteriophages (Hess *et al.* 2001; Dufresne *et al.* 2003; Palenik *et al.* 2003; Rocap *et al.* 2003; Lindell *et al.* 2004). The divergence in the host/pathogen gene organization may reflect multiple, independent acquisitions of the genes for photosynthesis (Lindell *et al.* 2004; Millard *et al.* 2004).

In regard to the fitness of 'hybrid' phages, Lindell *et al.* (2004) argued that the presence of highly conserved PSII reaction center and *hli* genes in three *different* phages infecting the cyanobacterium *Prochlorococcus* reflected positive natural selection. They argued—from the analogy of freshwater cyanobacteria, in which the production of bacterio-phage progeny is dependent upon continuous photosynthetic activity until just prior to lysis—that bacteriophages in the marine environment that con-tain functional photosynthesis genes might also facilitate their host's metabolism until just prior to lysis. This capability would thus provide a selective advantage over those bacteriophages that lacked the photosynthesis gene arrays (Lindell *et al.* 2004). However, as with lateral transfer between phage and host in general, the lateral transfer of photosyn-thesis genes is apparently not uni-directional. One example of phage-to-cyanobacterium exchange involves the evolution of the *hli* multigene family (Lindell *et al.* 2004). Indeed, the evolution of this gene family—via lateral transfer—is postulated to have caused adaptive evolution of the cyanobacte-ria as well (Partensky *et al.* 1999; Hess *et al.* 2001; Bhaya *et al.* 2002; Rocap *et al.* 2003). Genetic exchange has apparently altered the genomic com-position resulting in the adaptive evolution of both

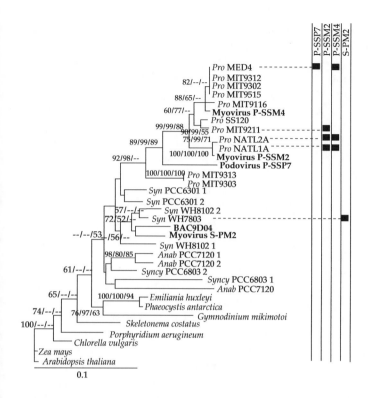

Figure 5.4 Phylogenetic tree of photosynthesis genes (PSII) found in bacteriophages (indicated by bold type) and their host species. The black boxes and lines indicate the cyanobacterium host strain infected by each bacteriophage strain. The numbers on the lineages indicate bootstrap values for distance and maximum parsimony and quartet puzzling values for maximum-likelihood analysis, respectively. Host designations are: *Pro, Prochlorococcus; Syn, Synechococcus; Anab, Anabaena; Syncy, Synechocystis* (from Lindell *et al.* 2004).

components of the cyanobacterium/bacteriophage system.

5.2.6 Yeast

It is now well appreciated that the genus *Saccharomyces* is a clear example of the process of reticulate evolution. Numerous strains are derivatives of crosses between various yeast species (e.g. Hansen and Kielland-Brandt 1994; Masneuf *et al.* 1998; Johnston *et al.* 2000; Torriani *et al.* 2004), with many of the hybrid lineages reflecting allopolyploidy (Sebastiani *et al.* 2002). However, yeast, like examples from the zoological literature (Arnold 1997), have been used to emphasize the lower fitness of some hybrid genotypes (Hunter *et al.* 1996; G. Fischer *et al.* 2000; Greig *et al.* 2001).

Many yeast hybrid genotypes—both naturally or experimentally generated—do indeed demonstrate very low levels of viability or fertility. However, as expected from crosses between related species, fitness estimates form a gradient from relatively low to relatively high (Marinoni *et al.* 1999; Johnston

et al. 2000; Greig *et al.* 2002; Sebastiani *et al.* 2002; Delneri *et al.* 2003). For example, Sebastiani *et al.* (2002) detected only 10^{-4} hybrid asci that were able to produce viable ascospores from crosses between *Saccharomyces cerevisiae* and *Saccharomyces bayanus*. Yet, the viable ascospores were capable of sporulation, and could mate successfully with either parental or hybrid yeast (Sebastiani *et al.* 2002). Similarly, Delneri *et al.* (2003) found a wide range of spore viabilities (from 1 to 96% viable) in crosses between yeast strains with or without differences in chromosomal rearrangements. Furthermore, Marinoni *et al.* (1999) found that crosses between *sensu stricto* species of *Saccharomyces* were much more likely to produce viable offspring than crosses between *sensu lato* species. Finally, Greig *et al.* (2002) tested for the potential of hybrid speciation in *Saccharomyces* by examining the fertility of later generation hybrids from crosses between *S. cerevisiae* and *Saccharomyces paradoxus*. These authors reported fertility estimates of 82 and 7.5% for crosses between hybrids and between the hybrids and either parent, respectively.

As is clear from the above studies, direct estimates of hybrid fitness from measures of viability/fertility reflect the expected outcome of a range of fitness values (Arnold and Hodges 1995; Arnold 1997). Yet hybrid fitness can be estimated indirectly from observations of population-genetic variation in natural isolates as well. For example, the detection of numerous, naturally occurring hybrid strains (Masneuf *et al.* 1998; Groth *et al.* 1999; Johnston *et al.* 2000; de Barros Lopes *et al.* 2002)—including allopolyploids, segmental allopolyploids, and various introgressants—provides support for the relatively high fitness of at least some naturally produced hybrid lineages. Thus, low hybrid fitness, though apparent for some hybrid *Saccharomyces* genotypes, is not universal and even when present does not diminish the potential for creative evolutionary outcomes from genetic exchange.

5.2.7 *Salmonella*

The final example of genetic exchange-mediated adaptive trait evolution, reflective of an increase in the fitness of a 'hybrid' microorganism, involves the genus *Salmonella*. Although some evidence exists suggesting the presence of certain restrictions for the horizontal transfer of some genomic regions among species of *Salmonella* (e.g. the *mutS* gene; Brown *et al.* 2003), there are ample data indicating the importance of genetic exchange in the evolution of these bacteria (Liu and Sanderson 1998; Marcus *et al.* 2000; Porwollik *et al.* 2002).

The World Health Organization reports that over 2500 *serotypes* of *Salmonella* were isolated in 2004 (www.who.int/mediacentre/factsheets/fs139/en/). Two of the most important serotypes are *Salmonella enterica serovar* Typhi (*Salmonella typhi*) and *S. enterica* serovar Typhimurium (*Salmonella typhimurium*). *S. typhi* and *S. enterica* account for millions of infections and numerous deaths worldwide each year (McClelland *et al.* 2001). For example, *S. typhi* causes approximately 15 million infections and 600 000 fatalities annually (Parkhill *et al.* 2001).

One of the greatest concerns regarding *S. typhi* (and indeed any bacterial pathogen) is the origin and spread of resistant strains, particularly those resistant to multiple classes of antibiotics (Murdoch *et al.* 1998; Parry *et al.* 1998). Parkhill *et al.* (2001) reported the entire genome sequence of one such multiple-drug-resistant strain, *S. typhi* CT18, and were able to test for the roles of such processes as horizontal transfer and gene silencing in the adaptive evolution of this human pathogen. In regard to gene silencing, these workers uncovered sequence data predicting 204 pseudogenes. Parkhill *et al.* (2001) concluded that this extensive array of inactivated genes was '. . . remarkable in a genome of an organism capable of growth both inside and outside the host.' Furthermore, they suggested that a large proportion of the phenotypic and host-range differences detected between *S. typhi* and *S. typhimurium* might be due to differences in cohorts of silenced genes (McClelland *et al.* 2001; Parkhill *et al.* 2001). Yet, horizontal transfer has also impacted greatly the evolution of adaptations seen, for example, in *S. typhi*.

Comparisons of the *S. typhi* and *E. coli* genomes detected a large degree of collinearity (Parkhill *et al.* 2001), even though these organisms separated from a common ancestor some 100 million YBP (Doolittle *et al.* 1996). Notwithstanding their similarity in gene order, there are major insertional differences between these species—both of longer and shorter length—generally involving gene systems that are key for survival in the host environment (Parkhill *et al.* 2001). Of interest for the present discussion, large gene arrays often include sequences with similarities to integrase- or transposase-encoding genes (Marcus *et al.* 2000) and are thought to reflect recent horizontal transfer events (Parkhill *et al.* 2001). Large insertions are, however, not the only type of DNA addition. In fact, there are hundreds of genes that may have been horizontally acquired by *S. typhi* in groups of five genes or fewer (Parkhill *et al.* 2001).

Of particular importance for the adaptation of *S. typhi* to the post-antibiotic stage of its evolutionary history was the apparent horizontal transfer of a plasmid (named pHCM1). This plasmid carries resistance genes for all of the '. . . first-line drugs used for the treatment of typhoid fever' (Parkhill *et al.* 2001). pHCM1 has an approximately 168-kb region that shares more than 99% sequence identity with a plasmid first isolated from *S. enterica*

(Sherburne *et al.* 2000; Parkhill *et al.* 2001). However, there are major sequence differences between these two plasmids, also due to horizontal transfers. It thus appears that pHCM1 is derived from the ancestral *S. enterica* plasmid, into which 46 coding sequences have been inserted. Significantly, 18 of these coding sequences contribute to the resistance of *S. typhi* to antimicrobials or heavy metals (Parkhill *et al.* 2001). As with the putative insertions into the *S. typhi* chromosome (see above), the additional coding sequences present in pHCM1 are surrounded by integrase- and transposase-like sequences. A number of the resistance genes were apparently acquired through multiple IS element-mediated events (Parkhill *et al.* 2001).

For each of the examples in this section, the evolution of adaptations has resulted, at least partly, from genetic exchanges. These novel adaptations most likely reflect an increase in fitness of the recombinant/hybrid form relative to the ancestral types. Thus the lateral transfer of DNA has allowed these microorganisms to invade novel habitats and demonstrate novel phenotypes. In the next two sections I will illustrate that, like lateral transfer in microorganisms, introgressive hybridization in plants and animals can produce a variety of genotypes that demonstrate a range of environment-dependent or -independent fitness estimates.

5.3 Genetic exchange and fitness: plants

5.3.1 *Ipomopsis*

In the genus *Ipomopsis*, pollinator behavior has played a significant role in the evolution of both floral form and reproductive isolation (Campbell *et al.* 1991, 1994; Wolf *et al.* 2001). Furthermore, of significance for the present topic, Campbell *et al.* (1997, 1998) have argued that *I. aggregata* × *I. tenuituba* natural hybrids were (i) the least fit when both of the normal pollen vectors (hawkmoths and hummingbirds) visited the hybrid zones or (ii) intermediate in fitness to the two parents when only hummingbirds were present (the normal situation). The first conclusion (that hybrids were uniformly unfit) was seen to be a result of hawkmoths favoring

I. tenuituba, and hummingbirds favoring *I. aggregata* (Campbell *et al.* 1997). In contrast, the hypothesis of a fitness gradient of *I. aggregata* → hybrids → *I. tenuituba* was based on these initial findings coupled with data from a second study in which hybrids had the most pollen transferred by hummingbirds. In the second analysis, the quantity of visitations favored *I. aggregata*, but the quality (i.e. the amount of pollen transferred) favored hybrid genotypes (Campbell *et al.* 1998). Considering all of the available data on pollinator-mediated selection, Campbell *et al.* (2002b) concluded, '. . . natural variation in pollination may produce spatiotemporal variation in hybridization and hybrid fitness.'

As reviewed in Chapter 4, a second component that can be used to estimate hybrid fitness is the relative competitive ability of hybrid and parental gametes. At the earliest stage of hybrid formation between *I. aggregata* and *I. tenuituba*, that of F_1 formation, Alarcón and Campbell (2000) detected no competitive advantage for conspecific pollen. These authors found no significant differences between the expected and observed proportion of F_1 seed formation on either species regardless of the ratio of conspecific and heterospecific pollen. Furthermore, pollinations of *I. aggregata* with either F_1 or F_2 pollen alone produced as many seeds as did conspecific pollinations (Campbell *et al.* 2003). However, when 50:50 mixtures of *I. aggregata* and hybrid (either F_1 or F_2) pollen were applied to *I. aggregata* stigmas, the hybrid pollen sired significantly fewer seeds than expected (Campbell *et al.* 2003), reflecting a lower fitness for the F_1 and F_2 hybrid 'classes', but not necessarily for individual hybrid genotypes (see Arnold and Hodges 1995 and Arnold 1997; see also Figure 1 in Campbell *et al.* 2003).

Once *Ipomopsis* progeny are formed, seed herbivory rates can demonstrate differential success in avoidance or resistance to such attacks by hybrids and parental genotypes, thus reflecting differential fitness. Campbell *et al.* (2002a) did indeed detect differential seed predation on *I. aggregata*, *I. tenuituba*, F_1, F_2, and natural hybrids. Specifically, they detected intermediate to higher oviposition levels on hybrid flowers and the lowest levels of oviposition on *I. tenuituba* flowers. Since the various classes did not differ in the damage caused by developing larvae (i.e. resistance), the quantity of oviposition reflected well differential fitness in terms of the likelihood of herbivore damage (Campbell *et al.* 2002a). This estimate then suggests an overall lower fitness for the various hybrid classes. Notwithstanding this observation, Campbell *et al.* (2002a) argued that F_1 hybrids and *I. aggregata* would, in fact, have the highest net reproductive rate, as indicated by the number of undamaged seeds produced under natural conditions. Consistent with this hypothesis, Campbell *et al.* (2002a) reviewed the findings of Melendez-Ackerman and Campbell (1998) and Campbell and Waser (2001) in which *I. aggregata* and *I. aggregata* / F_1 hybrids respectively produced the highest numbers of undamaged seeds. Thus fitness components estimated at the life-history stage of seed formation once again demonstrated that plant hybrids can possess high fitness relative to their parents.

Reciprocal transplant studies of *I. aggregata*, *I. tenuituba*, and experimental and natural hybrids also detected differential selection on various genotypes. In particular, some hybrid genotypes, in some ecological/environmental settings, possessed fitnesses that equaled or exceeded those seen for the parental taxa, while other hybrid genotype×environment combinations resulted in lower fitness estimates for hybrids. For example, an analysis of survivorship over a 5-year period detected equivalent survivorship of F_1 individuals compared to *I. aggregata* and *I. tenuituba* genotypes (Campbell and Waser 2001). However, the survivorship of F_1 individuals was asymmetric, with F_1 progeny from *I. aggregata* maternal plants surviving at a higher rate than those from the reciprocal F_1 class (Campbell and Waser 2001). Similarly, Campbell *et al.* (2005) found that F_1, F_2, and natural hybrid genotypes demonstrated high relative fitness (particularly the F_1 hybrids) compared to their parents for the fitness component of water-use efficiency. Unlike the findings of Campbell and Waser (2001), the latter analysis did not find evidence of inequality of hybrid fitness across reciprocal crosses, nor did they find evidence for hybrid breakdown in later-generation (post-F_1) hybrids (Campbell *et al.* 2005). In summary, the evolutionary importance of hybridization for this species complex, like estimates of hybrid fitness, demonstrates a spatiotemporal component, with it being of greater or lesser importance within certain environmental settings and time horizons.

5.3.2 Louisiana irises

As discussed in Chapter 1, a main focus for my group and associated colleagues has been the estimation of hybrid and parental fitness in various organisms in both experimental and natural habitats (e.g. Shoemaker *et al.* 1996; Williams *et al.* 1999; Promislow *et al.* 2001). In particular, we have drawn attention to the fact that hybrid genotypes can display a range of fitness estimates, sometimes due to environmental setting. The majority of the data used to illustrate this conclusion have come from studies of the Louisiana iris species *I. fulva*, *I. hexagona*, and *I. brevicaulis*. The studies have included experimental manipulations in both greenhouse and natural settings as well as from natural hybrid zones. Each of the studies has supported the hypothesis that hybrids vary in fitness and thus vary in the likelihood of their contributing to long-term evolutionary effects.

Though not as extensive as the studies involving *Ipomopsis*, we have also examined pollinator behavior as a means for inferring the fitness of parental and hybrid genotypes. In particular, Emms and Arnold (2000) and Wesselingh and Arnold (2000) examined pollinator interactions with *I. fulva*, *I. hexagona*, and experimental hybrids, and *I. fulva*, *I. brevicaulis* and experimental hybrids, respectively. The analysis of *I. fulva* paired with *I. hexagona* also included experimental F_1 hybrids (Emms and Arnold 2000). These authors examined pollinator behavior in two field sites, one containing a natural population of *I. fulva* and the other a population of *I. hexagona*. As expected, the major pollinator class at the '*I. fulva*' site was hummingbirds (the vector that prefers *I. fulva*) while the major pollinator found at the '*I. hexagona*' site consisted of bumblebees (the vector that prefers *I. hexagona*). The proportion of pollinator visitations was *I. fulva* $>F_1>$ *I. hexagona* at the *I. fulva* site, and *I. hexagona* $>F_1>$ *I. fulva* at the *I. hexagona* site (Emms and Arnold 2000). Because the F_1s were visited at an intermediate rate at both sites, their fitness would be estimated as less than the parental species that typified each site. It is, however, significant for the evolution of hybrid zones that the transitional flights by pollinators between F_1 individuals and their parents were significantly higher than expected (Emms and Arnold 2000). This indicates that the formation of advanced-generation hybrid classes would be facilitated. Thus less-fit hybrid genotypes are once again seen to play an important role, acting as a bridge for evolutionary change.

In contrast to the *I. fulva/I. hexagona* study, Wesselingh and Arnold (2000) found that among *I. fulva*, *I. brevicaulis*, and experimental F_1 and first-generation backcross plants, the pollinators favored the F_1 genotypes. In particular, the presence of large numbers of bumblebee workers (which preferred F_1 flowers over other types) resulted in the F_1 individuals having a higher relative fitness over all other classes. Indeed, the fact that this study was carried out later in the season than that undertaken by Emms and Arnold (2000) might explain the presence of many more bumblebee workers (from established nests) and thus the differential pollinator selection that favored the F_1 genotypes (Wesselingh and Arnold 2000).

Estimates of gamete-competition-derived selection in Louisiana irises have only been determined directly for interspecific pollen and conspecific pollen (Arnold *et al.* 1993; Carney *et al.* 1994, 1996; Emms *et al.* 1996; Carney and Arnold 1997). For a majority of these studies, conspecific pollen outcompeted heterospecific pollen. Thus the formation of F_1 hybrids was selected against. However, it is possible to also infer indirectly the levels of selection for F_1 pollen in competition with parental pollen, on both F_1 and parental species' stigmas. This inference can be deduced from data collected by Hodges *et al.* (1996) regarding patterns of seed siring by *I. fulva*, *I. hexagona*, and F_1 genotypes under natural pollinations. During the flowering season of this study, pollen from only 34 F_1 flowers—compared to more than 1500 parental flowers—resulted in backcross frequencies on fruits from *I. hexagona* and *I. fulva* flowers of 6.9 and 1.7%, respectively. This is in contrast to the frequency of F_1 offspring formed by these same *I. hexagona* and *I. fulva* plants of 0.74 and 0.03%, respectively (Hodges *et al.* 1996). Although F_1 pollen grains would have been at an extreme numerical and proportional disadvantage, compared to parental species' pollen

grains, the F_1 gametes displayed high competitive ability as reflected by siring ability.

A series of studies, both greenhouse- and field-based analyses, allow additional inferences of relative hybrid fitness based on various fitness components (Emms and Arnold 1997; Burke *et al.* 1998a, b; Johnston *et al.* 2001, 2003, 2004; Bouck *et al.* 2005; Martin *et al.* 2005, 2006). Included in the category of 'fitness components' are not only variables from sexual reproduction, but also asexual reproduction and vegetative growth. Louisiana irises produce an enormous amount of vegetative biomass during each growing season (Bennett and Grace 1990; Arnold and Bennett 1993). Much of this growth is directly associated with new, clonal copies that go on to reproduce sexually. The clonal copies also act as a potential buffer against a genotype's idiosyncratic loss from the population. For long-lived, clonally reproducing individuals, estimates of vegetative growth and clonal reproduction thus often correlate well with other fitness components such as survivorship and sexual reproduction.

The general finding from the many studies of fitness components of Louisiana irises is that some hybrid genotypes have low fitness, others intermediate fitness, and others high fitness relative to parental species' genotypes. One example of this finding comes from the work of Johnston *et al.* (2004). In this study, *I. brevicaulis*, *I. fulva*, F_1 hybrids, and reciprocal, first-generation backcross hybrids were exposed to two different environmental settings, wet and flood. In the wet treatment all three hybrid classes (the F_1 and both backcross categories) demonstrated significantly greater biomass production than either parent (which did not differ significantly from one another in either treatment). For the flood treatment the backcross toward *I. brevicaulis* and the F_1 class once again demonstrated significantly greater biomass (Johnston *et al.* 2004). For this clonal reproduction-associated fitness component the hybrid classes demonstrated a higher inferred fitness than either *I. fulva* or *I. brevicaulis*. Intriguingly, Johnston *et al.* (2004) found that the hybrid classes demonstrated significantly greater sexual reproductive output (as estimated by numbers of flower stalks and numbers of flowers) than *I. brevicaulis*, but equivalent output to *I. fulva*.

It appears that, in general, the hybrid genotypes combined the large amount of asexual reproduction characteristic of both species with the higher sexual reproduction of *I. fulva*. The combination of these life-history traits, characteristic for the two parental species, led to a greater cumulative fitness for these hybrid genotypes (Johnston *et al.* 2004). Burke *et al.* (1998a) made a similar observation in a study of asexual and sexual reproduction in *I. fulva*, *I. hexagona*, and experimental F₁ hybrids. In this study, the F₁ hybrids combined the greater asexual reproductive capacity of *I. hexagona* and the greater sexual reproductive output of *I. fulva*, resulting in a higher combined fitness than either parent (Burke *et al.* 1998a). Overall, Louisiana iris hybrid genotypes display high relative fitnesses, although some genotypes do display characteristics of later-generation hybrid breakdown (e.g. Cruzan and Arnold 1994; Burke *et al.* 1998b; Bouck *et al.* 2005; Martin *et al.* 2005).

5.3.3 *Helianthus*

As with the Louisiana irises, studies of annual sunflowers have provided a wealth of data concerning the evolutionary role of hybridization (e.g. Heiser 1951b, 1958; Rieseberg 1991; Rieseberg *et al.* 1999; Gross *et al.* 2004). In particular, recent analyses have resulted in fitness estimates for a variety of experimental hybrids and their parents, in both greenhouse and field environments. These estimates have then been utilized to predict the course of past evolutionary change resulting in adaptive trait introgression and/or hybrid speciation.

Adaptive trait introgression from *Helianthus debilis* var. *cucumerifolius* into *Helianthus annuus* (now recognized as ssp *texanus*) was hypothesized by Heiser (1951b) to account for (i) the expansion of *H. annuus* into the novel habitats of east Texas and (ii) the morphological mosaicism of east Texas *H. annuus*. In terms of the ecological expansion by *H. annuus*, he suggested that '*Helianthus annuus* . . . may have been poorly adapted and hence there might have been some selective premium placed on those hybrid forms which contained genes from *H. debilis* var *cucumerifolius*, a species already well adapted to this area' (Heiser 1951b). With regard to the unique *H. debilis* var. *cucumerifolius*-like

morphological characters demonstrated by *H. annuus* ssp *texanus* he argued 'It seems likely then that gene flow into *H. annuus* from *H. debilis* var *cucumerifolius* accounts for the many peculiarities observed in eastern Texas populations of *H. annuus*' (Heiser 1951b).

Recently, Kim and Rieseberg (1999, 2001) reevaluated Heiser's hypothesis concerning adaptive trait introgression from *H. debilis* var. *cucumerifolius* into *H. annuus* ssp *texanus*. A set of quantitative trait locus (QTL) analyses revealed the genetic architecture underlying the species-specific traits found in *H. annuus* and *H. debilis* ssp *cucumerifolius*. In particular, Kim and Rieseberg (1999, 2001) found that some QTLs were associated with regions that would not be predicted to introgress due to closely linked sterility loci. However, no restriction for introgression of the majority (45 of 56) of the morphological QTLs was detected (Kim and Rieseberg 1999). Furthermore, the pattern of segregation distortion found in the backcross population used for the mapping experiments indicated an overrepresentation of greater than 50% of the distorted loci. Kim and Rieseberg (1999) suggested that these *H. debilis* alleles might confer a fitness advantage for the backcross hybrids, or they might reflect segregation distorter loci. It is interesting that Rieseberg *et al.* (1999) also found greater than expected introgression of certain genomic regions in three different natural hybrid zones between *H. annuus* and *Helianthus petiolaris*. In this case they found very similar patterns of introgression involving the same genomic regions in the three independent hybrid zones. This suggests that the introgression of these loci is favored (Rieseberg *et al.* 1999).

In their analyses of hybrid speciation, Rieseberg and his colleagues have also collected data that allow inferences of the fitness of hybrid genotypes. For example, Rieseberg *et al.* (1996) detected gene interactions in various hybrids that resulted in lower, equivalent, or higher fitness estimates. In addition, several analyses have indicated that hybridization has been a key step in the derivation of hybrid species with unique ecological adaptations (Rieseberg *et al.* 2003). Thus Gross *et al.* (2004), Ludwig *et al.* (2004), and Lexer *et al.* (2003a, b) detected hybrid phenotypes that resembled those found in the ecologically differentiated hybrid

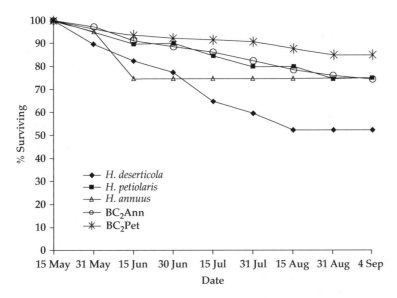

Figure 5.5 Survivorship of *Helianthus deserticola*, *H. petiolaris*, *H. annuus*, BC$_2$Ann (second-generation backcross toward *H. annuus*) and BC$_2$Pet (second-generation backcross toward *H. petiolaris*) samples transplanted into a common garden near naturally occurring *H. deserticola* (from Gross *et al.* 2004).

species. Furthermore, some experimental hybrid classes possessed a higher fitness than *H. annuus* and *H. petiolaris* under experimental and natural settings, including the unique environmental settings occupied by the hybrid species (Figure 5.5; Gross *et al.* 2004; Lexer *et al.* 2003a, b).

5.3.4 *Artemisia*

Species of *Artemisia* (subgenus *Tridentatae*) dominate much of western North America. The genetic variation that allows this broad ecological adaptation is facilitated by hybridization and polyploidization. (McArthur *et al.* 1998)

With this statement, McArthur *et al.* (1998) reflected their hypothesis concerning the importance of introgressive hybridization (in concert with polyploidy) for producing a portion of the genetic variation necessary to cause members of *Artemisia* to be a dominant component of the western North America environment (see also McArthur *et al.* 1988). The extent of introgression among members of this group is reflected in the confusion generated by morphologically based taxonomic treatments, as well as the identification of numerous,

contemporary hybrid zones (e.g. Beetle 1971; Freeman *et al.* 1991). The evolutionary effects from introgression between *Artemisia* taxa include (i) the transfer and fixation of cpDNA (i.e. cpDNA 'capture'; Rieseberg and Soltis 1991), leading to additional confusion for systematic treatments based on this molecular marker (Kornkven *et al.* 1999) and (ii) the origin of novel taxa (McArthur and Sanderson 1999).

In addition to the evidence for the evolutionary effects from hybridization, Freeman, Graham, and McArthur and their colleagues have produced a series of studies that address the factor of hybrid fitness in the *Artemisia tridentata* ssp *tridentata* × *Artemisia tridentata* ssp *vaseyana* hybrid complex. One class of data derives from studies of environment × genotype associations. First, Freeman *et al.* (1999) demonstrated that hybrid and parental taxa occurred in diagnosably different habitats. Specifically, hybrids and parental subspecies were associated with different plant species complexes and different edaphic zones; the latter estimated by such things as substrate, litter composition, etc. (Freeman *et al.* 1999). Furthermore, the hybrid and parental genotypes possess characteristics indicative of local adaptation to their respective niches such as different capacities for elemental uptake (Wang *et al.* 1999).

Additional inferences concerning hybrid and parental fitness were also possible from direct estimates of fitness for (i) plants from naturally occurring hybrid zones and (ii) *A. tridentata* ssp *tridentata*, *A. tridentata* ssp *vaseyana*, and hybrid genotypes reciprocally transplanted into the parental and hybrid habitat types. As usual the calculation of various fitness components for naturally occurring hybrid and parental individuals resulted in a complex pattern of estimates (e.g. compare Messina *et al.* 1996 with H. Wang *et al.* 1997). For example, in one study hybrids were found to have decreased recruitment and increased damage from herbivores (Graham *et al.* 1995). However, overall herbivore loads on the hybrids and parental individuals did not differ significantly (Graham *et al.* 1995). Finally, these workers found that seed production and germination by the hybrid and parental individuals was not significantly different. Similarly, a study of developmental stability of naturally occurring hybrid and parental plants did not suggest the occurrence of reduced hybrid fitness. In fact, when lower developmental stability was detected it was almost always found in one of the two parents and not in the hybrids (Freeman *et al.* 1995).

In addition to the surveys of naturally occurring hybrid and parental plants, H. Wang *et al.* (1997), by reciprocally transplanting parental and hybrid seedlings into the habitats characteristic for each of the three classes, detected environment-dependent fitness responses. Specifically, these workers found that *A. tridentata* ssp *tridentata*, *A. tridentata* ssp *vaseyana*, and hybrid genotypes had the highest cumulative fitness when placed in their native habitat (Figure 5.6; H. Wang *et al.* 1997). This, as well as previous observations, led H. Wang *et al.* (1997) to conclude that the *Artemisia* hybrid zones were best described by the *bounded hybrid superiority model*. However, this model was not clearly supported when data from a period of 9 years was taken into account. Instead of the hybrid and parental genotypes performing best in their respective environments, the hybrid plants demonstrated '. . . the highest intrinsic rate of increase, not only in their native garden, but in the parental gardens as well' (Miglia *et al.* 2005).

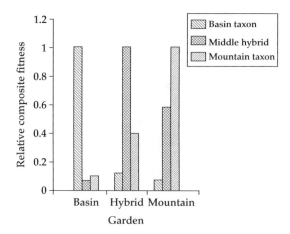

Figure 5.6 Relative fitness of basin (i.e. *A. tridentata* ssp *tridentata*), mountain (i.e. *A. tridentata* ssp *vaseyana*), and hybrid sagebrush genotypes transplanted into the respective habitats ('gardens') of the three classes (from H. Wang *et al.* 1997).

5.3.5 *Populus*

Many workers have presented data supporting the conclusion that introgressive hybridization between members of the genus *Populus* is widespread and has affected greatly the evolutionary history of both Old and New World taxa (Spies and Barnes 1981; Eckenwalder 1984a–c; Keim *et al.* 1989; Smith and Sytsma 1990; Floate 2004; Lexer *et al.* 2005). For example, the phylogenetic patterns defined by their analysis of both nuclear rDNA and cpDNA sequences led Hamzeh and Dayanandan (2004) to conclude, 'The incongruence between phylogenetic trees based on nuclear- and chloroplast-DNA sequence data suggests a reticulate evolution in the genus *Populus*.'

One research group that has utilized the *Populus* species complex as a model system for studying the evolutionary effects from hybridization is that of Whitham and his colleagues. An aspect they have examined in detail, both in North American *Populus* (i.e. cottonwood) species as well as in unrelated taxa such as species of *Eucalyptus*, is that of community-level effects from introgressive hybridization (Whitham *et al.* 1994, 1999, 2003; Floate and Whitham 1995; Dungey *et al.* 2000; Wimp and Whitham 2001; Lawrence *et al.* 2003; Bailey *et al.* 2004; Bangert *et al.* 2005; Wimp *et al.*

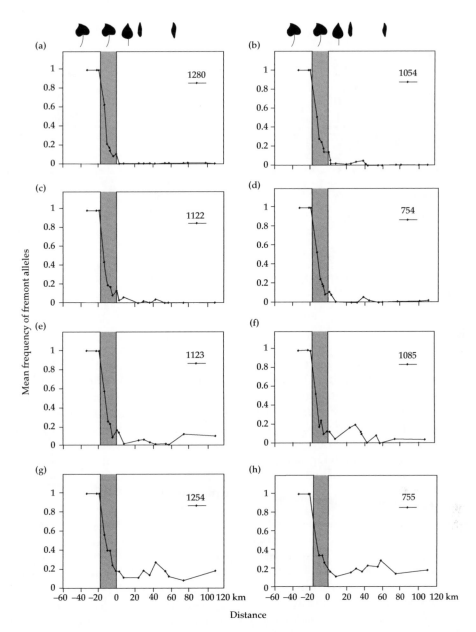

Figure 5.7 Frequencies of Fremont cottonwood (*P. fremontii*) alleles at eight markers. The frequencies indicate differential introgression (e.g. compare a with h) of the alleles from Fremont cottonwoods (indicated by a single, broad leaf above the graphs) across the current hybrid zone (indicated by three separate leaf shapes above the graphs) and into the range of narrowleaf cottonwood (*P. angustifolia*; indicated by a single, narrow leaf above the graphs; from Martinsen *et al.* 2001).

2005). In particular, Whitham and colleagues have tested hypotheses concerning the role that hybrid plants may play in determining the biodiversity of herbivores that utilize the trees. In general they have found a significant correlation with the genotype of the food source (i.e. tree) and the arthropod community feeding on the source. As one example, Wimp *et al.* (2005) detected no significant differences in arthropod species richness or abundance on individuals of *Populus fremontii*

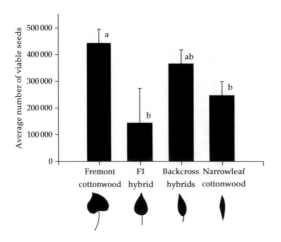

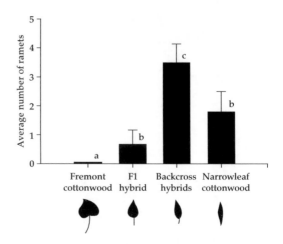

Figure 5.8 Viable seed production by Fremont (*P. fremontii*) and narrowleaf (*P. angustifolia*) cottonwoods, and their naturally occurring F$_1$ and backcross hybrids. Letters above each bar indicate statistical significance: different letters mean that the bars are significantly different (from Schweitzer *et al.* 2002).

Figure 5.9 Ramet production by Fremont (*P. fremontii*) and narrowleaf (*P. angustifolia*) cottonwoods, and experimental F$_1$ and backcross hybrids. Letters above each bar indicate statistical significance: different letters mean that the bars are significantly different (from Schweitzer *et al.* 2002).

(Fremont cottonwood), *Populus angustifolia* (narrowleaf cottonwood), or natural F$_1$ and backcross hybrid individuals. However, there were significant differences detected in the *composition* of the arthropod communities inhabiting the various genotypic classes. Specifically, hybrid genotypes housed a significantly different composition of herbivores than did individuals of the parental species (Wimp *et al.* 2005).

Data from herbivore analyses not only support the hypothesis of community-wide effects from introgression among cottonwood species, they also allow inferences concerning hybrid fitness. Indeed, Whitham (1989) used such data to describe hybrid zones as 'sinks for pests'. He argued that hybrid genotypes with lower fitness were attacked more often by pests and thus kept the pest species from encountering, and over evolutionary time adapting to, the parental genotypes (Whitham 1989). Regardless of the causes of the herbivore load in this hybrid zone (Floate *et al.* 1993), Whitham (1989) did record a higher incidence of pests on hybrid genotypes. This suggested that at least some hybrid genotypes possessed lower fitnesses than the parental individuals. Yet, as mentioned above, Wimp *et al.* (2005) did not find a similar quantitative difference in herbivores on hybrid (F$_1$ and backcross)

versus *P. fremontii* and *P. angustifolia* individuals. This indicates, at least with regard to resistance to herbivores, that the hybrids are not less fit than their parents. Similarly, Martinsen *et al.* (2001) discovered that while 80% of Fremont nuclear markers did not introgress across the hybrid zone, the remaining 20% of the nuclear markers along with the Fremont cpDNA and mtDNA were found in narrowleaf-like hybrids up to 100 km from the present-day zone of overlap (Figure 5.7). Martinsen *et al.* (2001) concluded that their findings provided '. . . evidence for positive, negative, and neutral effects of introgression.' Evidence of a range of fitnesses was also discovered by Schweitzer *et al.* (2002) in an analysis of sexual and asexual reproductive components of *P. fremontii*, *P. angustifolia*, and natural (F$_1$ and 'backcross') and experimental (F$_1$, BC$_1$, and BC$_2$) hybrids. These authors found that while F$_1$ and BC$_1$ hybrids demonstrated relatively low fitnesses, the naturally occurring backcrosses and the experimentally produced BC$_2$ hybrids demonstrated high relative fitness (Figures 5.8 and 5.9). As with the data for Louisiana irises (see section 5.3.2), some of the hybrid classes not only possessed high fitness, they also combined the greater sexual and asexual reproductive output of their parents (Schweitzer *et al.* 2002).

5.3.6 Hawaiian silversword complex

It has been argued that species of the Hawaiian genera *Argyroxiphium*, *Dubautia*, and *Wilkesia* constitute '. . . one of the most remarkable examples of adaptive radiation known to science' (Carr and Kyhos 1981). Carr and Kyhos (1981) substantiate their claim by pointing to the 28 species belonging to these three genera that (i) have bauplans ranging from large basal rosettes that form a single inflorescence before dying, to woody shrubs that annually produce many flowers, to trees, to vines and (ii) occur in widely varying terrestrial habitats including those near sea level to timberline, to regions receiving less than 45 cm annual precipitation, to an area characterized by approx. 1300 cm of rainfall per year, to very recent lava flows, to mature rain forests. The recognition of this group as an exemplar of a plant adaptive radiation has been repeatedly supported by a variety of studies (Witter and Carr 1988; Robichaux *et al.* 1990; Baldwin and Sanderson 1998).

In spite, or maybe reflective, of its remarkable diversity, and like other examples of adaptive radiation (e.g. members of the Cichlidae), the silversword alliance species remain capable of forming natural hybrids. Indeed, some 35 different combinations of intergeneric, intersubgeneric, and interspecific natural hybridization have been recorded (Carr and Kyhos 1981; Carr *et al.* 1989). Furthermore, depending largely on one's definition, either hybrid speciation or widespread introgression among various members of the silversword genera is necessary to account for patterns of phylogenetic discordance (Baldwin *et al.* 1990). It is perhaps then not surprising that the Hawaiian silversword genera arose from a hybridization event between North American species of tarweeds. Specifically, the Hawaiian forms reflect an allopolyploid speciation event caused by hybridization between members of the genus *Raillardiopsis*, and possibly also species of *Madia* (Baldwin *et al.* 1991; Barrier *et al.* 1999).

Data for the Hawaiian silversword alliance also allow inferences germane to the current discussion of hybrid fitness. First, a study by Robichaux (1984) found that *Dubatia ciliolata* and *Dubatia scabra* differed significantly in the water potentials of their tissues. The natural hybrid formed by these species demonstrates intermediacy for the same characteristics. This would suggest that the hybrid might have lower fitness relative to its parents in the respective parental habitats. However, the hybrid grows in the same habitat/site as *D. scabra* (younger lava flow; Robichaux 1984). This suggests that the natural hybrid genotypes are less fit and thus excluded from the *D. ciliolata* ecological setting, but do not suffer from lower fitness in the *D. scabra* habitat. A second series of studies also suggests that hybridization has contributed to adaptive diversification, and indeed the adaptive radiation in the Hawaiian silversword alliance. Specifically, Barrier *et al.* (2001) and Lawton-Rauh *et al.* (2003) found evidence consistent with the hypothesis that the original allopolyploid event released variation at floral regulatory genes, thus facilitating the adaptive radiation. Though possibly also reflective of selection and adaptive divergence, the finding of '. . . significantly accelerated rates of regulatory gene evolution in the Hawaiian species may reflect the influence of allopolyploidy . . . ' (Barrier *et al.* 2001). I will pursue this topic more fully in the next chapter, but it is significant to note that such whole-genome duplication is viewed as a major facilitator of organismic evolution (e.g. Furlong and Holland 2004).

5.3.7 *Salix*

Hybridization and introgression in the genus *Salix* (willows) has been recognized at least since the first half of the twentieth century. This is illustrated by the following quote from Lotsy (1916, pp. 122–123): 'Among plants, hybrids between different *Linneons* are quite common, and certain peculiarities in flower structure or behavior . . . even favor such crossfertilization among Linneons. An example of this is furnished by the willows.' The effects of hybridization among willow taxa can also be seen in taxonomic treatments, such as that of Wilkinson (1944), in which natural hybrids were recognized. Finally, the recent application of molecular markers to genotype plants from areas of spatial overlap between different willow species has confirmed genetic exchange through introgressive hybridization (e.g. Hardig *et al.* 2000).

Numerous analyses have collected data relevant to inferences of hybrid and parental fitnesses in willows. Like other asexually and sexually reproducing organisms, fitness estimates for willow genotypes must incorporate components for both reproductive avenues. Salik and Pfeffer (1999) pointed out that clonal reproduction could play a very significant role in allowing partially fertile, hybrid genotypes expanded opportunities for sexual reproduction. Indeed, their analyses identified one of four F_1 hybrid cross types—the two reciprocal F_1s from crosses of *Salix eriocephala* × *Salix exigua* and the two reciprocal F_1s from crosses of *S. eriocephala* × *Salix petiolaris*—that reproduced sexually significantly less than either its parents or the other three F_1 classes (Salik and Pfeffer 1999). Yet, all four of the F_1 cross types equaled or exceeded their parents in asexual reproduction. This suggests that the partially fertile F_1 class would have repeated opportunities to contribute to introgression. It is also important to note, however, that the other three F_1 classes equaled or exceeded the parental genotypes for sexual reproductive components as well (Salik and Pfeffer 1999). This reflects a higher composite fitness (based on asexual and sexual components) for at least three of the four hybrid classes relative to the parental genotypes.

Fritz and his colleagues have undertaken a series of studies to test, among other hypotheses, the finesses of various hybrid willow genotypes. In a study of the effect of water availability, Orians *et al.* (1999) estimated the fitness of *S. eriocephala*, *Salix sericea*, and their F_1 hybrids. These authors determined that the F_1 hybrids had a greater fitness (i.e. demonstrated heterosis) when water availability was not limited. However, when drought conditions were imposed the F_1 genotypes did relatively poorly compared with the parental genotypic classes. This result again indicates that, like almost all genotypes, rather than possessing uniformly high or low relative fitnesses, hybrid genotypes often demonstrate genotype × environment dependent fitness estimates.

Another area of particular emphasis for Fritz *et al.* has been the dissection of the effects of hybridization on plant–herbivore interactions in general, and for the willow system in particular (Fritz *et al.* 1994, 1999, 2003; Fritz 1999; Czesak *et al.*

2004). The data from these analyses reveal a range of resistances/susceptibilities by hybrids to various herbivores, reflecting once again the complex genotype by environment-mediated determination of hybrid (and parental) fitness. For example, Czesak *et al.* (2004) examined the resistance of *S. eriocephala*, *S. sericea*, and F_1, BC_1, and F_2 experimental hybrids to both aphids and mites. The F_2 genotypic class used in this analysis reflects well the different fitness estimates garnered from different data; F_2s, as a class, were significantly more susceptible to aphid attack, but were significantly less susceptible to mite damage. These alternate findings suggest that F_2 genotypes would be as successful, on balance, as the parents; the species show inverse susceptibilities, compared to the F_2 class, to these herbivores. It is important to note once more that the fitness estimates for the F_2s reflect a heterogeneous pool of genotypes and thus fitness estimates for individual F_2 genotypes could be much higher or lower than the average for F_2s (Arnold 1997).

The findings from each of the plant systems reviewed above lead to the same conclusion—hybrid genotypes demonstrate a variety of fitnesses, often in an environment-dependent fashion. In the last section, I address the same issue for animal species complexes.

5.4   Genetic exchange and fitness: animals

5.4.1   *Bombina*

In Chapter 4 I argued that the fire-bellied toad species *B. bombina* and *B. variegata* represent an excellent example of some of the evolutionary effects of hybridization. Also, as pointed out in the previous chapter, much of the work on this species complex, particularly by Szymura, Barton, and their colleagues, has emphasized the reduced fitness of some hybrid genotypes (e.g. Szymura and Barton 1986, 1991). In particular, the coincident and concordant step-clines for many unlinked markers have been used as evidence for an overwhelming influence of (i) selection against hybrids and (ii) continual immigration into regions of overlap, for the establishment and preservation of the hybrid zones (Barton and Hewitt 1985).

Although many analyses of *B. bombina* and *B. variegata* have detected step-clines suggestive of tension hybrid zones (i.e. environment-independent hybrid zones, in regard to hybrid fitness), recent studies have found a preponderance of evidence suggesting instead that the *Bombina* interactions form mosaic hybrid zones. This latter conclusion is reflective of the detection of habitat preferences for each of the species (MacCallum *et al.* 1998). In addition, unlike the prediction of the tension zone model, these zones are not characterized by hybrids of uniformly low fitness. For example, Kruuk *et al.* (1999) analyzed fitness by examining the development of eggs collected from a natural hybrid zone. On average, the central (or hybrid) populations demonstrated significantly higher mortality (Table 5.1). This indicates that hybrid fitness, on average, was significantly lower than that observed for the parental forms (Table 5.1; Kruuk *et al.* 1999). In addition to the overall lower fitness, eggs from the center of the hybrid zone also showed the greatest level of variation in fitness (compare the standard deviation for the 0.21–0.40 hybrid sample from the center of the zone with those of the 'pure' parental samples, 0.00–0.20 and 0.81–1.00; Table 5.1; Kruuk *et al.* 1999). Significantly, the increase in the variation of fitness estimates reflects not only hybrids possessing the lower range of values, but also hybrids that possess fitnesses equivalent to that of the parental genotypes (Kruuk *et al.* 1999).

Another indication of the role that variation in hybrid fitness can play in the evolution of *B. bombina* and *B. variegata* comes from a study of the genetic variation present within a hybrid zone near Apahida in Romania. The pattern of microsatellite variation detected by Vines *et al.* (2003) led them to conclude

Table 5.1 Total mortality across a Croatian hybrid zone between *B. bombina* and *B. variegata*. 0.00–0.20, *B. bombina*; 0.81–1.00, *B. variegata* (Kruuk *et al.* 1999).

Genotypic ranges	Mortality (%)	SD	Number of hatches
0.00–0.20	1.00	3.3	28
0.21–0.40	13.52	23.14	35
0.41–0.60	20.69	15.77	45
0.61–0.80	8.27	11.37	27
0.81–1.00	3.34	6.48	33

that introgression within this hybrid zone was likely causing a genetic admixture that would lead to the species gene pools being '. . . homogenized at all but the selected loci. . . ' However, selection *against* the transfer of certain traits is not reflective of the only type of selection that is apparently operating in this zone of overlap. Instead, the data for the fire-bellied toads indicates that hybrid genotypes in animals, like microorganisms and plants, can and do demonstrate a range of fitness estimates, including those that are equivalent (or greater) than their parents.

5.4.2 *Rana*

The frog genus *Rana* has been used to examine numerous evolutionary concepts and processes; among these are reproductive isolation and evolutionary diversification. Indeed, some of the *Rana* species complexes are excellent examples of the detection of numerous, previously unrecognized, evolutionary lineages. Such definitions have come from experimental crosses, surveys of natural mating call variation and genotypic distributions—each data set was utilized to estimate levels of reproductive isolation (e.g. Littlejohn and Oldham 1968; Cuellar 1971; Frost and Bagnara 1976; Santucci *et al.* 1996). In addition to its use in defining many fundamentally important evolutionary processes, *Rana* has also provided numerous examples of hybridization and introgression. Included among these latter studies are those of the North American leopard frogs known collectively as the *Rana pipiens* complex (e.g. Frost and Platz 1983; Kocher and Sage 1986).

Both earlier studies of hybridization and introgression (Kruse and Dunlap 1976; Sage and Selander 1979) and a set of recent analyses by Parris and colleagues (Parris 1999, 2000, 2001a, b; Parris *et al.* 1999) have shed light on the pattern of hybrid fitness in the *R. pipiens* complex. Parris *et al.* examined fitness components for the species *Rana blairi* and *Rana sphenocephala* and various experimental hybrids in both experimental and natural ecological settings. These analyses detected a complex set of genotype × environment fitness effects in the parental and hybrid classes. Two of these studies can serve as exemplars of the general findings. The first involved the rearing of *R. blairi*, *R. sphenocephala*, and experimental F_1, BC_1, and BC_2

larvae in experimental ponds (Parris 1999). The development (until metamorphosis) of each class was monitored in both single-genotype and two-way (i.e. larvae from both parents or from one parent and one hybrid class) mixtures. Parris (1999) summarized his findings in the following way: 'On average, primary-generation (F_1) hybrids experienced either increases or decreases in larval fitness component values relative to parental species, whereas advanced-generation (BC_1 and BC_2) hybrids had equivalent or higher values for most fitness components. . . ' A second analysis by Parris (2001a) considered the fitness of parental species and F_1 genotypes under natural conditions. In this case, the three classes of genotypes (both species and the F_1 hybrid) were transplanted into enclosures within natural ponds in three different habitats—prairie, woodland, and river floodplain. Parris (2001a) found no evidence for lower fitness (relative to the parents), in any of the environments, for the F_1 hybrids. Indeed, for one of the fitness components (body mass at metamorphosis), the F_1 hybrid exceeded that of *R. sphenocephala*, and was equivalent to *R. blairi* in the river floodplain habitat (where all three genotypic classes had their maximum survivorship and metamorphosis; Figure 5.10; Parris 2001a). It is significant that the F_1 demonstrated a high fitness in the floodplain habitat (Figure 5.10) since this is a primary site in which the species overlap and hybridize (Parris 2001a). This suggests that the initial hybrid generation would act as a very effective bridge for further gene exchange. In general then, the evolutionary role of introgression in the *R. pipiens* complex is likely affected by differential, often environmentally dependent, selection.

5.4.3 Cichlids

Members of the New and Old World clades of cichlid fish are reproductively isolated by at least the prezygotic mechanism of female mate choice (see section 4.3.2). Indeed, Smith *et al.* (2003) argued that male coloration (which affects female mate choice) '. . . is the central component maintaining separation of [cichlid] gene pools.' In spite of the reproductive isolation afforded by mate choice, introgressive hybridization has now been

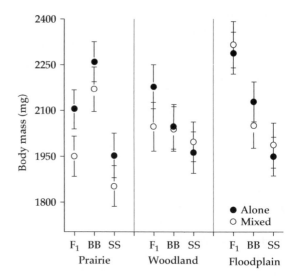

Figure 5.10 Body mass of F_1 hybrids, *R. blairi* (BB), and *R. sphenocephala* (SS) at the time of metamorphosis in three habitat settings (from Parris 2001a).

implicated as a major factor in the diversification of the African cichlids. This conclusion is supported by (i) the application of models of divergence that suggest gene flow between species, since the time of divergence from a common ancestor (Hey *et al.* 2004; Won *et al.* 2005), (ii) the population-genetic structure of sympatric species (Rüber *et al.* 2001), (iii) the origin of novel male coloration in hybrid populations (Smith *et al.* 2003), and (iv) the origin of new, stabilized species via introgressive hybridization (Figure 4.11; Salzburger *et al.* 2002). Indeed, Seehausen and his colleagues have hypothesized not only ongoing introgression and hybrid speciation for limited numbers of cichlid taxa, but for the entire species flock (Seehausen *et al.* 2002; Seehausen 2004; Joyce *et al.* 2005). Specifically, Seehausen (2004) has invoked a model of adaptive radiation of the cichlids from hybrid swarms, as defined by Anderson (1949).

Although it is now apparent that introgressive hybridization is one of the evolutionary forces affecting the genetic structure of cichlid populations, we still must ask whether the data are useful for addressing directly the issue of hybrid fitness. All of the introgression detected by the various studies of cichlid populations could reflect the diffusion of neutral alleles (Barton and Hewitt 1985). Several findings, however, belie this explanation. For example, Smith *et al.* (2003) detected a morphologically unique cichlid population from Lake Malawi. This population, belonging to the genus *Metriaclima* (i.e. *Pseodotropheus*), demonstrated a range of phenotypes intermediate between *Metriaclima thapsinogen* (possessing red dorsal fins) and *Metriaclima zebra* (possessing blue-black dorsal fins; Smith *et al.* 2003). Because male coloration is such a key factor in cichlid reproductive isolation, the unique pattern found in this population suggested the origin of a new adaptation. Indeed, reflecting the presence of the hybrid coloration, Smith *et al.* (2003) nominally referred to fish from this area as *Metriaclima* sp. Smith *et al.* (2003) also argued that *Metriaclima* sp. provided evidence for the role of introgression in replenishing genetic variation— and presumably producing hybrids with increased fitness—in cichlid species that have gone through population bottlenecks. Repeated (and in some cases ongoing) hybridization among previously allopatric taxa '. . . could thus explain the maintenance of high genetic variation within populations of Lake Malawi cichlids' (Smith *et al.* 2003). Consistent with the findings of the *Metriaclima* study, the observation of introgressive hybridization resulting in new species, and possibly the radiation of entire clades of African cichlids, also argues for the origin of adaptations as a result of hybrid, genetic mosaicism (Salzburger *et al.* 2002; Seehausen *et al.* 2002; Allender *et al.* 2003; Seehausen 2004; Joyce *et al.* 2005).

5.4.4 Whitefish, redfish, and charr

The work of Bernatchez and colleagues has added much to our understanding of adaptive evolution, particularly with regard to the role of ecological adaptations in evolutionary diversification. A main emphasis of Bernatchez's group has been to study the occurrence and causes of parallel evolutionary diversification in fish species complexes (e.g. Bernatchez *et al.* 1996; Turgeon *et al.* 1999; Rogers and Bernatchez 2005). For example, Pigeon *et al.* (1997) used phylogenetic evidence to test for the multiple independent origins of 'dwarf' whitefish

(i.e. *Coregonus clupeaformis*) phenotypes from the 'normal' morphotype. Pidgeon *et al.* (1997) found that the similar morphological transitions (i.e. normal→dwarf) had indeed arisen from independent lineages. This result provided excellent evidence of (i) the action of repeated bouts of natural selection and (ii) the role of a common selective regime that produced a common evolutionary product. In this case, the action of similar ecological pressures—as the cause of the parallel changes— was an obvious candidate (Pigeon *et al.* 1997). However, other studies have suggested that the origin of the whitefish lineages reflects a complex set of underlying mechanisms, including multiple invasions, ecological selection, and introgression among morphotypes (Bernatchez *et al.* 1996; Turgeon *et al.* 1999).

Studies of the whitefish radiation have also resulted in data that indicate the effects of differential selection on hybrid and parental genotypes. In this context, Campbell and Bernatchez (2004) and Rogers and Bernatchez (2005) suggested that only a small number of genomic regions (as little as 1.2% of the genome) were necessary to account for the divergent adaptations expressed by the various morphotypes. It was argued, however, that recombination within these regions (in hybrids) was likely to give rise to unfit genotypes, thus suppressing gene flow (Rogers and Bernatchez 2005). Of course, one corollary of this finding was that a large proportion of the whitefish genome should be permeable to gene flow, and available for the evolutionarily creative outcomes provided by such introgression. Intriguingly, the permeability of whitefish genomes is well substantiated, with some lakes demonstrating populations characterized as completely introgressed (Lu *et al.* 2001). Although the detection of such hybrid swarms has been attributed to a lack of differentiated niches into which new morphotypes could expand, and in which hybrids would be at a disadvantage (Lu *et al.* 2001), it seems just as plausible that high levels of genetic exchange reflect positive selection for some hybrid genotypes (Arnold 1997).

Additional analyses by Bernatchez and his colleagues, of both marine redfish (genus *Sebastes*) and freshwater charr (genus *Salvelinus*) species, reflected the potential role for positive selection on hybrid genotypes (Roques *et al.* 2001; Doiron *et al.* 2002). For the redfish, hybridization between *Sebastes fasciatus* and *Sebastes mentella* resulted in extensive introgression (Roques *et al.* 2001). Most significant for the present topic, the pattern and apparent stability of the introgressive hybridization (the formation of two reciprocal, species-like, introgressed forms; Figure 5.11) led to the conclusion that selection was favoring some hybrid genotypes in the intermediate ecological settings (Roques *et al.* 2001). Likewise, in a study of mtDNA variation in brook and arctic charr, Glémet *et al.* (1998) detected the introgression and substitution of the mtDNA from brook charr (a temperate-adapted species) with that from arctic charr (an arctic-adapted species) in northern populations of the brook charr. This result suggested the hypothesis that the introgression of arctic charr mtDNA, into brook charr, reflected an adaptive trait transfer event facilitating a fitness increase in northern, introgressed brook charr populations (Glémet *et al.* 1998; Doiron *et al.* 2002).

5.4.5 Flycatchers

The collared and pied flycatchers (*Ficedula albicollis* and *Ficedula hypoleuca*, respectively) have been used as a model system for studying reproductive isolation and biological speciation (e.g. Sætre *et al.* 2001). In terms of reproductive isolation, both pre-zygotic (i.e. female mate choice determined by male plumage) and post-zygotic (i.e. hybrid sterility) barriers have been examined (Tegelström and Gelter 1990; Sætre *et al.* 1997). In the case of mate choice, this species pair is an example of the occurrence of character displacement and putative reinforcement of premating reproductive barriers. Thus sympatric populations demonstrate a greater degree of male coloration and song divergence than allopatric populations (Sætre *et al.* 1997; Haavie *et al.* 2004). Yet natural hybrid zones between the flycatcher species are characterized by introgression (Tegelström and Gelter 1990; Sætre *et al.* 2001, 2003). For example, population-genetic variation in the maternally transmitted mtDNA

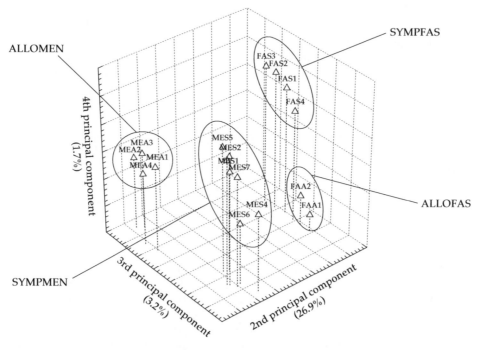

Figure 5.11 Principal components analysis (in a three-axis, multidimensional space) based on microsatellite variation of 17 redfish samples. The respective contributions of the axes are given in percentages (%). The four groups correspond to the sympatric and allopatric *S. mentella* and *S. fasciatus* samples: SYMPMEN, ALLOMEN, SYMPFAS, and ALLOFAS, respectively (from Roques *et al.* 2001).

and bi-parentally-inherited nuclear genes caused Tegelström and Gelter (1990) to conclude that sterile female hybrids (the heterogametic sex and thus reflective of Haldane's rule) were limiting mtDNA introgression, but that fertile males were facilitating nuclear gene exchange. Sætre *et al.* (2003) discovered the stereotypical pattern of differential permeability of genomic regions in detecting higher levels of introgression for autosomal versus sex-chromosome loci. Thus the introgression between these species once again results in mosaic genomes characterized by many introgressed regions as well as some regions that reflect limited genetic exchange.

The fitness of *F. albicollis* × *F. hypoleuca* hybrids has been estimated in a number of ways. As mentioned above, F_1 hybrids between these species demonstrate Haldane's rule, with the heterogametic sex (i.e. females) being sterile (Tegelström and Gelter 1990). This indicates that there is strong selection against the F_1 class as a whole, but not against both sexes of F_1 hybrids. Another data set that provides evidence for the occurrence of a range of fitnesses in hybrids involves the observation of differential introgression of autosomal compared with sex-chromosome markers (Sætre *et al.* 2003). In this regard, Sætre *et al.* (2003) found that not only did loci on the sex chromosomes contribute to hybrid sterility, but that the sex chromosomes also contained genes that affected traits for species recognition. This then suggested a cause for their dual observations of reinforcement of reproductive isolation for the sex chromosome loci and '. . . rather extensive introgression and recombination of autosomal genes.' The large-scale transfer of some genes is, as in other systems (e.g. *Mus*, Darwin's finches, Louisiana irises, and HIV) likely reflective of both neutral diffusion and positive selection. Indeed, support for the hypothesis of adaptive introgression between the flycatcher species comes from an analysis of rates of introgression among nine autosomal and four Z-linked (i.e. sex-chromosome) loci (Borge *et al.* 2005). Specifically, these authors found that the *F. hypoleuca* Alasy locus allele was found to have introgressed at very high frequencies into *F. albicollis*. Reminiscent of similar findings and conclusions concerning differential introgression in other

species complexes (e.g. Payseur *et al.* 2004), Borge *et al.* (2005) hypothesized that the pied flycatcher Alasy allele was linked to a locus positively selected in the hybrid background. The high-frequency introgression of the pied into the collared flycatcher background would thus reflect the fact that '. . . the collared flycatcher has inherited an adaptive allelic state from the pied flycatcher through introgressive hybridization. . . ' (Borge *et al.* 2005).

Another study that examined the fitness consequences of hybridization between the flycatcher species came to an intriguing conclusion. As is usual for many animal evolutionary biologists, Veen *et al.* (2001) emphasized that hybrids with lower fitness were produced when the two flycatcher species hybridized. The unique conclusion from their study was that fitness compensations occurred when there was a deficit of conspecific males (Veen *et al.* 2001). They thus surmised that in such demographic situations it would benefit females to mate with heterospecific males rather than to not mate at all. In Veen *et al.*'s (2001) words, '. . . hybrid pairing may sometimes represent adaptive mate choice, because the cost-reducing mechanisms we identify compensate for the fitness that would otherwise be lost owing to producing unfit hybrids.'

The data for the patterns and processes associated with *F. albicollis* × *F. hypoleuca* hybrid zones apparently reflect a range of selection coefficients for hybrid genotypes. In the last example of this chapter I again discuss findings from analyses of introgression between avian taxa, the manakins.

5.4.6   Manakins

The avian species *Manacus candei* (white-collared manakin) and *Manacus vitellinus* (golden-collared manakin) form a hybrid zone in the Bocas del Toro province of Panama (Parsons *et al.* 1993). Within this area of overlap, a highly discordant pattern of morphological and genetic character replacement occurs. The striking aspect of the manakin variation is reflected by offset, non-coincident, clinal changeover for molecular markers versus male plumage characteristics (Figure 5.12; Parsons *et al.* 1993; Brumfield *et al.* 2001). In particular, the male plumage characteristics from the golden-collared

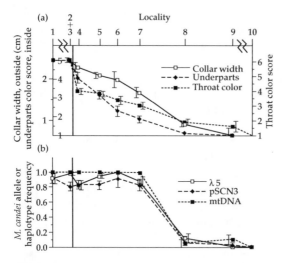

Figure 5.12 Patterns of frequency changeover for (a) male plumage and (b) genetic markers (mtDNA and two nuclear loci, λ5 and pSCN3) across the hybrid zone between the white-collared (*M. candei*) and golden-collared (*M. vitellinus*) manakins in Panama. All population samples in the rectangle to the left of each graph have white-collared phenotypes. All other birds outside of this rectangle have the golden-collared phenotype. Numbers along the *y* axis in (a) are: outside left, collar width (in cm); inside left, underparts index score; outside right, throat color index score. The cline for male plumage characteristics indicates introgression approximately 40 km further west than the molecular markers (from Parsons *et al.* 1993).

species appear to have introgressed into populations of the white-collared phenotype some 40 km further than any of the DNA markers (Figure 5.12; Parsons *et al.* 1993). Parsons *et al.* (1993) hypothesized that introgression of the male plumage characteristics from the golden- into the white-collared species reflected favorable natural selection. In particular, it was postulated (Parsons *et al.* 1993; Brumfield *et al.* 2001) that hybrid males exhibiting the golden-collared phenotype had a selective advantage relative to *M. candei* (i.e. the white-collared manakin). Parsons *et al.* (1993) argued that this advantage was most likely due to highly skewed male mating success at the lek sites of *Manacus*.

Strong support for the hypothesis of sexual selection driven introgression of the golden-collared traits came from a study of the male behavior of three classes of *Manacus* (McDonald *et al.* 2001). The three classes were (i) white-collared, (ii) golden-collared, and (iii) 'lemon-collared'. These

phenotypic/genotypic classes reflected *M. candei*, *M. vitellinus*, and natural hybrids, respectively. McDonald *et al.* (2001) hypothesized that sexual selection-derived introgression (i.e. adaptive trait introgression) would be facilitated by higher levels of aggressiveness by the golden- and lemon-collared males, relative to the white-collared form. This was indeed their finding. First, the golden and lemon phenotypes attacked the taxidermy-mounted specimens presented to them significantly more often than the white-collared birds. Second, the lemon-colored hybrid males demonstrated more vocalizations than either of the parental species (McDonald *et al.* 2001). In sum, these data argued again for adaptive trait transfer, and reflected elevated hybrid fitness.

5.5 Summary and conclusions

My goal for this chapter was to demonstrate the truism that hybrid genotypes vary in fitness. Often this variation is environment-dependent, but hybrids that have low or high fitness across many habitats (i.e. their fitness appears environment-independent) are also found. I have emphasized two different signatures that can indicate the relative fitness of hybrids. The first involves evidence that there has been a transfer of traits that are adaptive in the hybrids. Such a transfer can be detected by a novel phenotype or by the introgression of markers or traits at a significantly higher than expected frequency in natural or experimental hybrid populations. The second method for estimating hybrid fitness comes from studies of fitness components in experimental or natural environments. These latter studies provide a detailed, composite fitness not available from the more correlative analyses of natural populations.

As I stated at the outset, my choice of examples for this chapter was designed to point to the fact that hybrid fitness is conceptually and empirically the same measurement as that for any non-hybrid genotype. Furthermore, whether a hybrid is the result of viral recombination, lateral gene transfer, or introgressive hybridization, the patterns of hybrid fitness found in natural and laboratory populations are concordant.

CHAPTER 6

Gene duplication

The merger of two genomes with different evolutionary histories in a common nucleus appears to offer unique avenues for phenotypic response to selection.

(Jiang *et al.* 1998)

The patterns of hybridization and floral morphology in *Lepidium* support the hypothesis that the preponderance of species with reduced floral forms could have been produced by . . . allopolyploid speciation.

(Lee *et al.* 2002)

Manipulation of such molecular mechanisms would be important under conditions where the degree of ploidy impacts both the level and profile of gene expression and ultimately, the cellular physiology.

(Ravid *et al.* 2002)

Homoeologous copies of ASAP1 and ASAP3/TM6 show differing levels and patterns of nucleotide polymorphism . . . These results suggest that differing evolutionary forces can affect duplicate loci arising from allopolyploidization.

(Lawton-Rauh *et al.* 2003)

We conclude that a WGD [whole-genome duplication] event occurred in the *Saccharomyces* lineage . . . either by endo-duplication (auto-polyploidy) or fusion of two close relatives (allo-polyploidy).

(Kellis *et al.* 2004)

. . . we have apparently caught concerted evolution homogenizing rDNA units between parental genomes in the act.

(Kovarik *et al.* 2005)

6.1 Gene duplication and evolution

The central goal for this chapter is to illustrate some of the evolutionary effects of gene duplication arising from genetic exchange. Though I will discuss several cases in which the processes leading to gene duplication have led to what are recognized as novel taxa (or even entire clades), I will not discuss in detail the formation of new evolutionary lineages via genetic exchange. Instead, this latter topic will be reviewed in Chapter 7 in the context of other classes of hybrid lineage formation. In the present chapter, I will limit my discussion to the possible effects from partial genome duplication and whole-genome duplication (WGD)—arising from genetic exchange—in terms of the evolution of (i) whole genomes, (ii) genes and gene families (including the origin of novel adaptations), and (iii) whole clades of organisms (i.e. adaptive radiations). It has been postulated that the evolution of organismal complexity and new adaptations is highly correlated to the origin of novel gene functions fed by duplication events (e.g. see Force *et al.* 1999; Adams and Wendel 2005; Comai 2005; Davis and Petrov 2005; De Bodt *et al.* 2005; Panopoulou and Poustka 2005; Crow *et al.* 2006 for discussions). Duplication events may arise along a single chromosome or through chromosome doubling, or through a combination of both processes (e.g. the ANTP *homeobox genes*; Castro and Holland 2003). Furthermore, WGD events can reflect either allopolyploidy (i.e. from crosses between divergent evolutionary lineages) or *autopolyploidy* (i.e. from crosses between individuals from the same or similar evolutionary lineages). For the present discussion I will not concern myself with whether or not WGDs fall within one or other of these categories. There are three reasons for this decision: (i) in general, patterns of gene and genome evolution following duplication are similar whether the original event involved allopolyploidy, autopolyploidy, introgression, lateral gene transfer, or viral recombination; (ii) in terms of WGD events, the definitions of auto- and allopolyploidy overlap to a great degree (Stebbins 1947); and (iii) I emphasize throughout this book that the degree of differentiation between lineages that exchange genes is largely irrelevant (e.g. see the definition of 'natural hybrid' given in the Glossary) with regard to the potential evolutionary effects.

Another goal for this chapter is to highlight the fact that the evolution of genomes, gene functions, adaptive traits and adaptive radiations are not discrete events and processes; each of the topics is highly correlated with the remaining topics. One example of this is reflected by the interrelationship of DNA methylation, gene expression and transposable element activation in plant allopolyploids. In this case the addition of cytosine methylation can affect the expression of individual genes (e.g. Wang *et al.* 2004) and gene families and may also lead to the reactivation of transposons (Riddle and Birchler 2003). Such correlations will lead inevitably to examples placed in one section of this chapter that can also be used to exemplify other processes. I emphasize this not as an apology or caveat, but rather to indicate the interdependence of the various evolutionary processes. This key aspect of interdependent effects is becoming more evident as additional genomic, phenotypic, phylogenetic and

fitness data become available for organisms affected by genetic exchange-mediated duplications.

6.2 Genetic exchange: genomewide evolution following duplication

6.2.1 Genomewide effects—epigenetic changes through methylation

Following duplication events many processes may affect smaller or larger proportions of the genome. In the last section I will consider examples of duplication events that may have been causal in organismic evolution, leading to radiations of entire clades. Yet, the basis of these radiations begins with the molecular events that affect the functioning of the genomic elements that have been duplicated. When considered in the context of whole genome evolution, genetic transfer-induced duplications may lead to widespread changes resulting from *epigenetic* modifications (e.g. *methylation*), the activation of transposable elements, *genome downsizing* and chromosome rearrangements (e.g. Riddle and Birchler 2003; Soltis *et al.* 2003; Leitch and Bennett 2004; Levy and Feldman 2004; Pires *et al.* 2004; Salmon *et al.* 2005). At the level of individual genes or gene families (see Section 6.3), some of the possible molecular evolutionary processes include *nonfunctionalization* (i.e. *silencing*), *sub-functionalization*, *neo-functionalization* or *concerted evolution* (Prince and Pickett 2002; Hartwell *et al.* 2004; de Souza *et al.* 2005; Duarte *et al.* 2006).

As indicated in the Glossary definition, epigenetic alterations are genomic modifications that can be passed from one generation to another, but that are not caused by DNA-based (i.e. mutational) changes. Soltis *et al.* (2003) reflected this definition when they stated, 'Epigenetic changes, such as DNA methylation, histone modification, RNA interference, and dosage compensation, may alter gene expression without a change in DNA sequence. . .' Furthermore, as observed by Riddle and Birchler (2003) in a review of some of the effects from hybridization and allopolyploidy (particularly in plants), 'Epigenetic changes, especially altered cytosine methylation patterns, are assumed to be responsible for altered gene expression states, and for the reactivation of transposable elements.' As an exemplar for discussing the occurrence and possible affects from epigenetic changes that follow duplications, I will consider patterns of methylation. In the following sections, I will review data that are consistent with the hypothesis that partial or whole-genome duplications cause changes in patterns of methylation that can lead to alterations in gene expression and the reactivation of transposons. However, it is first necessary to address whether or not the distribution of methylated bases is different in the hybrid/duplicated genome relative to the progenitor genotypes. I will illustrate the answer to this question with four examples of allopolyploidy in plants—cotton, wheat, *Brassica*, and *Spartina*.

I wish to begin the discussion of epigenetic modifications, and methylation per se, with an example that presents a '... striking difference from other synthetic allopolyploids. . .' (Riddle and Birchler 2003). My rationale for discussing an 'exception' first is to highlight the processes encountered in the majority of examples. This exception to the rule involves members of the cotton genus *Gossypium*. Not all aspects of cotton allopolyploid evolution demonstrate significant differences relative to other polyploid systems. For example, repeated-sequence copy-number evolution, and the activation of transposable elements, has been inferred for cotton as it has been for other allopolyploids (Zhao *et al.* 1998a, b). Furthermore, concerted evolution of rRNA genes has resulted in a homogeneous array of members of this repeat family (Wendel *et al.* 1995). Finally, patterns of gene expression show significant differences between the progenitor species and the allopolyploid cotton derivatives (Adams *et al.* 2003; Adams and Wendel 2004). Notwithstanding these similarities, cotton allopolyploids do present a significant difference in their genomic stability relative to other species complexes. In regard to the present topic, the alteration of gene expression and transposon activity were not attributable to changes in the distribution of methylation: methylation patterns were not changed between the parental genotypes and their allopolyploid offspring (Liu *et al.* 2001). Thus for *Gossypium* it appears that the (normally) methylation-driven changes are caused by other factors (Riddle and Birchler 2003).

In contrast to *Gossypium*, studies of wheat allopolyploids have detected the entire series of events (i.e. gene inactivation, genome downsizing, transposon activation, and chromosomal rearrangements) that are likely caused, at least partially, by the initial repatterning of cytosine methylation. For example, Shaked *et al.* (2001) analyzed methylation patterns in F_1 hybrids and their allopolyploid derivatives from crosses involving *Aegilops* and *Triticum* (both members of the wheat clade). These investigators utilized the 'methylation-sensitive amplification polymorphism' methodology (Reyna-López *et al.* 1997 as modified by Xiong *et al.* 1999) to assay methylated and non-methylated loci. The series of analyses by Shaked *et al.* (2001) revealed that: (i) 13% of the loci demonstrated alterations in the methylation patterns between the diploid parents and their allopolyploid progeny (cytosine base pairs in the allopolyploids were methylated or demethylated relative to the parents); (ii) methylation changes affected both low-copy number and repetitive fractions of the allopolyploids' genomes; and (iii) repetitive, retro-element members demonstrated changes in the distribution of methylation. A second study (Kashkush *et al.* 2002) also found evidence for methylation-induced changes in the genomes of a newly synthesized wheat allopolyploid, in this case from a cross between *Aegilops sharonensis* and *Triticum monococcum*. Furthermore, Kashkush *et al.* (2002) estimated that 1–5% of the genes in the allopolyploid had been silenced. Of the silenced genes, a large proportion had altered methylation relative to their parents (Kashkush *et al.* 2002). Overall, the alterations in cytosine methylation in the wheat allopolyploids were extensive and likely affected further genomic alterations.

The genus *Brassica* represents another model system developed to decipher genomic changes following allopolyploidization. A number of results suggest the role played by methylation-mediated epigenetic changes in allopolyploid lineages. For example, Chen and Pikaard (1997) detected *nucleolar dominance* in *Brassica* allopolyploids. This epigenetic modification, resulting in the transcription of only one of the parental species' rRNA gene arrays, is likely partially reflective of differential methylation. Indeed, Frieman *et al.* (1999) have argued that although other factors are obviously involved in the initial inactivation of one set of parental rRNA genes, 'Silencing is enforced by cytosine methylation and histone deacetylation.' In addition to nucleolar dominance, Song *et al.* (1995) found evidence for genome-wide effects from alterations in the methylation of synthetic allopolyploids formed from crosses between *Brassica rapa*, *Brassica nigra*, and *Brassica oleracea*. Though these authors argued for a minimal effect on genome functioning from methylation subsequent to allopolyploidization, two of nine (i.e. 22%) molecular probes detected modified methylation patterns.

The final example of the occurrence of epigenetic affects due to modifications in methylated bases comes from the plant genus *Spartina*. Of particular interest is *Spartina anglica*, an allopolyploid species that originated in England at the end of the nineteenth century from natural hybridization between the native species *Spartina maritima* and the introduced North American species *Spartina alterniflora*. One advantage—for tests of hypotheses concerning genomic evolution—realized from this neo-species is the provision of data concerning genomic alterations during the earliest stages of allopolyploidization. With regard to epigenetic alterations, Ainouche *et al.* (2004) and Salmon *et al.* (2005) detected major changes in the distribution of methylated bases in the parental species (*S. maritima, S. alterniflora*), natural F_1 hybrids (e.g. *Spartina* × *townsendii*), and the natural allopolyploid species *S. anglica* (Figure 6.1). For example, Salmon *et al.* (2005) reported methylation changes in both natural hybrid classes that affected almost 30% of the parental fragments (Figure 6.1). The *Spartina* data reflect the highest rates of alterations in methylation patterns in plants seen thus far (Salmon *et al.* 2005). However, as discussed above, alteration in the genomic distribution of methylated bases is the rule following genomic duplications.

6.2.2   Genomewide effects—activation of transposable elements

In discussing the role of transposable-element reactivation in the evolution of gene function and genome structure, Feschotte *et al.* (2002) made the following observation, 'The recent demonstration that *Arabidopsis* TEs [transposable elements] can

be reactivated in genetic backgrounds that are deficient in aspects of epigenetic regulation brings this story full-circle. . .' The 'story' referred to by these authors was that written largely by Barbara McClintock, including her description of unstable mutations, some of which caused the 'breakage-fusion-bridge cycle' (McClintock 1984). Regarding the subject of this chapter, it is also evident that the conclusion of Feschotte *et al.* (2002) is reflective of possible effects from genomic duplications as well. Thus numerous authors speak of the 'genomic shock' resulting from the bringing together of divergent genomes, the effects of which may include the reactivation of transposable elements (Kashkush *et al.* 2003). Such reactivations may cause additional phenomena (e.g. chromosome rearrangements and the modification of gene expression) that will be discussed below. However, as with putative methylation-induced genomic alterations, it is first necessary to define whether or not genome duplications do indeed cause transposon reactivation.

As with DNA methylation, not all cases of allopolyploidy provide evidence of elevated transposon activation. In regard to the *Spartina* example discussed above, it is particularly noteworthy that Ainouche *et al.* (2004) concluded the following: 'No burst of retroelements has been encountered in the [natural] F_1 hybrid or in the allopolyploid, suggesting a "structural genome stasis" rather than "rapid genomic changes" '. These authors did, however, find evidence for a significant difference in the pattern of methylation in the progenitor species and their hybrid derivatives (see also Salmon *et al.* 2005). Though methylation differences in *Spartina* co-occur with chromosome reorganization and were also suggested as a potential facilitator of adaptive changes in the allopolyploid (Salmon *et al.* 2005), they are not seen to have affected a detectable reactivation of transposable elements.

In contrast to *Spartina*, genome duplication via allopolyploidy in wheat produces an apparent transcriptionally mediated reactivation of transposable elements. Specifically, changes in cytosine methylation in the hybrid offspring were found to affect both retro-element family members (and other repetitive sequences as well) and low-copy DNA loci at approximately equal frequencies.

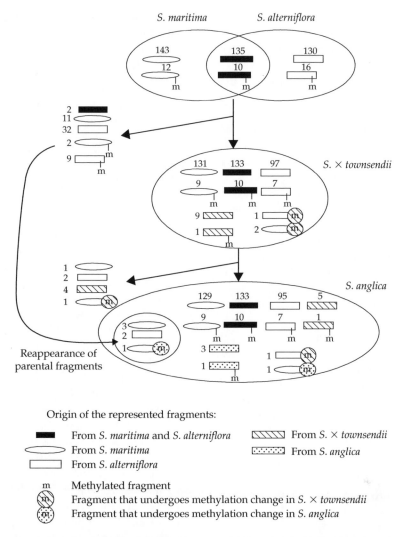

Figure 6.1   Patterns of methylation found in the parental diploid *Spartina* species (*S. maritima* and *S. alterniflora*), their natural F₁ hybrid (*S.* × *townsendii*), and the allopolyploid derivative (*S. anglica*). Large ovals represent the various genomes with the markers for each species present within the respective ovals. Absent DNA fragments are placed outside of the ovals (e.g. *S. maritima* and *S. alterniflora* fragments 2, 11, 32, 2 and 9 are not found in *S.* × *townsendii*, but some reappear in the allopolyploid, *S. anglica*; from Salmon *et al.* 2005).

Furthermore, in one study the only activated genes with definable functions were retro-elements (Kashkush *et al.* 2002). From a pool of 12 transcripts activated in the synthetic allopolyploid, the WIS 2–1A retroelement was isolated and characterized further. Kashkush *et al.* (2002) determined that this element was '. . . strongly activated in the amphiploid whereas a transcript was barely detectable in the diploid parents. . . ' Yet these authors did not detect evidence for new sites of

insertion for this activated retroelement. They concluded from this that although there was a significant alteration in transcriptional activity for this and other retro-elements in the allopolyploid derivatives relative to the diploid progenitors, the activations had not led to detectable transposition events (Kashkush *et al.* 2002; Levy and Feldman 2004).

Unlike *Spartina* and wheat, studies of tobacco and *Arabidopsis* have detected evidence for the activation *and* insertion of transposable elements in the allopolyploid derivatives. First, Melayah *et al.* (2004) reported that in *Nicotiana tabacum* the copy number of the retro-element Tnt1 was only 67% of the expectation derived from the numbers found in the diploid progenitors. This indicates the loss of elements from this family since the polyploid event that formed tobacco. However, the distribution of insertion sites for Tnt1 reflects a high level of transpositional activity in the tobacco lineage. Specifically, Melayah *et al.* (2004) found that two-thirds of the insertion sites in *N. tabacum* matched those in the parental species. The remaining sites were unique for the polyploid—suggesting their origin since the allopolyploid event that resulted in this species. Similarly, Madlung *et al.* (2005) detected insertional activity of the *Sunfish* retro-element in the autotetraploid lineages used to synthesize an *Arabidopsis* allopolyploid (Figure 6.2). For example, these workers found the excision of the *Sunfish* elements from sites on chromosome 5 in two of 20 autotetraploids. These transposition events were also suggested to be causal for the chromosomal rearrangements seen in these synthetic hybrids (Madlung *et al.* 2005). Thus, in all but *Spartina*, there is evidence of hybridization-/allopolyploid-mediated activation of retro-elements.

6.2.3 Genomewide effects—genome downsizing

The loss of genomic elements has apparently been a common factor in the evolution of many, but not all (e.g. *Gossypium*; Rong *et al.* 2005), polyploids. One indication of this can be seen by the lack of agreement between the ploidy levels of organisms and the amount of DNA they contain. As expressed by Leitch and Bennett (2004), 'All else being equal polyploids are expected to have larger C-values (amount of DNA in the unreplicated gametic nucleus) than their diploid progenitors, increasing in direct proportion with ploidy.' An analysis of DNA amounts for Angiosperm taxa did detect lineages in which this expectation was met. Yet these authors also found groups in which there was a greater than additive increase as ploidy increased, and other lineages in which there was a decrease in DNA amount as ploidy level increased. In contrast to their expectation, Leitch and Bennett (2004) found overall that (i) the mean DNA amount among the Angiosperm taxa did not increase in direct proportion with ploidy level and (ii) '... the mean DNA amount per basic genome (calculated by dividing the 2C value by ploidy) tended to decrease with increasing ploidy' (Figures 6.3 and 6.4; Leitch and Bennett 2004). The widespread genome downsizing detected in the Angiosperm data set is also reflected by examinations of specific clades and lineages (Leitch and Bennett 2004). For example, since the hypothesized polyploidization event leading to the human lineage (Spring 1997; Furlong and Holland 2002, 2004), approximately

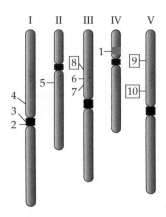

Figure 6.2 Genomic locations of 10 members of the *Arabidopsis* retro-element family *Sunfish*. Boxes reflect positions of *Sunfish* elements in the original diploid *A. thaliana* parent used to construct the experimental allopolyploid, which are not found in the allopolyploid derivative. Analysis of *Sunfish* element 9 confirmed its excision during the construction of the allopolyploid lineage (from Madlung *et al.* 2005).

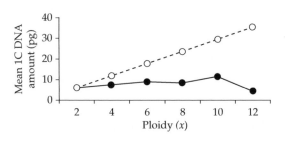

Figure 6.3 Mean 1C DNA amounts for various Angiosperm taxa demonstrating different levels of polyploidy. ●, Observed values; ○, values expected from an additive model of increase due to increases in ploidy (from Leitch and Bennett 2004).

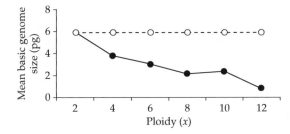

Figure 6.4   The mean DNA amount per basic genome (calculated by dividing the 2C value by ploidy) for various Angiosperm taxa demonstrating different levels of polyploidy. ●, Observed values; ○, values expected from an additive model of increase due to increases in ploidy (from Leitch and Bennett 2004).

50% of the duplicated loci have been deleted (Leitch and Bennett 2004).

Levy and Feldman (2004) arrived at the same conclusion drawn by Leitch and Bennett (2004) concerning the loss of DNA sequences during allopolyploidization, in their case within the wheat species complex (see also Dvorak and Akhunov 2005). For example, Shaked *et al.* (2001) utilized an AFLP methodology to assay 3661 anonymous loci. These authors concluded that downsizing was a major factor in the evolution of the genomes of newly created allopolyploids. Specifically, they detected the elimination of 5–14% of all loci (Shaked *et al.* 2001). Furthermore, Levy and Feldman (2004) summarized data for sequences in wheat that showed either a genome- or chromosome-specific distribution (Figure 6.5). Though the limited distribution of these sequences could have resulted from an origin subsequent to the polyploidization event(s), it was more parsimonious to infer their loss from certain genomic sites following hybridization/polyploidization (Levy and Feldman (2004). Like previous findings related to wheat genomic evolution, Han *et al.* (2005) detected large-scale deletions of the pGc1R-1 repeated-element family in newly synthesized polyploids. Thus both fluorescence *in situ* hybridization and DNA gel-blot analysis detected the elimination of 70–90% of the members of this sequence family within two or three generations (Han *et al.* 2005). Further examples of such widespread—genomically and taxonomically—sequence elimination are also reflected by data for the lineages that include rice, *Arabidopsis*,

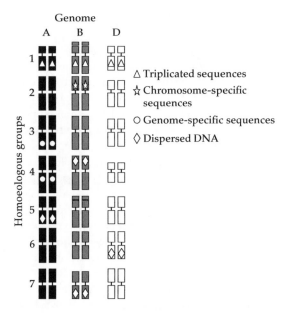

Figure 6.5   A schematic representation of the wheat genome/karyotype. The chromosomal elements in hexaploid wheat are identified as originating from three separate progenitors (the A, B, and D genomes). The chromosomes originating from the three $2n = 14$ progenitors are arranged into seven homoeologous groupings. In this schematic, sequences that are triplicated (have representative sequences from all three progenitors still present in the hexaploid), chromosome-specific (found on only one chromosome pair from one genome), genome-specific (found on non-homologous chromosomes, but only from a single genome), and dispersed (found in multiple locations within and among the three genomes) are indicated (from Levy and Feldman 2004).

and *S. cerevisiae*. In these species, the proportions of duplicated genes lost have been estimated to be approx. 30–65, 70, and 90%, respectively (Wolfe and Shields 1997; Bowers *et al.* 2003; Kellis *et al.* 2004; Paterson *et al.* 2005; X. Wang *et al.* 2005).

6.2.4   Genomewide effects—chromosome rearrangements

Barbara McClintock described the effect of externally mediated challenges to the genomes of organisms as 'shocks'. In particular, she held that 'There are "shocks" that a genome must face repeatedly, and for which it is prepared to respond in a programmed manner. Examples are the "heat shock" responses in eukaryotic organisms and the "SOS" responses in bacteria' (McClintock 1984). Yet she also recognized the very important set of responses to challenges that were not hard-wired into the genomic makeup of organisms. In this regard she observed that 'An experiment conducted in the mid-1940's prepared me to expect unusual responses of a genome to challenges that the genome is unprepared to meet in an orderly, programmed manner.' McClintock suggested that one such response could result from the genomic shock associated with crosses between divergent evolutionary lineages (e.g. Y.-M. Wang *et al.* 2005). Specifically with regard to the present topic she stated, 'Major restructuring of chromosome components may arise in a hybrid plant and continue to arise in its progeny, sometimes over successive plant generations. The restructuring may range from apparently simple to obviously complex' (McClintock 1984).

It is now understood that McClintock was, as usual, extremely insightful in her assertion that chromosome rearrangements caused by hybridization were frequent and important. Indeed, such changes may reflect some of the rearrangements seen as causal in some models of evolutionary diversification (i.e. speciation models)—for example 'stasipatric' (White 1978) and 'recombinational' (Grant 1981). However, it is likely that McClintock was incorrect in her assertion that the chromosome rearrangements were uniformly unpredictable. Thus it appears that the same rearrangements may occur, and become fixed, repeatedly in response to the genomic challenges from mixing divergent genomic components (e.g. Shaw *et al.* 1983; Rieseberg *et al.* 1996). In the case of genome-duplication events, and particularly instances of allopolyploidy, chromosome rearrangements (both intra- and intergenomic rearrangements) are encountered frequently. For the present discussion I will use three genera (*Arabidopsis, Nicotiana,* and *Brassica*) to illustrate this common outcome from allopolyploidization.

In addition to their detection of expression and insertional activation of the *Sunfish* retro-elements in the experimental *Arabidopsis* polyploids (see above), Madlung *et al.* (2005) were also able to test the hypothesis that 'Even limited transposon activity can have profound effects on genome structure if it leads to chromosome breaks, fusions and translocations' (i.e. reminiscent of McClintock's breakage-fusion-bridge cycle). These authors did

indeed find numerous cytological data pointing to elevated frequencies of chromosome rearrangements in the allopolyploids relative to the parental lineages. Specifically, Madlung *et al.* (2005) detected approximately 30% abnormal meioses in the synthetic allopolyploids. Consistent with allopolyploidy-caused chromosome rearrangements are data collected from analyses of both the naturally occurring and experimentally produced allotetraploid species, *Arabidopsis suecica* (Pontes *et al.* 2004). Specifically, Pontes *et al.* (2004) detected deletions of entire nucleolar organizing and 5 S regions (i.e. rRNA genes) in both the natural and synthetic allopolyploids, as well as translocations (or transpositions) in the synthetic lineages (Figure 6.6).

Like *Arabidopsis,* natural and artificial allopolyploid derivatives in *Nicotiana* and *Brassica* also revealed evidence of chromosomal structural changes in the polyploid, versus diploid, lineages (Lim *et al.* 2004; Pires *et al.* 2004; Skalická *et al.* 2005). For example, in *Nicotiana* similar intergenomic translocation alterations were found in the naturally occurring *N. tabacum* samples, and in two of three synthetically produced tobacco plants (Lim *et al.* 2004). This is not only evidence for genome duplication-mediated changes, but also for the repeated origin of chromosome rearrangements in divergent lineages. Likewise, the deletion of the same quantitative trait loci, through non-reciprocal transposition events, was detected in an artificially produced, allopolyploid lineage resembling *Brassica napus* (Pires *et al.* 2004). In this case, the synthetic hybrid was produced through crosses of the diploid parents of the naturally occurring *B. napus, B. rapa,* and *B. oleracea.* Significantly, not only did the synthetic hybrid lineages demonstrate large-scale chromosomal rearrangements, the alterations were associated with the fitness trait of flowering time (Pires *et al.* 2004).

6.3 Genetic exchange: gene and gene family evolution following duplication

6.3.1 Genes and gene families: concerted evolution

Dover and Tautz (1986) described both the process of concerted evolution (a term coined by Zimmer *et al.* 1980, and later described as 'molecular drive'

(a) Natural *A. suecica*

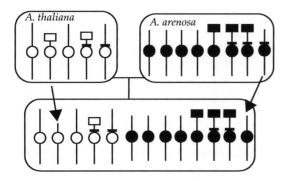

(b) Synthetic *A. suecica* (F3 1459a)

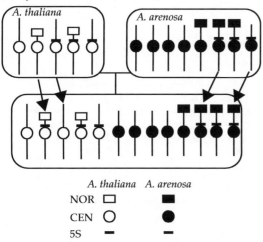

	A. thaliana	A. arenosa
NOR	□	■
CEN	O	●
5S	▬	▬

Figure 6.6 Diagrammatic representation of chromosome rearrangements involving the nucleolar organizing and 5 S genic regions in the naturally occurring (a) and experimentally produced (b) allopolyploid *A. suecica* (from Pontes *et al.* 2004).

by Dover 1982), and its outcome, in the following manner: 'Knowledge of the rates, units and biases of widespread mechanisms of non-reciprocal DNA exchange, in particular within multigene families, provides alternative explanations for conservation and divergence . . . Such mechanisms of DNA turnover cause continual fluctuations in the copy-number of variant genes in an individual and, hence, promote the gradual and cohesive spread of a variant gene throughout a family (homogenization) and throughout a population (fixation).' Thus when a new mutation arises in one member of a gene or repeated DNA family it can be, and often

is, transferred to other members by concerted evolution (e.g. Figure 6.7; Hall *et al.* 2005; Pröschold *et al.* 2005). The mechanisms recognized as potentially affecting the process of concerted evolution include unequal crossing over, *gene conversion* and the movement of transposable elements (e.g. Smith 1976; Slightom *et al.* 1980; Kidwell and Lisch 2000; Skalická *et al.* 2003; Johannesson *et al.* 2005; Sugino and Innan 2005). In the context of genome duplication, concerted evolution is seen as a major factor in gene family evolution. In contrast to instances of within-genome concerted evolution that is fed by newly arisen mutations, allopolyploid formation brings together two or more divergent genomes through hybridization—genomes that likely differ by many mutational changes. Yet, for different instances of allopolyploidy the pattern of transfer of genetic information between the divergent genomes can be quite dissimilar. It is probable that the differences are reflective of (i) inherent barriers to genomic exchange for some allopolyploid taxa, (ii) the effects of differential levels of gene expression (i.e. highly expressed *paralogous genes*, in the case of within-genome duplications, may demonstrate less sequence divergence from one another than duplicates that are expressed at lower levels; Drummond *et al.* 2005; Pyne *et al.* 2005), and (iii) the time since formation of the polyploid lineage (i.e. the number of generations available to facilitate the concerted events). Two plant allopolyploid systems—*Tragopogon* and *Nicotiana*—can be used to illustrate the apparent idiosyncratic expression of concerted evolutionary processes.

The genus *Tragopogon* reflects a rich example of the various processes associated with the evolution of allopolyploid taxa. Furthermore, members of this genus that are found in North America illustrate the outcome of concerted evolution of divergent genomes brought together through natural crosses. The number of generations over which the concerted events occurred is known because the North American allopolyploids originated within the past 150 years, subsequent to the introduction of the diploid, Eurasian natives, *Tragopogon dubius*, *Tragopogon porrifolius*, and *Tragopogon pratensis* (Soltis *et al.* 2004). From morphological and chromosomal data Ownbey (1950) described the two allopolyploid derivatives, *Tragopogon mirus* and

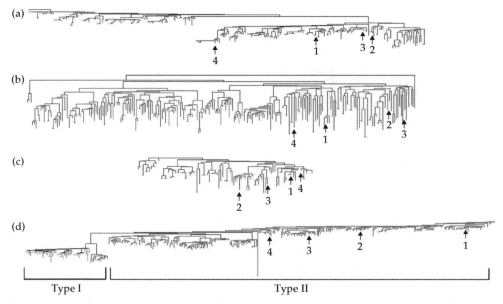

Figure 6.7 Phylogenies derived from repeated DNA sequences from the following four plant species belonging to the family Brassicaceae: (a) *Arabidopsis arenosa*; (b) *Olimarabidopsis pumila*; (c) *Capsella rubella*; and (d) *Sisymbrium irio*. The colors in each phylogeny reflect the bacterial artificial chromosome (BAC) from which the repeat sequences were derived. Thus repeat sequences (within a phylogeny) with the same color likely came from the same genomic region. The overall pattern of clustering of repeats from the same genomic region, in clades divergent from repeats from other BAC clones, suggests that concerted evolution affects most efficiently those repeat elements that are located in the same genomic regions (from Hall *et al.* 2005). See also Plate 4.

Tragopogon miscellus as having formed from hybridization between *T. dubius/T. porrifolius* and *T. dubius/T. pratensis*, respectively. Though analyses of the rRNA gene families present in *T. mirus* and *T. miscellus* have detected the presence of the gene families from both diploid progenitors, these families are not found in the equivalent frequencies expected from an additive model (Soltis *et al.* 2004). Indeed, Kovarik *et al.* (2005) found° that the rRNA genes of *T. dubius*, the common parent for both allopolyploid species, were in a significant minority (i.e. making up only 5% of the family members in some plants) in all but two of the natural populations of the allopolyploids, *T. mirus* and *T. miscellus*. However, in the two exceptions (both *T. mirus* populations), the repeats of *T. dubius* made up the majority of rRNA repeat elements. These results reflect an asymmetric (though bi-directional; Kovarik *et al.* 2005) pattern for the concerted evolution of ribosomal repeats within the *Tragopogon* allopolyploids. Specifically, concerted changes result most often in the preferential loss of *T. dubius* repeats.

Studies of rRNA gene sequences in three natural *Nicotiana* allopolyploids also detected the effects from concerted evolution following hybrid lineage formation. In particular, Kovarik *et al.* (2004) examined the ratios of parental repeat variants in *N. tabacum*, *Nicotiana rustica* and *Nicotiana arentsii*. As in *Tragopogon*, the three *Nicotiana* allopolyploid derivatives demonstrated an asymmetric exchange of rDNA repeat variation that resulted in the predominance of one parent's rRNA genes (Table 6.1). However, concerted evolution has not been uniform in its effect on the resulting rDNA variation in *N. tabacum*, *N. rustica*, and *N. arentsii*. Kovarik *et al.* (2004) concluded that *N. tabacum* and *N. rustica* demonstrated the outcome of concerted evolution (including both intra- and interlocus gene conversion) that has partially replaced one parental rDNA genotype with the other, while *N. arentsii* demonstrated a pattern indicative of complete replacement. The effect of concerted evolutionary processes is thus seen to have impacted greatly the distribution of parental rRNA genes in allopolyploid derivatives of both *Tragopogon* and *Nicotiana*.

Table 6.1 The percentage of the rDNA repeat units originating from each diploid progenitor for three natural *Nicotiana* allopolyploids (Kovarik *et al.* 2004).

Allopolyploid	Parental species	Percentage of rDNA repeats	
		from P-1	from P-2
Nicotiana tabacum	*Nicotiana sylvestris* (P-1)	2–25%	75–98%
	Nicotiana tomentosiformis (P-2)		
Nicotiana rustica	*Nicotiana paniculata* (P-1)	20%	80%
	Nicotiana undulata (P-2)		
Nicotiana arentsii	*Nicotiana undulata* (P-1)	>99%	<1%
	Nicotiana wigandioides (P-2)		

6.3.2 Genes and gene families: changes in gene expression patterns and function

An excellent example for illustrating gene expression/function evolution—following WGD—comes from the yeast genus *Saccharomyces*. Comparisons of sequences from its genome with other fungal species, as well as with more distantly related organisms, have yielded a wealth of information concerning the evolution of expression for duplicate genes (Li *et al.* 2005). For example, Kellis *et al.* (2004) compared the whole genome sequence from the related yeast species *Kluyveromyces waltii* with the previously sequenced genome of *S. cerevisiae*. Their analysis allowed the confirmation of a WGD event in the lineage leading to *Saccharomyces*. In particular, comparison of the *K. waltii* and *S. cerevisiae* genomes suggested that the polyploidization occurred along the *Saccharomyces* lineage following its divergence from the lineage that gave rise to *Kluyveromyces* (Kellis *et al.* 2004). Subsequent to this WGD, members of numerous duplicated gene pairs diverged from one another. Thus Kellis *et al.* (2004) concluded (from their own analyses and from a literature survey) the following: (i) 17% of the *S. cerevisiae* gene pairs demonstrated accelerated protein evolution relative to that found in *K. waltii*; (ii) in 95% of the cases of gene pairs that demonstrated accelerated evolution, the elevated level of change was in only one member of the pair; (iii) the slower evolving member of a gene pair is more likely to have retained the ancestral function while the rapidly evolving member possesses a novel function; and (iv) in 18% of the cases of accelerated evolution, deletion of the slower-evolving (i.e. the gene coding

for the ancestral function) member was lethal, but in no case was lethality obtained by deleting the faster-evolving gene.

Pyne *et al.* (2005) and Gu *et al.* (2005) also tested the hypothesis that gene duplication is the major cause of genetic (and thus organismal) novelty. Both groups gathered data concerning the evolution of expression and function of duplicated genes following the WGD event in the *Saccharomyces* lineage. The major conclusion of Pyne *et al.* (2005) concerning the evolutionary pattern for different pairs of duplicates was that '. . . poorly expressed genes diverge rapidly from their paralog, while highly expressed genes diverge little, if at all.' Likewise, Gu *et al.* (2005) detected divergence of gene expression patterns following the *Saccharomyces* WGD. Indeed, like Kellis *et al.* (2004) this latter analysis identified a trend in which rapid evolutionary change (in this case reflected in divergent gene expression patterns) was found in only one of the copies of the duplicated genes. In addition, Gu *et al.* (2005) were able to infer that the initial rate of expression and regulatory network evolution was > 10-fold higher than seen prior to the duplication event (Figure 6.8). From this observation, they concluded that the increase in complexity of the *Saccharomyces* regulatory network was affected greatly by rapid evolution subsequent and proximate to the WGD event (Gu *et al.* 2005). Taken together the studies discussed above (and those reviewed by these authors) indicate that, 'Compared to multiple independent duplications and divergence of individual genes or segments, WGD may be more efficient and may offer great opportunities for coordinated evolution' (Kellis *et al.* 2004).

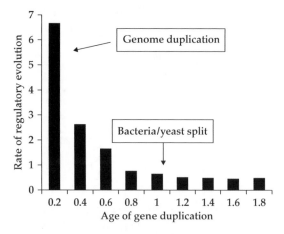

Figure 6.8 Rate of regulatory evolution before and after the WGD in the evolutionary lineage leading to the yeast genus *Saccharomyces*. The time units are calibrated by the estimated divergence time of 2 billion YBP for the 'bacteria/yeast split' (from Gu *et al.* 2005).

6.3.3 Genes and gene families: evolution of adaptations

A common theme throughout this book is the effect genetic exchange may have on adaptations—both the transfer of existing adaptations between lineages and the *de novo* formation of adaptations in hybrid/reassortant genotypes. In the current section I wish to highlight the outcomes from gene/genome duplication resulting from genetic exchange in terms of the formation of novel adaptations. To accomplish this, I will use three examples, one involving parthenogenetic geckos of the genus *Heteronotia* and two from the yeast genus *Saccharomyces*. In all three cases, adaptations reflect apparent increased fitness under environmentally mediated selection. This common result reflects again the reflective and predictive power of the web-of-life metaphor; that is, genetic exchange plays a role in the evolutionary history of much of biological life leaving a creative signature along many lineages (Arnold and Larson 2004).

Australian species of the gecko genus *Heteronotia* reflect a classic example of the formation of allopolyploid-based *parthenogenesis* (Moritz 1983; Moritz *et al.* 1989a; Strasburg and Kearney 2005). Occurring throughout a large portion of the central arid region of Australia, the allopolyploid parthenogenetic lineages (placed, along with the

diploid forms, in the species *Heteronotia binoei*) were formed multiple times from natural hybridization among the diploid races (Moritz *et al.* 1989a). Indeed, this pattern of multiple origins is a common characteristic for allopolyploid lineages (Soltis and Soltis 1993). It has been postulated that cyclical climatic alterations led to repeated range expansions and contractions that resulted in hybridization between the diploid races (Strasburg and Kearney 2005), thus leading to multiple, independent formations of triploid *H. binoei* parthenogenetic lineages.

Not only have research findings identified *Heteronotia* as an exemplar for understanding the formation of allopolyploid and asexual lineages, they have also suggested the production of novel adaptations in the allopolyploid races. For example, with regard to aerobically sustained locomotion, Kearney *et al.* (2005) detected a fitness advantage in the parthenogens relative to their diploid progenitors. Specifically, in lower temperature environments they found (i) significantly greater endurance, (ii) higher maximum oxygen-consumption rates, (iii) higher maximum aerobic speeds, and (iv) greater levels of voluntary activity in the parthenogenetic female geckos relative to diploid females. This led them to conclude that 'Parthenogenetic lineages of *Heteronotia* thus have an advantage over sexual lineages in being capable of greater aerobic activity.' It thus appears that genomic duplication (or in this case triplication) has facilitated the origin of phenotypic traits that reflect adaptive novelty (see Johnson 2005 for similar findings in parthenogenetic snails of the genus *Campeloma*).

As discussed above, *Saccharomyces* has been used as a model system for defining the effects of genome duplications in the evolution of new gene functions. Indeed, these new gene functions are obvious candidates for defining WGD-based origins of novel adaptations. However, in addition to the effects from the polyploid event leading to *S. cerevisiae* (see section 6.3.2), allopolyploidy continues to enrich the biological/species composition of *Saccharomyces*. One example of apparent allopolyploidy-derived adaptations is reflected in the *Saccharomyces* species used in the production of the two categories of beer known as lager and ale.

The production of these two types of beer is quite distinct, with the lager yeast being known as a 'bottom-dwelling' strain. This adaptation for the propensity to grow in this different environment is seen to be the result of an allopolyploidy event that gave rise to the species *Saccharomyces pastorianus* (synonymous with *Saccharomyces carlsbergensis*; Masneuf *et al.* 1998). However, *S. pastorianus* does not appear to be an isolated case of genome duplication-facilitated adaptation in *Saccharomyces*. For example, Masneuf *et al.* (1998) reported hybrid strains that were both cryophilic (i.e. possessed a higher growth rate and an optimum fermentability at low temperatures) and had the capacity to produce wines with higher amounts of flavor-active esters. These novel adaptations were correlated with their hybrid (i.e. allopolyploid) constitution. Thus, like *Heteronotia*, the various allopolyploid *Saccharomyces* species possess novel adaptations not found in their diploid progenitors.

6.4 Genetic exchange: genome duplication and adaptive radiations

6.4.1 Duplication and adaptive radiation: the vertebrate lineage

Because the divergence of regulatory genes is being considered necessary to bring about phenotypic variation and increase in biological complexity, it is tempting to conclude that such large-scale gene duplication events have indeed been of major importance for evolution in general . . .

(Maere et al. 2005)

This conclusion reflects a paradigm that is well-represented in the evolutionary literature and yet is still controversial (e.g. Gu *et al.* 2002; Castro *et al.* 2004; Furlong and Holland 2004; Mulley and Holland 2004; Donoghue and Purnell 2005). As with genetic exchange in general, there are a plethora of data sets (some discussed above) that have as their conceptual framework an assumption of a causal role for duplication events in evolutionary/developmental transitions and adaptive radiations. Within the animal assemblage, data from invertebrate and vertebrate taxa have led to the inference of multiple WGDs—both in the

lineage leading to vertebrates and within the vertebrate clade as well. Such inferences are now possible because of the availability of whole-genome data sets for invertebrate and vertebrate species. Comparisons of these data sets lent support to the hypothesis that vertebrate genomes contained more complexity than those of invertebrates (Donoghue and Purnell 2005). These comparisons further indicated that a large measure of the difference in complexity between invertebrate and vertebrate taxa was due to a series of duplications.

In regard to the hypothesis of WGD-facilitated adaptive radiations, it is seen as significant that a duplication event occurred coincidentally with the origin of vertebrates (Spring 1997; Furlong and Holland 2002, 2004; Gu *et al.* 2002). That a WGD event occurred near the point of origin of the vertebrate lineage is apparent from comparisons of vertebrate genomes with that of their closest living invertebrate relative, amphioxus (Spring 1997; Furlong and Holland 2002; Donoghue and Purnell 2005). Indeed, two rounds of tetraploidy (leading to an octoploid derivative) were hypothesized to be temporally associated with the timing of the separation of the vertebrate and invertebrate lineages (Figure 6.9; Furlong and Holland 2002; but see Panopoulou and Poustka 2005). Therefore, as in viral, bacterial, fungal, and plant clades, genetic

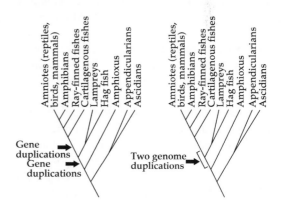

Figure 6.9 Alternative phylogenetic placement of the two hypothesized polyploid events in the lineage leading to the vertebrate clade. Both duplications have been suggested to involve tetraploidy thus leading to an octoploid genomic condition (from Furlong and Holland 2002).

exchange is correlated with the explosive, adaptive radiation resulting in the vertebrate clade.

6.4.2 Duplication and adaptive radiation: the fish clade

The final examples of potentially duplication-driven adaptive radiations come from fish. Not only do these examples illustrate the possibility of an important role for WGD in the radiation of entire clades, they also reflect the difficulty of testing for the causes of such radiations. Thus Le Comber and Smith (2004) considered examples primarily from the fish families Cyprinidae and Salmonidae. They compared the phylogenetic placement of polyploidy with the number of species present in clades rooted or not rooted in WGDs. Le Comber and Smith (2004) also took into account the ecological, morphological, behavioral, physiological, and life-history diversity of fish in general. Their analysis led to the conclusion that '. . . polyploidy may have been of considerable importance in the evolution of fishes' (Le Comber and Smith 2004). Further they recognized that, in general, the groups in which WGDs had occurred were also those reported to have the most cases of contemporary hybridization. This resulted in the additional inference that the cases of polyploidy were most likely due to hybridization (i.e. reflected allopolyploidy; Le Comber and Smith 2004).

In contrast to the above study, Donoghue and Purnell (2005) included both living and fossil forms in a phylogenetic assessment designed to test for associations of diversification with WGD. One of the events on which they focused was the duplication hypothesized as causal in radiation of the teleosts (ray-finned fishes). As emphasized by Donoghue and Purnell (2005), this group is the most diverse of any vertebrate clade. Furthermore, the cause of teleost diversity, in terms of phenotype and numbers of species, has been ascribed to a WGD at the base of the clade. Yet when extinct clades are taken into account, the timing of the event becomes less well resolved. In particular, Donoghue and Purnell's (2005) analysis suggests that the WGD is just as likely to have occurred subsequent, rather than prior, to the origin of extinct teleost lineages. This finding, along with a similar observation for the evolution of gnathostomes (jawed vertebrates), casts doubt on the possibility of WGD-derived adaptive radiations in these groups (Donoghue and Purnell 2005). However, these authors concluded that definitional constraints prevented a rigorous test of the hypothesis that genomic duplication caused an increase in phenotypic and taxonomic diversity. Overall then, the hypothesis that duplications of the genome (either whole or partial) have led to adaptive radiations, though intriguing, remains controversial.

6.5 Summary and conclusions

The examples discussed above reflect the diversity of effects hypothesized from genetic exchange-mediated genomic duplications. Most, but not all, of these examples reflect polyploidy. Yet partial genome duplications are seen as having the same potential for spurring evolutionary innovations. The importance of duplications is thus considered to reside in the availability of the duplicated material to acquire changes. These changes can be epigenetic or they can reflect mutations. In either case the changes would likely be deleterious for an organism that had only one copy of a particular region of functional DNA. The malleability of the genomes of organisms containing duplications is easily demonstrated. What is less easily documented is any causal affect on evolutionary change, thus leading to such responses as new adaptations and adaptive radiations. However, for many examples the comparison of whole-genome data sets with biological and phylogenetic characteristics, suggests such causal links. It is likely that future data sets will identify additional examples of the association of genome duplications with evolutionary innovation and diversification.

CHAPTER 7

Origin of new evolutionary lineages

Linnaeus and Kerner were the outstanding early exponents of the idea that natural hybrids can be the starting points of new species.

(Grant 1981, p. 245)

... *Poa labradorica* is of intergeneric hybrid origin between *Dupontia fisheri* subsp. *psilosantha* ... and *Poa eminens* ... Incapable of sexual reproduction, *P. labradorica* dominates large areas of coastal marsh through vigorous production of rhizomes.

(Darbyshire *et al.* 1992)

This observation allows the strong inference that the formation of *I. nelsonii* did indeed involve hybridization between *I. fulva*, *I. hexagona*, and *I. brevicaulis*.

(Arnold 1993)

Comparison of mtDNA and Y-chromosome topologies suggests a possible hybrid origin of *M. arctoides* from a cross between proto-*M. fascicularis* and an early [*Macaca*] *sinica* group population ...

(Tosi *et al.* 2003)

... the differential distribution of large mobile elements carrying virulence and drug-resistance determinants may be responsible for the clinically important phenotypic differences in these strains.

(Holden *et al.* 2004a)

We hypothesize that two or more hybridization events involving the parental species *Clitarchus hookeri* and an unknown taxon probably resulted in the formation of the parthenogenetic genus *Acanthoxyla*.

(Morgan-Richards and Trewick 2005)

7.1 Viral recombination, lateral transfer, natural hybridization, and evolutionary diversification

Genetic exchange not only creates the opportunity for evolutionary novelty, it also opens the possibility for evolutionary diversification (e.g. see Anderson and Stebbins 1954; Grant 1981; Dowling and DeMarais 1993; Arnold 1997, 2004a; Rieseberg 1997; Holmes 2004; Gevers *et al.* 2005). In particular, all mechanisms of gene transfer have the potential to form the genetic foundation for new evolutionary lineages. In some cases these novel lineages will be given a formal taxonomic designation. In others they will be referred to simply as races or strains of one of the original taxa. As emphasized in previous chapters, the presence or absence of such taxonomic recognition, though non-trivial for recognizing evolutionary pattern and process, is nonetheless not necessarily indicative of the evolutionary/ ecological/epidemiological significance of viral recombination, lateral transfer, and natural hybridization. Thus, whether a newly formed lineage of organisms is deemed by researchers as deserving of a taxonomic label, or is instead termed a hybrid, recombinant, or reassortant population, the question remains as to whether or not it reflects evolutionary diversification. For example, the evolutionary longevity of hybrid taxa that reproduce exclusively (or nearly so) through asexual means has been termed '. . . but an evening gone' (Maynard Smith 1992), reflecting a viewpoint that such lineages are inconsequential due to the relative brevity of their existence compared to sexually reproducing lineages (see also Griffiths and Butlin 1995). However, across widely varying clades of plants and animals, the same asexual lineages are repeatedly formed. Therefore, although the product of a single origin event may be (relatively) short-lived, the recurrent formation of lineages through natural hybridization between the same sexual forms facilitates the evolutionary diversity and longevity of the asexual taxa. This is likely a partial explanation for why so-called *agamic species complexes* and *clonal microspecies* abound in plant and animal complexes, forming important constituents of certain ecosystems (e.g. Grant and Grant 1971; Darbyshire *et al*. 1992; King 1993; Vollmer and Palumbi 2002; Schranz *et al*. 2005).

The major goal for this chapter is to illustrate that hybrid lineage formation (*sensu lato*) occurs throughout biological life forms and evolutionary time. The following discussion includes numerous examples taken from what is normally termed the speciation literature. Many authors, including myself, often place discussions concerning the derivation of recombinant lineages into the context of the formation of new, hybrid species. However, I will sample from a broad landscape of examples that also reflect the formation of lineages that, though not termed new species, still form the foundation for evolutionary innovation and diversity. To continue the subject begun in Chapter 6, I will first consider examples involving WGD events. The categories under which I will discuss the WGD-associated diversifications include plants and animals. For plants, I have selected a diverse array of non-angiosperm (i.e. ferns and bryophytes) and angiosperm clades. Some allopolyploid derivatives

reproduce only sexually, others by a mixture of sexual and asexual reproduction, while others reproduce exclusively (or nearly so) through asexual means. I will discuss representative examples for each of these three broad categories. Like plant allopolyploid lineages, under the category of animals I will present data from analyses of asexually and sexually reproducing polyploid derivatives. I will illustrate the evolutionary outcome seen in asexual, allopolyploid animal lineages by considering examples of parthenogenesis and *gynogenesis*. Although sexually reproducing animal allopolyploids are relatively rare, I will illustrate well-defined examples of this outcome from genetic exchange as well.

In plants and animals, one can also illustrate lineage formation through genetic exchange using examples not involving WGD. Such hybrid lineage formation results in taxa that possess the same, or nearly the same, chromosome number as the progenitor individuals. Once again, those who work on such 'homoploid' hybrid formation often discuss these lineages in terms of speciation. Furthermore, homoploid hybrids are normally defined as demonstrating sexual reproduction (see Rieseberg 1997 for a discussion). Thus diploid hybrid lineages that reproduce via asexuality are generally placed in other categories—such as clonal microspecies (Grant and Grant 1971; Grant 1981). In this chapter I will break with this tradition and discuss the formation of homoploid hybrid lineages, regardless of their mode of reproduction (e.g. via *hybridogenesis*). I believe that this emphasizes the most important element (in the context of this book), that of genetic exchange, rather than focusing on mechanisms of reproductive isolation (Grant 1981; Rieseberg 1997).

The discussion of homoploid formation is biased toward plant examples. This is due to the relative scarcity of homoploid animal examples reported thus far. However, it is important to emphasize again that this is due somewhat to definitional constraints. Specifically, the number of microorganism, animal, and plant 'hybrid' lineages is legion. If we divorce ourselves from the requirement that a hybrid derivative be assigned a species name before it is considered to be of evolutionary importance, many more examples of homoploid hybrid lineage formation become apparent in the zoological literature as well (e.g. see Llopart *et al.* 2005 for a description of a hybrid lineage that would fall within this conceptual framework).

The final topic explored in this chapter will be the role of viral recombination, lateral gene transfer, and introgressive hybridization in the formation of microorganisms. In this section I will illustrate again the outcome of genetic exchange-mediated evolutionary diversification. For this category, I will draw examples from bacterial, bacteriophage and protozoan taxa. I will use each of these cases to draw attention to the observation that in microorganisms, as in the examples from higher eukaryotes, genetic exchange can lead to evolutionary diversification.

7.2 Natural hybridization, allopolyploidy, and evolutionary diversification in non-flowering plants

7.2.1 Ferns

Natural hybridization, introgression, polyploidy, and lateral gene transfer have been detected in numerous fern assemblages and related clades (e.g. Wagner 1987; Rabe and Haufler 1992; Kentner and Mesler 2000; Hoot *et al.* 2004; Davis *et al.* 2005). For example, WGD characterizes the fern genus *Asplenium*, and indeed the entire family Aspleniaceae. Van den Heede *et al.* (2003) summarized this conclusion by observing that 'Both auto- and allopolyploidy are common driving forces of evolution in Aspleniaceae . . . in the well-known Holarctic fern flora, *Asplenium* consists of about 53% (ancestral) diploids, 24% autopolyploids, 20% allopolyploids, and 3% *apogamous taxa*. Polyploidy is at least equally common in the tropical taxa, but their status and ancestry remain largely unexplored.' In their analysis, Van den Heede *et al.* (2003) examined the phylogenetic relationships of 20 taxa from the *Asplenium* subgenus, *Ceterach*. Most germane to the present discussion, their results supported the existence of numerous allopolyploid (and autopolyploid) lineages. Furthermore, they found concordance between geographic region and subclades within *Ceterach*. Specifically, they detected species complexes in particular geographic regions consisting of

diploid, autopolyploid, and allopolyploid taxa. It is clear that for *Asplenium*, subgenus *Ceterach*, reticulate evolution has generated a large proportion of its evolutionary diversity.

Old World *Asplenium* taxa are not the only clades that reflect evolutionary diversification via genetic exchange. One of the classical examples of reticulate evolution within *Asplenium* involves the New World species located in the Appalachian Mountains of North America. Wagner (1954) and Werth *et al.* (1985) used morphological, cytological, and allozyme data to define a complex of three diploid progenitors (*Asplenium platyneuron*, *Asplenium montanum*, and *Asplenium rhizophyllum*), three allopolyploid derivatives (*Asplenium ebenoides* (*A. platyneuron* × *A. rhizophyllum*), *Asplenium pinnatifidum* (*A. montanum* × *A. rhizophyllum*) and *Asplenium bradleyi* (*A. montanum* × *A. platyneuron*)) plus several other sterile, hybrid taxa (Figure 7.1). Though the diploid taxa are differentiated morphologically and allozymically (Wagner 1954; Werth *et al.* 1985), the allopolyploid taxa overlap morphologically with (and show combinations of the allozymes characteristic for) their various progenitor species. Like the Old World *Asplenium* complexes, the Appalachian species demonstrate the effects of multiple reticulate evolutionary events.

A second example of the effects of genetic exchange, and particularly allopolyploidy, on the evolutionary diversification of fern species is seen for the *Polypodium vulgare* complex. As with *Asplenium*, numerous hybridization events have resulted in a series of ancestral/diploid and derivative/allopolyploid taxa. Haufler *et al.* (1995a, b) deciphered the web-like evolutionary history of this group using both allozyme and cpDNA data. These authors were able to use the nuclear (i.e. allozyme) and cytoplasmic (i.e. cpDNA) data to infer the parentage of the seven allopolyploid species (Figure 7.2; Haufler *et al.* 1995a, b). Furthermore, the discovery of both parental cpDNA haplotypes in populations of the various allopolyploids indicated reciprocal, and thus recurrent, formation of the polyploid lineages (Haufler *et al.* 1995a). This latter process—multiple origins for a single allopolyploid taxon—is commonly found for many plant and animal polyploid complexes (e.g. Hedrén 2003; Pongratz *et al.* 2003).

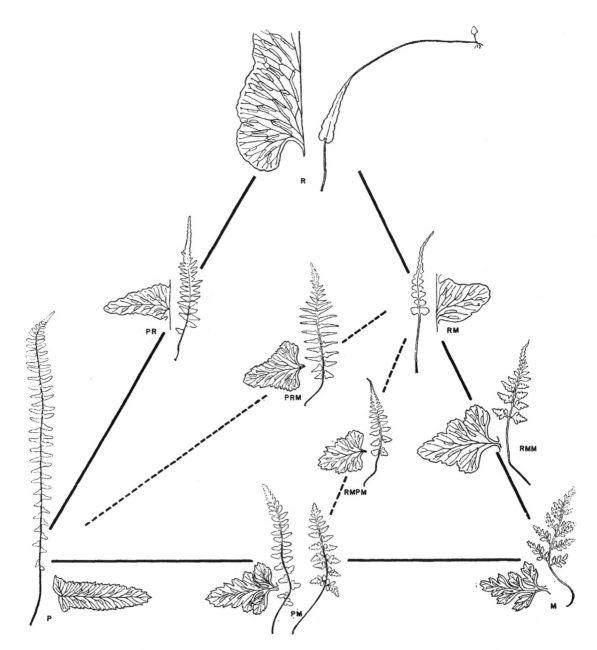

Figure 7.1 Evolutionary derivation of hybrid populations and allopolyploid taxa of Appalachian *Asplenium*. The three diploid progenitors are found at the corners of the triangle: R, *A. rhizophyllum*; P, *A. platyneuron*; M, *A. montanum*. The three fertile allopolyploid derivatives are found along the major axes: PR, *A. ebenoides*; RM, *A. pinnatifidum*; PM, *A. bradleyi*. One of the sterile, allopolyploid taxa (RMM, *A. trudellii*) is found along the axis with *A. pinnatifidum*. The other two sterile, allopolyploid taxa (PRM and RMPM, *A. kentuckiense* and *A. gravesii*, respectively) are found within the triangle. Intersecting lines indicate the progenitors for the various fertile and sterile allopolyploids (from Wagner 1954).

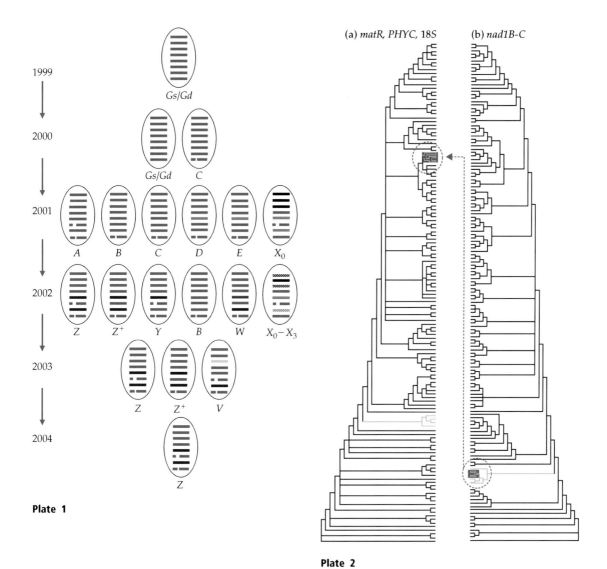

Plate 1

Plate 2

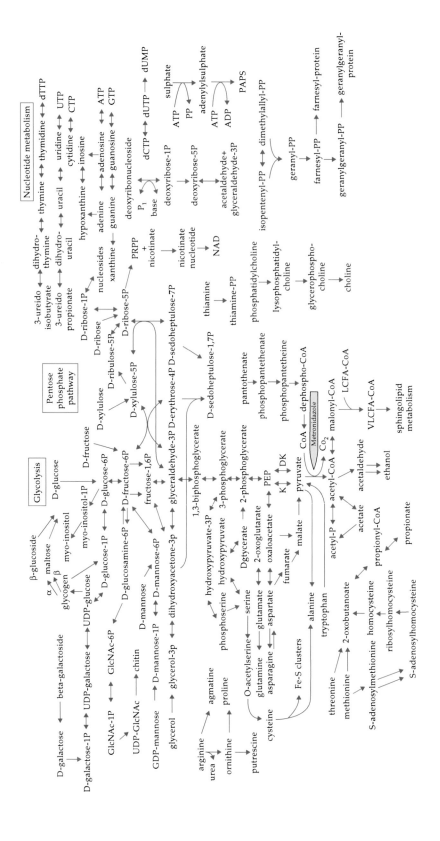

Plate 3

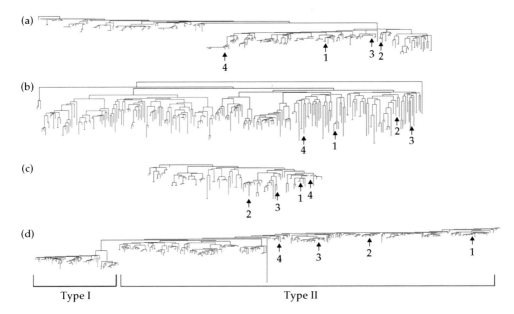

(a)

(b)

(c)

(d)

Type I   Type II

Plate 4

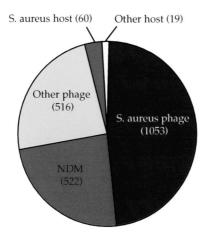

S. aureus host (60)   Other host (19)

Other phage
(516)

S. aureus phage
(1053)

NDM
(522)

Plate 5

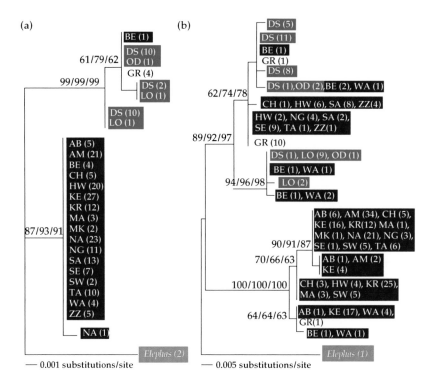

(a)

BE (1)
DS (10)
OD (1)
GR (4)
DS (2)
LO (1)

61/79/62

99/99/99

DS (10)
LO (1)

AB (5)
AM (21)
BE (4)
CH (5)
HW (20)
KE (27)
KR (12)
MA (3)
MK (2)
NA (23)
NG (11)
SA (13)
SE (7)
SW (2)
TA (10)
WA (4)
ZZ (5)

87/93/91

NA (1)

Eleplus (2)

— 0.001 substitutions/site

(b)

DS (5)
DS (11)
BE (1)
GR (1)
DS (8)
DS (1),OD (2),BE (2), WA (1)
CH (1), HW (6), SA (8), ZZ(4)
HW (2), NG (4), SA (2),
SE (9), TA (1), ZZ(1)
GR (10)
DS (1), LO (9), OD (1)
BE (1), WA (1)
LO (2)
BE (1), WA (2)

62/74/78

89/92/97

94/96/98

AB (6), AM (34), CH (5),
KE (16), KR(12) MA (1),
MK (1), NA (21), NG (3),
SE (1), SW (5), TA (6)
AB (1), AM (2)
KE (4)
CH (3), HW (4), KR (25),
MA (3), SW (5)
AB (1), KE (17), WA (4),
GR(1)
BE (1), WA (1)

90/91/87

70/66/63

100/100/100

64/64/63

Eleplus (1)

— 0.005 substitutions/site

Plate 6

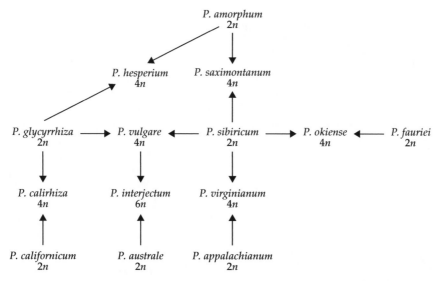

Figure 7.2   Pattern of reticulate evolution for the *P. vulgare* species complex. The diploid (2*n*) progenitors of allopolyploid (4*n* and 6*n*) taxa are indicated at the bases of the various arrows. The allopolyploids are found at the tips of the intersecting arrowheads (from Haufler *et al.* 1995b).

7.2.2   Bryophytes

Byologists...have rarely accepted hybridization among bryophyte species as anything more than an ephemeral and evolutionary insignificant phenomenon...a kind of 'evolutionary noise'.

(Natcheva and Cronberg 2004)

With this reflection, Natcheva and Cronberg (2004) introduced a review of data allowing a test of the importance of natural hybridization in the evolutionary history of bryophyte clades. One conclusion drawn by these authors was that genetic exchange, in the form of both introgression and allopolyploidization, had played a significant role in the evolutionary trajectories of some bryophyte lineages. It is particularly significant that the earlier assumption of bryophyte polyploidy having arisen from within evolutionary lineages (i.e. autopolyploidy) is now understood to have been incorrect (Natcheva and Cronberg 2004). Instead, recent analyses involving the application of molecular methodologies have demonstrated '... that allopolyploidy is the rule rather than the exception, and is common in both mosses and hepatics...' (i.e. liverworts; Natcheva and Cronberg 2004).

Wyatt and his colleagues produced a classic series of studies demonstrating that polyploidy in bryophytes resulted from hybridization between divergent evolutionary lineages (e.g. Wyatt *et al.* 1988, 1992). These investigations highlighted the resolving power of molecular data (in this case mainly allozyme analyses) for deciphering (i) the mechanism of polyploidy (allo- versus auto-), (ii) the progenitors of the various allopolyploid taxa, and (iii) whether specific allopolyploid taxa had been derived from single or multiple hybridization events. The moss genus examined by Wyatt *et al.* (1988, 1992) was *Plagiomnium* and the taxon of interest was *Plagiomnium medium*. Previous studies had designated *P. medium*—the only polyploid of the seven *Plagiomnium* species—to be of autopolyploid derivation (Wyatt *et al.* 1988). In contrast to these assumptions, Wyatt *et al.* (1988, 1992) found fixed heterozygosity at a number of allozyme loci. This fixed heterozygosity reflected combinations of species-specific allelic variation found in the two haploid species, *Plagiomnium ellipticum* and *Plagiomnium insigne*. Thus the allozyme loci indicated that *P. medium*, rather than deriving from only one of these lineages, originated instead from hybridization between the lineages (Wyatt *et al.* 1988, 1992). In addition, the genetic variation within this allopolyploid indicated that it had formed from multiple hybridization events (Wyatt *et al.* 1988, 1992). Finally, using maternally inherited cpDNA markers, Wyatt *et al.* (1988) were also able to infer that *P. insigne* acted as the most common (or sole) maternal parent in the crosses that originated *P. medium*.

Another clade of mosses known to demonstrate the effects of genetic exchange is that containing the peatmosses (genus *Sphagnum*; Shaw *et al.* 2005). For example, Shaw and Goffinet (2000) detected incongruent phylogenetic placements of *Sphagnum mendocinum, Sphagnum cuculliforme, Sphagnum falcatulum*, and *Sphagnum ehyalinum* in evolutionary trees derived from nuclear and plastid DNA sequences. This result supported the hypothesis that these species had received the plastid and nuclear genes from different species of *Sphagnum* belonging to divergent sections. Their findings also led to the general caution that 'Future systematic studies in the peatmosses cannot ignore possible hybridization as a factor underlying ecological, morphological, or molecular variation' (Shaw and Goffinet 2000). This latter conclusion was made in the context of *Sphagnum* taxa being 'notoriously difficult' to delineate at the species level (Shaw and Goffinet 2000). Såstad *et al.* (2001) also documented the effects of reticulate events involving species of peatmoss. In their analysis, they detected the role of allopolyploid formation in the origin of the narrow endemic, *Sphagnum troendelagicum*. The parental taxa in the allopolyploidy event were inferred to be *Sphagnum tenellum* and *Sphagnum balticum* (Såstad *et al.* 2001). As already mentioned, recurrent formation of the same allopolyploid taxon is the rule. Intriguingly, though *S. troendelagicum* had an extremely restricted distribution (only five known populations from central Norway), it also possessed genetic variation indicating its origin through multiple bouts of hybridization. Thus its restricted range was not the result of a single hybridization episode followed by limited geographic expansion (Såstad *et al.* 2001).

The final example of WGD-mediated evolution in bryophytes (and in non-flowering plants) involves the liverwort genus *Targionia*. Boisselier-Dubayle and Bischler (1999) defined the genetic variation in the *Targionia hypophylla—Targionia lorbeeriana* species complex. This species complex has a worldwide distribution and contains both haploid and triploid cytotypes. Boisselier-Dubayle and Bischler (1999) focused their study on haploid and triploid lineages from Europe and the Atlantic islands (i.e. the Azores, the Canary Islands, and Madeira). Interestingly, these authors argued for the involvement of both autopolyploidy and allopolyploidy in the derivation of the triploid lineages. Thus an auto-duplicated genome was hypothesized to have joined with a second, divergent genome through hybridization (Boisselier-Dubayle and Bischler 1999). This additional evolutionary step—the formation of these lineages from an auto- *and* allopolyploid event—did not deter recurrent formation of the triploid lineage, as indicated by the separate origins of the European-Atlantic and African-Australasian polyploids (Boisselier-Dubayle and Bischler 1999).

7.3   Natural hybridization, allopolyploidy, and evolutionary diversification in flowering plants

7.3.1   *Draba*

Allopolyploidy has been recognized for decades as a major evolutionary process in Angiosperms (Stebbins 1947, 1950; Grant 1981; Soltis and Soltis 1993; Masterson 1994). Various estimates have led to the conclusion that the majority of flowering plants have polyploid ancestors somewhere in their phylogenetic history. Since it is generally accepted that allopolyploidy is the most common form of WGD in angiosperms (Stebbins 1947), most species of flowering plants can be said to have a hybridization event somewhere in their evolutionary lineage. Many of the cases of hybrid (i.e. allopolyploid) lineage formation were ancient and predate the diversification of specific clades.

In addition to instances of WGD that predate the diversification of entire complexes, there are also numerous examples of Angiosperm clades in which allopolyploidization has been recurrent throughout their evolutionary history, resulting in the derivation of multiple lineages (Soltis and Soltis 1993, 1995). One flora that is recognized as a resource for studying the process of polyploidy in general, and allopolyploidy in particular (Brochmann *et al*. 2004), is that found in the Arctic. A component of this flora that has been studied extensively—largely by Brochmann and his colleagues—is the genus *Draba* (Brochmann 1992, 1993; Brochmann *et al*. 1992a–d, 1993; Widmer and Baltisberger 1999a, b; Scheen *et al*. 2002). This genus reflects well the conclusion that the Arctic is '. . . one of the Earth's most polyploid-rich areas, in particular of high-level and recently evolved polyploids' (Brochmann *et al*. 2004). In a study using experimental crosses and cytological characters Brochmann *et al*. (1993) were able to assign 101 population samples into 16 species and three sections of *Draba*: sections *Draba*, *Chrysodraba*, and *Drabella*. Of the 16 species, three were diploid and 13 were allopolyploid derivatives demonstrating 4×, 6×, 8×, 10×, and 16× ploidy levels.

Additional analyses of *Draba* taxa have further refined the extensive role that allopolyploidy has played in the evolutionary diversification of this genus. For example, Widmer and Baltisberger (1999b) used nuclear ITS and cpDNA sequences in a phylogenetic analysis of the tetraploid species, *Draba ladina*, along with the putative diploid parental species *Draba aizoides*, *Draba dubia*, and *Draba tomentosa*. The phylogenetic analyses resolved the allopolyploid nature of *D. ladina* and its progenitors. Thus *D. ladina* demonstrated the ITS haplotypes of both *D. aizoides* and *D. tomentosa* (Figure 7.3), but only a cpDNA haplotype characteristic of the latter species (Figure 7.4). These phylogenetic results indicated that *D. aizoides* and *D. tomentosa* were the likely progenitors of this allopolyploid taxon, and that *D. tomentosa* had acted as the maternal (cpDNA-donor) species (Widmer and Baltisberger (1999b). Interestingly, the pattern of genetic variation in *D. ladina* also resulted in the inference that it had arisen through a single hybridization event. This species is thus an exception to the rule that specific allopolyploids are formed repeatedly. However, the pattern within the genus *Draba* as a whole is for the recurrent formation of individual allopolyploid taxa (e.g. Brochmann *et al*. 1992b, c). Indeed, such recurrent formations along with gene flow between taxa of different ploidy groups may explain a portion of the taxonomic complexity encountered within this genus (Brochmann *et al*. 1992d).

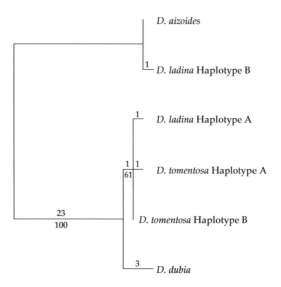

Figure 7.3 Phylogenetic hypothesis, based on nuclear ITS rDNA sequences, for the allotetraploid species *D. ladina* and its possible diploid progenitors (*D. aizoides*, *D. tomentosa*, and *D. dubia*). These results demonstrate that *D. ladina* has both haplotypes found in *D. aizoides* and *D. tomentosa*, thus indicating their role in its formation. Numbers above the lines indicate the number of base-pair substitutions; numbers below the lines are bootstrap values (from Widmer and Baltisberger 1999b).

7.3.2 *Paeonia*

The genus *Paeonia* is a classic model system for understanding factors and processes associated with genetic exchange, including both diploid- and allopolyploid-level hybrid diversification. This species complex has been developed into such a model system through the accumulation of morphological, cytogenetic, artificial hybridization, and—most recently—molecular phylogenetic findings (e.g. Stebbins 1938; Sang *et al*. 1995, 1997a–b, 2004; Sang and Zhang 1999; Ferguson and Sang 2001; Hong *et al*. 2001). In particular, it is now understood that this genus is characterized by reticulate evolution leading to new evolutionary lineages. In some cases, the formation of the hybrid lineages involved homoploidy. Interestingly, as pointed out by Ferguson and Sang (2001), such 'homoploid' diversification can involve hybridization between individuals belonging to different

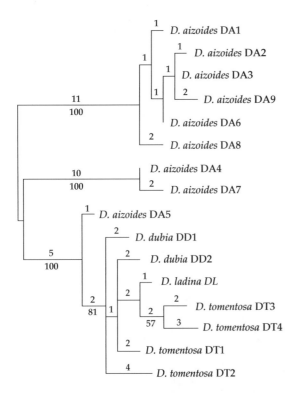

Figure 7.4 Phylogenetic hypothesis derived from the cpDNA haplotype diversity found in the allotetraploid species *D. ladina* and its possible diploid progenitors (*D. aizoides, D. tomentosa,* and *D. dubia*). The results indicate that the most likely maternal (i.e. cpDNA-contributing) parent in the allopolyploid formation of *D. ladina* was *D. tomentosa*. Numbers above the lines indicate the number of base-pair substitutions; numbers below the lines are bootstrap values (from Widmer and Baltisberger 1999b).

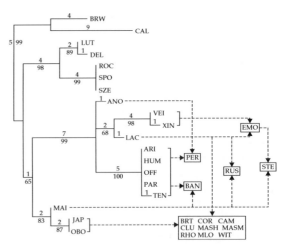

Figure 7.5 Nuclear ITS rDNA phylogeny for taxa of the plant genus *Paeonia*. Each acronym reflects a separate taxon. Boxed taxa are inferred to be of hybrid origin (EMO, BRT, CAM, RHO, and MLO are diploid hybrids; PER, RUS, BAN, COR, MASH, MASM, and WIT are allotetraploids; CLU contains both diploid and tetraploid lineages; the ploidy of STE is unknown). The dashed lines lead to the hybrid taxa from their presumed parents. Numbers above the lines indicate the number of base-pair substitutions; numbers below the lines are bootstrap values (from Sang *et al.* 1997a).

allopolyploid lineages resulting in new taxa with the same number of chromosomes possessed by the (allopolyploid) progenitors.

Although homoploid hybrid formation is frequent in *Paeonia*, the most common means of origin for hybrid lineages is through allopolyploidy (Figure 7.5; Sang *et al.* 1997a, 2004). For example, of the 25 species from the section *Paeonia* examined phylogenetically, 19 were defined as 'hybrid' (i.e. six diploid, 10 tetraploid, and three with both diploid and tetraploid populations; Sang *et al.* 1997a). The extensive allopolyploidization (as well as homoploid lineage formation) has been inferred most strongly using tests for (i) additivity of nuclear markers fixed in various diploid lineages and (ii) discordance between phylogenies constructed from cytoplasmic

and nuclear markers. Thus Sang *et al.* (1995, 1997a) and Sang and Zhang (1999) concluded that additivity for nuclear, ITS, and *Adh* loci was most parsimoniously interpreted as resulting from separate contributions by progenitor species (Figure 7.5). Furthermore, comparisons of phylogenetic interpretations drawn from the uniparentally (i.e. maternally) inherited cpDNA with those arrived at from the biparentally inherited nuclear loci detected significant discordances. The pattern of genetic additivity and the phylogenetic complexity were thus used as indicators of web-like processes, resulting in the taxonomic recognition of hybrid subspecies or species (e.g. see Hong *et al.* 2001 and Sang *et al.* 2004).

The molecular marker data for *Paeonia* has also been used to inform the discussion concerning the process of polyploidy in general. In this case, Sang *et al.* (2004) found evidence for the role of hybrid lineage formation that could not be sharply defined as being either auto- or allopolyploidy. Specifically, *Paeonia obovata* had been termed an autopolyploid derivative of the morphologically identical diploid species of the same name. However, Sang *et al.*

(2004) found *Adh* variation in the tetraploids that did not match with the geographically adjacent diploid populations, indicating that the polyploid plants were likely derivatives of crosses between divergent diploid *P. obovata* lineages. Sang *et al.* (2004) concluded that the origin of the *P. obovata* tetraploid populations demonstrated intermediacy between auto- and allopolyploidy. Such intermediacy reflected the continuum of diploid>autopolyploid>allopolyploid derivation of hybrid lineages. These authors recognized the complexity of reticulate patterns in *Paeonia* by arguing for the importance of studying hybrid derivations '. . . along the continuous range of genomic divergence.'

7.3.3 *Glycine*

The evolution (both organismic and molecular) and systematics of the soybean genus *Glycine* has been the focus of attention for multiple research groups. Of particular importance for the current discussion of WGD and the formation of hybrid taxa, analyses of the subgenus *Glycine* (the sister clade to the progenitor of the cultivated soybean and to the cultivar itself) have shown that it consists of both diploid and allopolyploid taxa (e.g. Doyle and Brown 1985; Doyle and Brown 1989; Doyle *et al.* 1990, 2000, 2002, 2003; Rauscher *et al.* 2004). Furthermore, each of the individual studies have contributed to the general conclusion that the '. . . various polyploid taxa known from the subgenus are all part of a single large allopolyploid complex, linked by shared diploid genomes' (Figure 7.6; Doyle *et al.* 2004).

The evidence for the extensive, allopolyploidy-mediated web relationships (Figure 7.6) in *Glycine* subgenus *Glycine* comes from a diverse array of data sets. An earlier analysis by Doyle and Brown (1985) involved an examination of isozyme variation. From these data they concluded that each of their tetraploid samples was closely related to a maximum of two of the possible diploid progenitor groups. More recent molecular marker studies—consisting of cytoplasmic and nuclear loci—have further defined the complex, but interrelated, connections among the diploid and polyploid lineages. In particular, the work of JJ Doyle and his colleagues has resulted in the inference of (i) multiple maternal lineages in the founding of the allopolyploids

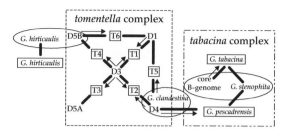

Figure 7.6 The *G. tomentella, G. tabacina*, and *G. hirticaulis* diploid/polyploid complexes. The dashed and dashed/dotted boxes enclose the *G. tomentella* and *G. tabacina* complexes, respectively. The *G. hirticaulis* complex is shown to the left of the *G. tomentella* box. The letter/number designations within the *G. tomentella* complex are races. Those not enclosed by a box are diploid; allopolyploid races are boxed. Within the *G. tabacina* complex, the boxes indicate allopolyploid taxa. Similarly, the diploid and allopolyploid *G. hirticaulis* complex races are indicated by the absence or presence of boxes. Ovals surround diploid taxa that possess the same diploid genome. Solid lines connect allopolyploids with their diploid progenitors and arrows indicate the cpDNA donor species (from Doyle *et al.* 2004).

(i.e. from cpDNA variation; Doyle *et al.* 1990), (ii) multiple polyploid races within systematically recognized taxa (Doyle *et al.* 2000), (iii) recurrent and, in some cases, bi-directional formations of the same allopolyploid taxon (Figure 7.7; Doyle *et al* 2002, 2003; Rauscher *et al.* 2004), (iv) concerted evolution that affects ITS homoeologous loci (Rauscher *et al.* 2004), and (v) introgressive hybridization among the diploid progenitors of the allopolyploid taxa (Doyle *et al.* 2003). For example, the six allopolyploid 'races' within the *Glycine tomentella* complex demonstrate patterns of cpDNA and nuclear genetic variation indicative of multiple, and relatively recent, origins with subsequent introgression between some races (Figures 7.6 and 7.7; Doyle *et al.* 2002, 2004). The recurrent nature of allopolyploid formation is well-illustrated by the *T3* polyploid race. This race was found to be polymorphic at the homoeologous histone H3-D loci. Since the diploid progenitors possessed only a single H3-D locus, the finding of three allele classes among members of the *T3* race (for one of the homoeologous loci) indicated a minimum of two separate origins (Figure 7.7; Doyle *et al.* 2002).

Significantly, the reticulate evolution detected in *G. tomentella* is not unique. Thus, the results from studies of this diploid/polyploid complex reflect a similar web-like pattern found in the sister

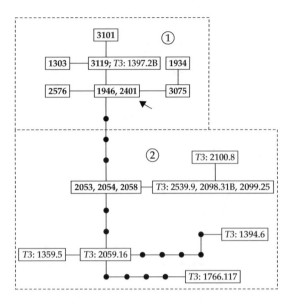

Figure 7.7   Network of histone H3-D alleles present in the diploid (bold numbers) races and the *T3* allopolyploid race of *G. tomentella*. The finding of multiple allelic forms for this one homoeologous locus (*D5A*) indicates a minimum of two separate formations for the *T3* polyploid (from Doyle *et al.* 2002).

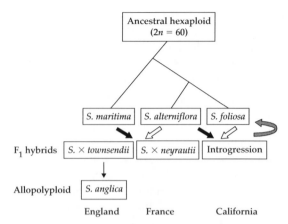

Figure 7.8   Introgressive hybridization and allopolyploidy in the hexaploid lineage of *Spartina*. Each of the events is due to the introductions (into Europe and the west coast of North America) of the eastern, North American species *S. alterniflora*. Arrows indicate hybridization between the taxa. The unfilled arrows indicate the maternal genome donors in the various hybridization events (from Ainouche *et al.* 2003).

complex, *Glycine tabacina*—with which *G. tomentella* shares a diploid genome (Figure 7.7; Doyle *et al.* 2004). As with *Draba* and *Paeonia*, *Glycine* highlights the pervasive effect of genetic exchange, and particularly allopolyploidization, on the evolution of a plant species complex.

7.3.4   *Spartina*

Evolutionary studies of *Spartina* have identified this genus as a paradigm for understanding the fundamentally important effects from hybridization, introgression, and allopolyploidy. In particular, this genus is typified by the formation of natural hybrids that have been the basis for the origin of numerous allopolyploid lineages (Thompson 1991; Thompson *et al.* 1991a–c; Ferris *et al.* 1997; Baumel *et al.* 2001, 2002, 2003; Ainouche *et al.* 2003). The genus is largely New World (13 taxa), with only a few forms native to Western Europe (four taxa; Baumel *et al.* 2002). Within the genus there are two anciently diverged clades (as defined by DNA sequences; Baumel *et al.* 2002)—clade I contains three hexaploid species, *Spartina foliosa*, *Spartina alterniflora*, and

Spartina maritima, with one tetraploid species, *Spartina argentinensis*, basal to the hexaploids; Clade II contains only tetraploid taxa. Overlain onto this cladogenetic patterning are the phylogenetic signatures of introgression and allopolyploidization (Figure 7.8; Ainouche *et al.* 2003). For example, introgressive hybridization is extensive in areas of California where the invading *S. alterniflora* (a North American east-coast species) is forming hybrid swarms with the native *S. foliosa* (Ainouche *et al.* 2003). In addition, molecular data collected for progenitor and derivative taxa demonstrate (i) additivity for nuclear genes and (ii) discordance between the nuclear and cpDNA phylogenies for recently formed hybrid diploid and allopolyploid taxa. The discordance, as with other such examples, results from the hybrid taxa containing multiple nuclear lineages, but only a single cpDNA lineage deriving from one of the parents (Ferris *et al.* 1997; Baumel *et al.* 2001).

The effects of genetic exchange among *Spartina* taxa are also well exemplified by the clade that includes the invasive allopolyploid *Spartina anglica*. It is now understood that *S. anglica* arose through hybridization between *S. alterniflora* and *S. maritima* (Figure 7.8; Thompson 1991). The hybridization was a consequence of the introduction of *S. alterniflora* from North America into the United Kingdom in the early part of the nineteenth century. The first step in the formation of *S. anglica* was the origin of the sterile, F_1 hybrid, *Spartina* × *townsendii* (Figure 7.8; Thompson 1991). A similar, but independently formed, F_1 hybrid between these species (*Spartina* × *neyrautii*) is known from southwest France (Figure 7.8; Baumel *et al.* 2003). Both of these sterile F_1 hybrids have the cpDNA haplotype of *S. alterniflora*, indicating that this species was the maternal parent in the origin of these F_1 lineages (Ferris *et al.* 1997; Baumel *et al.* 2003). The final stage in the formation of *S. anglica* involved chromosome doubling in *S.* × *townsendii*. This event restored fertility and also resulted in a highly fit allopolyploid that was able to invade large areas of aquatic habitat (Thompson 1991). Interestingly, this WGD event was not followed by significant genomic changes (Ainouche *et al.* 2003).

In each of the above examples of allopolyploidy in plants, the process of bringing genomes together has led to further evolution. This has included not only the formation of multiple, polyploid lineages, but also introgression between diploids and polyploids and between the allopolyploid lineages as well. This additional genetic exchange has, in many cases, resulted in further evolutionary change and lineage formation. In the next section I will extend the discussion of WGD-mediated diversification to animal taxa.

7.4 Natural hybridization, allopolyploidy, and evolutionary diversification in animals

7.4.1 Parthenogenesis

Parthenogenesis is reproduction '. . . in which the female's nuclear genome is transmitted intact to the egg, which then develops into an offspring genetically identical to the mother . . .' (Figure 7.9; Avise *et al.* 1992). Importantly, allopolyploidy leading to animal lineages that reproduce by parthenogenesis (or other modes of asexual or sexual reproduction; see below) shares important similarities with allopolyploidy in microorganisms and plants. For example, as is the rule in

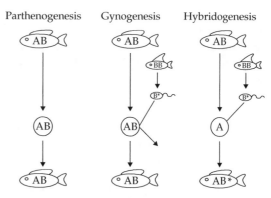

Figure 7.9 Schematic illustration indicating three mechanisms of all-female reproduction in animals. AB indicates the female's genotype. For gynogenesis and hybridogenesis, the BB genotype is that belonging to a male. In gynogenetic reproduction, a sperm is required to stimulate egg development. During each generation of hybridogenetic reproduction, the paternal chromosomes (indicated by B) are lost (indicated by the egg containing only the A genome), but then replaced again through fertilization with a related sexual species (from Avise *et al.* 1992).

other clades, allopolyploid animal lineages are formed repeatedly (Theisen *et al.* 1995; Adams *et al.* 2003).

In the context of this book, I will focus on cases in which the parthenogenetic mode of reproduction is a consequence of hybridization. In this regard, Kearney (2005) has argued that hybridization may be the reason parthenogenetic organisms are often successful in colonizing new/available habitats like those produced by glacial cycling during the late Pleistocene era. Furthermore, in the current section, I will limit my focus to include only those examples in which hybridization *and* WGD events form the basis of the parthenogenetic lineages. Placing examples of hybridization-associated parthenogenesis into polyploid or diploid categories does not, however, reflect the total complexity for all of the groups discussed. I will thus use as one of my examples triploid, parthenogenetic species of whiptail lizards (genus *Cnemidophorus*) that have been formed through hybridization. Yet this genus also contains hybrid, parthenogenetic lineages that possess a diploid chromosome number (Dessauer and Cole 1989). Examples of the formation of lineages containing diploid, hybrid parthenogens will be discussed in detail in section 7.5.2.

Heteronotia binoei

The Australian gecko species complex—placed under the scientific name of *Heteronotia binoei*—represents a classic example of hybridization→ polyploidy→parthenogenesis. Beginning with the work of Moritz *et al.* and more recently Kearney and his colleagues, this diploid—polyploid complex has been dissected in terms of the mode of origins, evolutionary and ecological history, molecular evolution, and the relative fitness of the sexual and parthenogenetic lineages (e.g. Moritz 1983, 1991; Moritz *et al.* 1989a; Hillis *et al.* 1991; Kearney 2003; Kearney and Shine 2004a, b; Strasburg and Kearney 2005).

Like so many allopolyploid (and diploid; e.g. see Webb *et al.* 1978; Honeycutt and Wilkinson 1989) parthenogenetic forms, triploid *Heteronotia* parthenogens had recurrent origins (Moritz 1983, 1991; Moritz *et al.* 1989a). The parthenogenetic *Heteronotia* originated from reciprocal crosses between two of three diploid, sexual races (Figure 7.10; Moritz 1983), with backcrossing between the initial, diploid parthenogenetic form and the sexual diploids leading to the formation of the triploid populations (Figure 7.10; Strasburg and Kearney 2005). Furthermore, it has been postulated that gradual aridification and, more recently, cyclical arid and mesic periods, altered the geographic ranges of the CA6 and SM6 sexual races (Strasburg and Kearney 2005). The triploid lineages overlap with both of their progenitors and with a third sexual, diploid race (Figure 7.11). Findings indicate that this latter race was not involved in the hybridization/ parthenogenetic events (Strasburg and Kearney 2005). The putative backcrosses (Figure 7.10) were suggested to explain not only the triploid, parthenogen formation, but also the high level of nuclear-marker polymorphism demonstrated by the parthenogenetic lineages (Moritz 1983; Moritz *et al.* 1989a).

The parthenogenetic *H. binoei* occur in the arid region of the Australian continent. Indeed, Kearney (2003) has suggested that the observation of numerous parthenogenetic animal and plant forms in the arid region of Australia is suggestive of a cause-and-effect relationship. Thus the parthenogens may be favored due to the uncertainty of finding mates in the harsh environments. However, Kearney (2003)

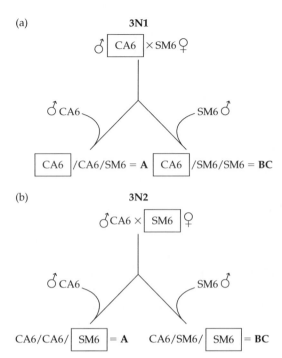

Figure 7.10 Reciprocal formations of the *H. binoei* triploid, parthenogenetic races (a) 3N1 and (b) 3N2. Backcrossing (indicated at the bottom of each panel) between putative diploid, hybrid parthenogens produced multiple types of clones (A and BC) within 3N1 and 3N2 (from Strasburg and Kearney 2005).

argued that this was not a sufficient explanation for the correlated patterns of aridity/parthenogenesis. He also argued that hybridization and polyploidy per se could give at least a short-term advantage to parthenogens through the bringing together of variation from two divergent evolutionary lineages. Consistent with this hypothesis, various analyses of fitness components detected higher, lower, and equivalent estimates for parthenogenetic *H. binoei* relative to their diploid progenitors (Kearney and Shine 2004a, b; Kearney *et al.* 2005).

Parthenogenetic *H. binoei* also exemplify post-WGD molecular evolutionary processes. In this regard, Hillis *et al.* (1991) documented the role of concerted evolution in the structuring of DNA variation among the ribosomal genes of the parthenogenetic lineages. Specifically, their analyses demonstrated the concerted replacement of the CA6 rDNA variants by allelic forms of the SM6 parent; these data were consistent with biased

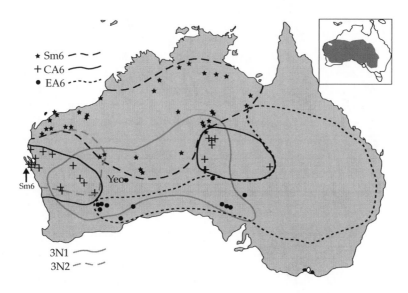

Figure 7.11   The geographic distribution of sexual (SM6, CA6, EA6) and parthenogenetic (3N1 and 3N2) lineages within the *H. binoei* complex. The arid zone of Australia is indicated by dark grey shading in the inset (from Strasburg and Kearney 2005).

gene-conversion-mediated molecular evolution (Hillis *et al.* 1991). Thus not only did the initial hybridization events—followed by rounds of back-crossing (Figure 7.10)—bring together highly divergent genomes, they also set off molecular mechanisms that affected the genetic variation found in the present-day parthenogenetic lineages.

Daphnia

Hebert and his co-workers have analyzed the cladoceran genus *Daphnia* extensively with regard to the evolutionary trajectory of its species complexes. Though not true for every *Daphnia* species complex (Taylor *et al.* 1998), in general this genus exemplifies the important effects from genetic exchange. In this regard, Taylor *et al.* (1998) stated, 'Past genetic studies on *Daphnia* biogeography have focused on syngameons such as the *longispina*, *pulex*, and *carinata* complexes. In these groups, there is incomplete reproductive isolation, resulting in regionally dominant hybrid clones and introgression' (Figure 7.12; Weider *et al.* 1999). Some of these hybridization events are also seen as the trigger for the origin of many of the parthenogenetic *Daphnia* lineages. Specifically, hybridization and allopolyploidy have led to the formation of numerous taxa that demonstrate either *cyclic* or *obligate*

parthenogenesis (Lynch 1984; Crease *et al.* 1989). For example, within the *Daphnia pulex* complex, all polyploid lineages are allopolyploid and partheno-genetic (Colbourne *et al.* 1998).

The *D. pulex* complex reflects the diversity of evolutionary phenomena detected within *Daphnia* as a whole. In particular, as stated above, genetic exchange is linked to WGD and parthenogenesis. The hybridization events have occurred repeatedly, resulting in the recurrent origin of various hybrid parthenogenetic lineages (Dufresne and Hebert 1994, 1997). The *D. pulex* species complex contains both diploid, parthenogenetic (often originating from within a single sexual species; Hebert and Finston 2001) and diploid, sexual species (Adamowicz *et al.* 2002). In North America and Europe, members of this species complex show a complex pattern of geographic distributions with (i) a preponderance of polyploids in the high Arctic, (ii) both diploid and polyploid lineages found in the low Arctic, and (iii) only diploids occurring in temperate regions (Adamowicz *et al.* 2002).

Natural selection favoring allopolyploids has been suggested as a partial explanation for their high frequency in the Arctic environments (Beaton and Hebert 1988). Similarly, diploids are viewed as having a competitive advantage—relative to

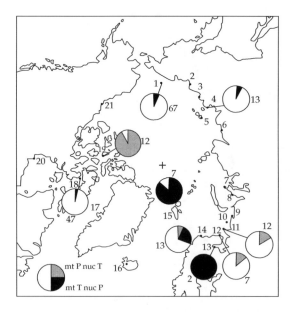

Figure 7.12 The frequency of introgressive hybridization between members of the *D. pulex* groups, *pulicaria* (P) and *tenebrosa* (T). The filled portions of each diagram reflect the proportion of parthenogenetic clones in a certain region that possessed the mtDNA of the *tenebrosa* lineage (T), but the nuclear (nuc), genetic markers of the *pulicaria* lineage (P). Shaded portions of each diagram reflect the reciprocal pattern of mtDNA and nuclear variation. Sample sizes are indicated next to each diagram (from Weider *et al*. 1999).

closely related allopolyploid taxa—in more temperate environments (Adamowicz *et al*. 2002). The combination of these opposing forces could have established a selective gradient leading to the clinal variation in the frequency of diploid/polyploid populations seen in North America and Europe. However, there is no apparent clinal distribution for the diploid and polyploid parthenogens (and sexual species) in the southern hemisphere. Specifically, Adamowicz *et al*. (2002) detected only allopolyploid, asexual lineages in samples from 'temperate or cool temperate' regions of Argentina. It was argued that the presence of the allopolyploids in atypical (i.e. temperate) environments and the absence of diploid *D. pulex* could be the result of the allopolyploids being 'fortunate founders' (Adamowicz *et al*. 2002, 2004). The asexual allopolyploids were hypothesized to have originated, and become established, in the absence of their diploid, sexual progenitors (Adamowicz *et al*. 2002).

Cnemidophorus

The genus *Cnemidophorus* is another well-studied group of animals that demonstrates the effects of hybridization and WGD on the evolutionary diversification of parthenogenetic lineages. Also known as whiptail lizards, this genus is replete with unisexual taxa. To quote Densmore *et al*. (1989b): 'The lizard genus *Cnemidophorus* consists of about 50 species, one-third of which are unisexual and consist exclusively of parthenogenetically reproducing females'. However, as mentioned above, not all of the unisexual taxa reflect WGD-mediated hybridization. For example, of the 20 species examined by Dessauer and Cole (1989), eight were listed as reproducing only through parthenogenesis, while one was found to have both bisexual and unisexual populations. Of these nine species, two had only diploid asexual populations, six had only triploid unisexual forms, and one (*Cnemidophorus tesselatus*) had both diploid and triploid parthenogenetic lineages.

C. tesselatus exemplifies well the evolutionary complexity possible in clades containing unisexual forms. In particular, this species illustrates how taxa thought to reproduce only asexually may also be involved in the production of additional parthenogenetic lineages during occasional bouts of sexual reproduction with bi-sexual forms. The diploid, unisexual *C. tesselatus* populations were derived from hybridization between the bi-sexual species *Cnemidophorus tigris* and *Cnemidophorus septemvittatus* (Parker and Selander 1976). The triploid, parthenogenetic *C. tesselatus* were subsequently derived from hybridization between diploid parthenogens of this species and a third bi-sexual species, *Cnemidophorus sexlineatus* (Parker and Selander 1976). Not only does this 'asexual species'— *C. tesselatus*—illustrate the formation of both diploid and polyploid parthenogenetic lineages through hybridization, it also indicates the permeability of the boundary between asexual and sexual reproduction in some parthenogenetic forms (see also Tinti and Scali 1996 and Scali *et al*. 2003 for a similar example from the stick insect genus *Bacillus*).

The large number of detailed studies of *Cnemidophorus* has allowed additional conclusions concerning the evolutionary steps leading to the origin of the various parthenogenetic lineages.

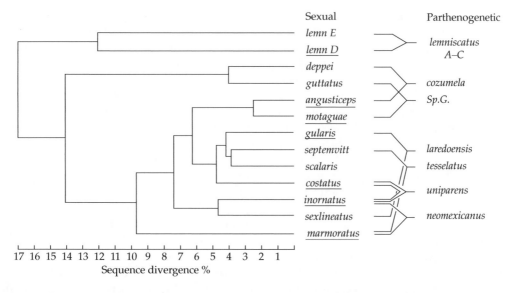

Figure 7.13 Evolutionary relationships among sexually and asexually reproducing species of whiptail lizards of the genus *Cnemidophorus*. The relationships were inferred using mtDNA restriction-site data. The crosses that produced the parthenogens are indicated with intersecting lines that run from each of the two bi-sexual parents and terminate to the left of the unisexual taxa. Underlining indicates the sexual species that acted as the maternal parent in each cross (from Moritz *et al.* 1992).

Studies of mtDNA not only provided estimates of the age of origin (see above) of unisexual taxa; they also led to the discovery that different unisexual taxa often shared a maternal ancestor. For example, in the case of nine *C. sexlineatus*-group parthenogenetic species, it was determined that the most likely maternal progenitor was *Cnemidophorus inornatus* (Densmore *et al.* 1989a). Specifically, the nine unisexual species were probably derived from crosses involving female *C. inornatus* belonging to the subspecies *arizonae* (Densmore *et al.* 1989a). Another inference made possible by the collection of mtDNA sequence data, and the application of phylogenetic methodologies to these data, concerned the evolutionary distance of the bi-sexual progenitors. In this regard, Moritz *et al.* (1992) detected apparent 'phylogenetic constraints' on the hybridization events that could give rise to asexual taxa (Figure 7.13). Moritz *et al.* (1992) analyzed diploid, rather than polyploid, parthenogens. However, the diploids act as the intermediate step to WGD events (e.g. see Dessauer and Cole 1989) resulting in the allopolyploid unisexuals. Thus any phylogenetic constraints on the origin of diploid, asexual taxa reflect constraints on the polyploids

as well. Specifically Moritz *et al.* (1992) found that, '. . . the combination of bi-sexual species that have resulted in parthenogenetic lineages are generally distantly related or genetically divergent. This . . . is consistent with the hypothesis that some minimal level of divergence is necessary to stimulate parthenogenetic reproduction in hybrids' (Moritz *et al.* 1992; Figure 7.13). Finally, the whiptail lizard clade reflects phenomena likely common to most asexually reproducing animal lineages. Some parthenogens are formed recurrently (Parker and Selander 1976; Densmore *et al.* 1989b) and parthenogenetic taxa are of apparent recent origin (Brown and Wright 1979; Densmore *et al.* 1989a; Moritz *et al.* 1989b).

7.4.2   Gynogenesis

The gynogenetic mode of reproduction (like hybridogenesis) can be considered an intermediate form of reproduction—containing some of the characteristics of both asexual and sexual reproduction (Schlupp 2005). Unlike parthenogenesis, for gynogenetic reproduction '. . . sperm from a related bi-sexual species is required to stimulate egg development' (Figure 7.9; Avise *et al.* 1992). This requirement means that the gynogenetic form must have a 'host' bi-sexual species with which it can interact reproductively. Without this host, the gynogen should become extinct (but see section 7.5.2 for an exception). Although lineages characterized by this form of reproduction might be less likely to persist than parthenogenetic taxa (due to the added requirement of the co-habitation of the gynogenetic and bi-sexual species) there are numerous excellent examples of evolutionary diversification in animals via hybridization→ WGD→gynogenesis.

Poecilia formosa

Poecilia formosa (known previously as *Mollienisia formosa*) was the first vertebrate species to be described as possessing unisexual reproduction (Hubbs and Hubbs 1932). The title of the landmark paper recording this finding, 'Apparent parthenogenesis in nature, in a form of fish of hybrid origin' (Hubbs and Hubbs 1932), also reflects the process of hybridization-mediated formation of an evolutionary lineage. Hubbs and Hubbs (1932) defined the morphology of *P. formosa* as being 'exactly intermediate'—reflecting its hybrid origin—between the presumed parents (i.e. *Mollienisia latipinna* and *Mollienisia sphenops*, now *Poecilia latipinna* and *Poecilia sphenops*; Hubbs and Hubbs 1932). Though *P. formosa* is indeed unisexual, 'parthenogenesis' is more accurately replaced by 'gynogenesis' in describing its mode of reproduction. Indeed, Hubbs and Hubbs (1932) gave an excellent description of the expected association of a gynogen and its bi-sexual progenitors in their observation, 'Where this hybrid form exists in nature solely as females, it occurs with only one of the parent species; never with neither.'

Subsequent to the original description in *Poecilia*, various analyses have addressed hypotheses concerning (i) the progenitor species involved in producing the gynogens, (ii) the number of genes involved in affecting the switch to a unisexual lifestyle, and (iii) whether the formation of triploid gynogenetic lineages was a single event or a series of recurrent events (Turner *et al.* 1980; Avise *et al.* 1991; Lampert *et al.* 2005). As for the question of

parentage, it is now accepted that *P. latipinna* was the most likely male parent in the cross that resulted in the gynogen, *P. formosa* (Turner *et al.* 1980; Avise *et al.* 1991). However, it is also now well accepted that instead of *P. sphenops*, *Poecilia mexicana*, a close relative of *P. sphenops*, acted as the maternal parent in the origin of the gynogenetic species (Turner *et al.* 1980; Avise *et al.* 1991). In regard to point (ii), Turner *et al.* (1980) argued that their failure to recreate the gynogen through experimental crosses of the progenitors might be best explained by the control of gynogenesis by a few genes that were polymorphic in the populations of the progenitors. Their failure was hypothesized to have resulted from the use of populations for their crosses that did not possess the correct genotypes at these few loci. Consistent with the conclusion of Turner *et al.* (1980), Lampert *et al.* (2005) argued for a single origin of the triploid, gynogenetic clones of *P. formosa* (Figure 7.14). Thus, unlike the majority of other unisexual lineages of animals, Lampert *et al.* (2005) argued for a lack of recurrent formation of this asexual form (Figure 7.14). However, similar to most other asexual animals this hypothesized single origin was inferred to be recent (Lampert *et al.* 2005).

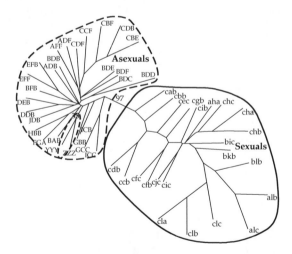

Figure 7.14 Phylogenetic analysis of gynogenetic *P. formosa* clones (asexuals) and sexual *P. formosa* samples (sexuals) using microsatellite data. The two triploid, gynogenetic clones are YYY and ZZZ (indicated by the small, dashed ellipse within the asexual assemblage). The remaining asexual lineages are diploid gynogens (from Lampert *et al.* 2005).

Cobitis

The European spined loaches—genus *Cobitis*—are a freshwater fish group that was originally thought to be a single species (*Cobitis taenia*) containing many subspecies (Bohlen and Ráb 2001). However, within the past decade they have come to be recognized as a species complex containing not only several sexual species, but also numerous unisexual, gynogenetic forms as well (Bohlen and Ráb 2001). Consistent with previous (and following) examples of unisexual animals, asexual *Cobitis* can be either diploid or polyploid (i.e. triploid; Janko *et al.* 2003). The triploid gynogenetic taxa apparently derived from '. . . the incorporation of a haploid sperm genome into unreduced ova produced by diploid hybrids' (Janko *et al.* 2003).

The unisexual *Cobitis* taxa have been shown to have arisen from hybridizations between (i) *Cobitis elongatoides* and *C. taenia*, (ii) *C. elongatoides* and *Cobitis tanaitica*, and (iii) *C. elongatoides* and an unnamed species (*C.* sp.; Janko *et al.* 2003, 2005). The various gynogenetic lineages have arisen recurrently through repeated hybridization events between the sexual forms (Janko *et al.* 2003). Within each hybridization complex, the directionality of crosses has apparently varied. mtDNA analyses indicated that *C. elongatoides* acted as the sole maternal progenitor for unisexual lineages stemming from its pairings with *C. tanaitica* (Janko *et al.* 2003). In contrast, gynogenetic forms originating between *C. elongatoides* and *C. taenia* indicated the occurrence of reciprocal crosses (Janko *et al.* 2003). It has been inferred that the recent hybridization events leading to the numerous unisexual taxa resulted from the expansion of the sexual species from various Pleistocene refugia (Janko *et al.* 2005). Subsequent to their formation, the clonal lineages spread into the ranges of the parental taxa (Janko *et al.* 2005). However, there is also evidence that, in addition to the recently formed gynogenetic taxa (Janko *et al.* 2003, 2005), a few ancient clonal lineages survived in some of the same refuges occupied by their parents (Janko *et al.* 2005). Once again, the data for *Cobitis* indicate that hybridization and polyploidy can result in evolutionary diversification through the formation of numerous asexual forms.

7.4.3 Sexually reproducing, allopolyploid animals

Numerous hypotheses have been posited that address the question of why polyploidy is relatively rare in animal taxa—relative to plants for example. Stebbins (1950) suggested that the barrier to establishment of WGD products in animals was caused by a more easily perturbed developmental system. More recently, Orr (1990) argued that polyploidy in animals was relatively rare due to '. . . the difficulty of establishing a tetraploid line in organisms with a genetically degenerate sex chromosome: although polyploid speciation does not necessarily disrupt sex determination in such species, it *does* invariably disrupt the balance of X chromosome relative to autosomal gene product normally maintained by dosage compensation.'

Notwithstanding the potential barriers to the formation of new, sexually reproducing lineages of animals, there is compelling evidence that polyploidy—and specifically allopolyploidy—has indeed played a significant role in the evolution of the animal clade. First, as discussed in Chapter 6, WGD events may be causal for the radiation of the entire vertebrate clade (Spring 1997; Furlong and Holland 2002; Donoghue and Purnell 2005) and for subclades (such as fish; Le Comber and Smith 2004) within vertebrates. Regardless of the controversy over whether such events cause the diversification of entire clades, the WGD events themselves are accepted as well substantiated. Second, recent WGD events leading to sexually reproducing animal taxa have also been detected. Thus polyploidization within lineages of animals has occurred from the earliest divergence of vertebrates through to recent radiations. In this section, I will focus on the latter category of WGD events that have led to the origin of bi-sexual (as opposed to asexual) lineages of animals. In particular, I will discuss vertebrate systems, since they reflect those lineages thought to be most susceptible to the problems caused by polyploidization, and thus the least likely to be involved in this process (Stebbins 1950; Orr 1990).

Octodontidae

Gallardo and his colleagues have produced a series of data sets consistent with the hypothesis that allopolyploidy events have taken place within the South American hystricognath rodents of the family Octodontidae (Gallardo *et al.* 1999, 2003, 2004; Gallardo and Kirsch 2001; Honeycutt *et al.* 2003). The first species for which they inferred WGD was the desert-adapted *Tympanoctomys barrerae* (Figure 7.15). Cytogenetic and DNA content analyses discovered the largest chromosome number (102) and a genome size (16.8 pg) twice that of *T. barrerae*'s closest living relatives (*Octodontomys gliroides* and *Octomys mimax* possess genome sizes of 8.2 and 7.6 pg; Gallardo *et al.* 1999). Indeed, the ancestral chromosome number ($2n = 46$–56) and genome size (8.2 pg) for the entire superfamily to which *T. barrerae* belongs suggests a 'quantum shift' (Gallardo *et al.* 2003) in the lineage leading to this species (Honeycutt *et al.* 2003). In addition, cell sizes (including sperm) were significantly larger in this species relative to its sister taxa (Gallardo *et al.* 1999, 2003). Finally, estimates of gene copy number—for loci that are single copy in diploid organisms—also indicated a WGD event in the origin of *T. barrerae* (Gallardo *et al.* 2004). All of these data sets were consistent with *T. barrerae* possessing a polyploid (i.e. tetraploid) genome. Analyses of meiotic pairing allowed the further inference that this species was most likely an allopolyploid derivative. In particular, 51 bivalents were present in the meiotic cells of this $4n = 102$ species. Such diploid-like

pairing is expected from a hybrid-derived, polyploid species given that the chromosomes from each progenitor species pair with only their own species' chromosomes.

Gallardo *et al.* (2004) have recently inferred that a second species from this clade also originated from allopolyploidy (Figure 7.15). Specifically, they found that *Pipanacoctomys aureus* possessed (i) 92 chromosomes, (ii) gene duplication, (iii) 46 bivalents in meiotic cells, (iv) significantly larger sperm cell dimensions (relative to diploid sister taxa), and (v) a genome size that is twice as large as the inferred ancestral genome size for hystricognath rodents (Gallardo *et al.* 2004). Phylogenetic reconstructions support the conclusion of a close evolutionary relationship between *T. barrerae* and *P. aureus* (Figure 7.15). Furthermore, Gallardo *et al.* (2004) concluded that, for *P. aureus*, 'Apparently, two ancestral lineages allied to *O. mimax* (but differing in chromosome number between them) may have donated its 92-chromosome complement.' Thus two of the sexually reproducing, desert-adapted species in this clade reflect evolutionary diversification by allopolyploidization.

Xenopus and Silurana

Members of the African clawed frog genera *Xenopus* and *Silurana* reflect the pervasive effects of genetic exchange. In particular, species from these genera demonstrate the role of WGD through allopolyploidization. These two genera are sister clades within the family Pipidae and subfamily Xenopodinae (De Sá and Hillis 1990; Evans *et al.* 2004). Phylogeographic and molecular clock analyses have suggested that extant *Silurana* and *Xenopus* taxa originated some 64 million YBP in the central and/or eastern portions of equatorial Africa (Evans *et al.* 2004).

The role of WGD in the evolution of the African clawed frogs is first indicated by the numbers of chromosomes possessed by the various species of *Silurana* and *Xenopus*. Specifically, *Silurana* contains species with either 20 or 40 chromosomes reflecting diploidy and tetraploidy, respectively (Evans *et al.* 2005). There are no known diploid *Xenopus* taxa; rather there are 10 tetraploid ($4n = 36$), five octoploid ($8n = 72$), and two dodecaploid ($12n = 108$) species (Evans *et al.* 2005). Additional evidence of

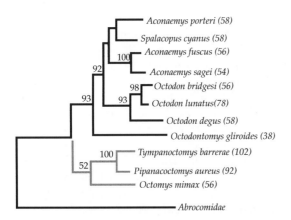

Figure 7.15 Phylogenetic hypothesis for species of the rodent family Octodontidae based on mtDNA sequence data. Bootstrap values are given above the lineages near the appropriate nodes. Chromosome numbers are indicated in parentheses. Gray lines indicate the desert-adapted taxa (from Gallardo *et al.* 2004).

the polyploid constituency of these species is reflected by the increasing series of DNA contents in the various chromosomal forms (Kobel and Du Pasquier 1986).

Several lines of evidence supported the hypothesis that WGD within *Xenopus* and *Silurana* was the result of allopolyploidization. First, comparative phylogenetic analyses suggested the frequent occurrence of hybridization events in the evolution of these species complexes. As discussed previously, phylogenetic discordance is one possible signature of past reticulations. Within *Xenopus* and *Silurana*, phylogenetic discordance has been detected in comparisons of phylogenetic inferences drawn from either multiple molecular data sets or molecular versus morphological data (e.g. Carr *et al.* 1987; Evans *et al.* 2005). Second, there is evidence that both natural and experimental hybrids can produce polyploid gametes (Kobel and Du Pasquier 1986). Third, there are several instances of natural hybridization between various species of *Xenopus* (e.g. Evans *et al.* 1997; W.J. Fischer *et al.* 2000).

All of the available data sets point to a fundamentally important role for genetic exchange in the evolution of the clawed frogs. Inferences from phylogenetic analyses, and associated simulations, supported the occurrence of one origin of allotetraploids in both *Xenopus* and *Silurana* (Evans *et al.* 2005). In contrast, estimates for the number of origins of the *Xenopus* octoploids and dodecaploids were five and two, respectively (Evans *et al.* 2005). Thus the *Xenopus/Silurana* clade reflects numerous instances of natural hybridization-mediated, evolutionary diversification.

7.5   Natural hybridization, homoploidy, and evolutionary diversification

Homoploid hybrid diversification has been most widely discussed in reference to bi-sexual plant taxa (see reviews by Grant 1981; Abbott 1992; Arnold 1997; Rieseberg 1997). One of the best-explored cases of such plant lineage formation is that involving the annual sunflowers. I have already detailed the earlier research findings of Heiser and the more recent ones of Rieseberg and his colleagues in Chapters 4 and 5. Thus I will not repeat the discussion of their many and varied findings in the

present chapter; rather I direct the reader to those two earlier chapters for reviews of the *Helianthus* work. Although sexually reproducing plants dominate the examples of hybrid lineages that possess the same, or nearly the same, chromosome number as their parents, similar processes have also been identified in a handful of animal clades (e.g. DeMarais *et al.* 1992; Franck *et al.* 2000; Tosi *et al.* 2003; Schwarz *et al.* 2005; Taylor *et al.* 2005). Many additional hypotheses of homoploid diversification in animals are made possible by a recognition that the numerous instances of 'introgressive hybridization' between animal lineages lead to similar genetic and phylogenetic signatures as those seen in 'homoploid taxa' (e.g. DeMarais *et al.* 1992; Dowling and DeMarais 1993). Furthermore, as mentioned above, I have chosen to highlight the origin of hybrid, unisexual—but diploid—animal lineages under the heading of homoploidy as well. In this case, I will discuss examples of the origin of both diploid parthenogenetic and hybridogenetic lineages resulting from hybridization between bi-sexual relatives.

7.5.1   Evolution by homoploid hybrid lineage formation in plants

Machaeranthera

The taxonomy of the genus *Machaeranthera* has had a checkered past, with much controversy surrounding the placement of the genus in relation to other genera, and the alignment of species within the genus itself (e.g. Cronquist and Keck 1957; Turner and Horne 1964; Morgan and Simpson 1992). One example of the controversy associated with the evolutionary/phylogenetic associations of *Machaeranthera* is reflected by the following quote from a paper by Cronquist and Keck (1957), entitled 'A reconstitution of the genus *Machaeranthera*': '. . . we are convinced that both *Machaeranthera* and *Xylorhiza* are more nearly related to *Haploppapus* than to *Aster*, and that their inclusion in *Aster* tends to destroy both the morphologic and the phylogenetic homogeneity of the genus [*Aster*].' Thus, instead of *Machaeranthera* belonging within the genus *Aster*, these authors held that it was deserving of its own genus-level recognition. Furthermore, confusion has also existed as to what species share the ancestor of the *Machaeranthera* lineage. For example, Turner and Horne (1964) proposed that members of the genus *Psilactis* were part of the radiation of the *Machaeranthera* clade.

Recent molecular systematic investigations by Morgan and his colleagues initially added to the confusion concerning the evolutionary history of *Machaeranthera*. In particular, Morgan and Simpson (1992) found conflicts between their cpDNA-derived phylogenies and taxonomic conclusions based on morphological characters. These authors gave two explanations for the disagreement among the data sets—(i) variable rates of evolutionary change for the morphological characters or (ii) cpDNA introgression among the species examined (Morgan and Simpson 1992). Several subsequent studies, utilizing a combination of cpDNA and nuclear rDNA sequences, have led to a clearer resolution of evolutionary relationships within *Machaeranthera*, and most importantly for the present discussion, resolved a fundamentally important role for homoploid hybrid diversification (Morgan 1993, 1997, 2003). Figure 7.16 (Morgan 2003) reflects a hypothesized evolutionary/phylogenetic scenario to explain discordances, particularly between phylogenies derived from either nuclear or cpDNA data sets. Specifically, Morgan (2003) proposed that there were a minimum of seven homoploid diversification events resulting in hybrid derivatives assigned to the genera *Machaeranthera*, *Pyrrocoma*, or *Oonopsis* (Figure 7.16). Aside from the phylogenetic discordance for different data sets, the hypothesis of extensive reticulate evolution within this clade is also supported by the observation that members of this genus readily form both natural and experimental hybrids (Morgan 1997). The occurrence of homoploid hybrid lineage formation in *Machaeranthera* has thus been a pervasive force in shaping the evolutionary trajectory of this clade.

Pinus densata

Occupying habitats on the Tibetan plateau and southwestern China, *Pinus densata* belongs to the Asian *Pinus* species complex, several members of which are thought to have arisen via hybridization (see references cited by Wang *et al.* 2001). In particular, the genetic exchange between *Pinus*

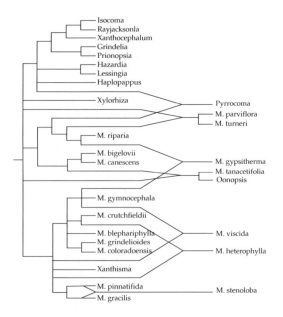

Figure 7.16 Phylogenetic hypothesis reflecting seven reticulate events leading to taxa of the plant genera *Machaeranthera*, *Pyrrocoma*, and *Oonopsis*. The phylogeny was constructed by (i) identifying and aligning species in evolutionary relationships supported by both cpDNA and nuclear DNA sequences, (ii) identifying and aligning species in evolutionary relationships supported by one data set that do not conflict with the other data set, and (iii) placing two lineages on the phylogenetic framework for those species for which there was a conflict and then joining the two lineages to indicate the homoploid events (from Morgan 2003).

tabuliformis (sometimes listed as *Pinus tabulaeformis*) and *Pinus yunnanensis* has been inferred in the origin of *P. densata*. Morphological analyses of *P. densata* have detected intermediacy between this species and its putative parents (Wang and Szmidt 1994). Furthermore, the observation that *P. densata*, *P. tabuliformis*, and *P. yunnanensis* each possess 24 chromosomes led to the inference of homoploid hybrid lineage formation (Wang and Szmidt 1994).

Although morphological data suggested the hybrid origin of *P. densata* (along with other Asian pine species), molecular marker data sets provided the increased resolution needed to test this hypothesis rigorously. These data have come from analyses of both cytoplasmic (cpDNA and mtDNA) and nuclear (allozymes and rDNA) components. In pines, mtDNA is maternally inherited, while cpDNA demonstrates paternal transmission (e.g. Neale and Sederoff 1989). Song *et al.* (2002, 2003) detected multiple mtDNA and cpDNA cytotypes in *P. densata* that were combinations of variants found in *P. tabuliformis* and *P. yunnanensis*. The cytoplasmic marker data were thus consistent with *P. densata*'s designation as a homoploid, hybrid species formed by hybridization between *P. tabuliformis* and *P. yunnanensis* (Song *et al.* 2002, 2003). Nuclear-marker analyses also supported this conclusion. Specifically, allozyme and rDNA (i.e. 5 S and 18 S–5.8 S–25 S) variation demonstrated additivity in *P. densata* for alternate markers found in the two putative parents (Wang *et al.* 2001; Liu *et al.* 2003a, b).

In addition to revealing homoploid hybrid speciation, the molecular markers also suggested added complexity in the evolutionary history of *P. densata*. For example, both nuclear and cytoplasmic data indicated that different *P. densata* populations have evolved independently from one another. This is reflected by the varying genetic constitution of populations from different geographic regions (Wang *et al.* 2001; Song *et al.* 2002, 2003). Furthermore, the paternal and maternal markers indicated that *P. tabuliformis* and *P. yunnanensis* had acted as both father and mother in the origin of the *P. densata* lineage (Song *et al.* 2003). Finally, a unique cpDNA haplotype (i.e. not found in *P. tabuliformis* and *P. yunnanensis*) was detected in the hybrid taxon (Wang and Szmidt 1994; Song *et al.* 2003). This variant could reflect a mutation unique to the *P. densata* lineage.

However, Wang and Szmidt (1994) concluded that the presence of the three haplotypes in *P. densata* instead reflected the contributions of not only *P. tabuliformis* and *P. yunnanensis* to the evolution of the homoploid hybrid, but a third species as well.

Scaevola

Although a majority of the species belonging to the plant genus *Scaevola* occur on the Australian continent, nearly one-third have resulted from radiations into the Pacific Basin (see Gillett 1966 and Howarth *et al.* 2003 for discussions). Analyses of species distributions and morphological and molecular variation have led to the conclusion that the adaptive radiation of the Hawaiian *Scaevola* clade was accomplished through three separate dispersal events from

Australia (Gillett 1966; Howarth *et al.* 2003). Of importance here, the Hawaiian archipelago radiation is thought also to have been an adaptive radiation that included introgressive hybridization and homoploid hybrid lineage derivation (Gillett 1966; Howarth and Baum 2002, 2005).

To test the hypothesis of homoploid hybrid formation, Howarth and Baum (2005) examined sequence variation for the ITS rDNA, *floricaula/leafy*, nitrate reductase, and glyceraldehyde-3-phosphate dehydrogenase loci. For the latter three genes, intron sequences were collected. For five of the seven species of Hawaiian *Scaevola* included in their study (*Scaevola coriacea, Scaevola gaudichaudii, Scaevola mollis, Scaevola gaudichaudiana,* and *Scaevola chamissoniana*), Howarth and Baum (2005) found concordant phylogenetic signals (Figure 7.17). In contrast, when sequence data for *Scaevola procera* and *Scaevola kilaueae* were included in the analysis, significant phylogenetic discordance was detected (Figure 7.17). It was concluded that the non-concordance was due to reticulate events between two different pairs of *Scaevola* species resulting in the two homoploid hybrid species (Figure 7.17; Howarth and Baum 2005). These findings led Howarth and Baum (2005) to a conclusion similar to that of Seehausen (2004); they argued that homoploid hybrid speciation might be an important contributor to adaptive radiations, particularly on islands like those of the Hawaiian chain.

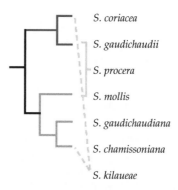

Figure 7.17 Phylogenetic hypothesis for seven species of Hawaiian *Scaevola*. Reticulate events that gave rise to *S. procera* and *S. kilaueae* are indicated by intersecting lines drawn from their respective progenitors (from Howarth and Baum 2005).

7.5.2   Evolution by homoploid hybrid lineage formation in animals—parthenogenetic and hybridogenetic taxa

Warramaba

An extensive series of studies by White and his colleagues defined the reproductive biology, population genetics, and evolutionary origins of the parthenogenetic Australian grasshopper, *Warramaba virgo* (e.g. White *et al.* 1963, 1977, 1980, 1982; Webb *et al.* 1978; White 1980; Dennis *et al.* 1981; White and Contreras 1982; Honeycutt and Wilkinson 1989). In terms of its reproductive biology, White (1980) defined the species as having '. . . a form of automictic parthenogenesis in which a premeiotic doubling of the karyotype in the oocyte . . . is followed by a synapsis which is restricted to sister chromosomes.' He concluded further that this mode of reproduction '. . . automatically ensures that all the offspring of a given female are genetically identical to one another and to their mother. No genetic recombination occurs, the chiasmata being without genetic consequences, since they are formed between chromosomes that are molecular copies of one another . . .'

Though *W. virgo* does indeed reproduce parthenogenetically, White's (1980) conclusions concerning a lack of genetic variation produced by recombination were modified substantially due to a subsequent study of allozyme variation (Honeycutt and Wilkinson 1989). The pattern of allozyme variation in the parthenogenetic lineage reflected significant bouts of recombination-generated variation (Honeycutt and Wilkinson 1989). Furthermore, as with asexual, *allopolyploid* lineages, cytogenetic and genetic variation in populations of this diploid parthenogen indicated its recurrent formation by hybridization between its bi-sexual progenitors (White *et al.* 1977; Webb *et al.* 1978; Dennis *et al.* 1981). The allozyme analysis by Honeycutt and Wilkinson (1989) also provided the greatest resolution for deciphering the origin and post-formation evolution of *W. virgo*. These data supported the following conclusions: (i) *W. virgo* consisted of many clonal lineages resulting from numerous hybridization events between the sexual forms known as P196 and P169; (ii) on the basis of its clonal diversity and geographically extensive distribution,

W. virgo appeared to be very fit relative to its diploid progenitors; and (iii) new variation had arisen since the formation of this parthenogen, indicative of it being a genetically dynamic lineage (Honeycutt and Wilkinson 1989).

Rana esculenta

As illustrated by Figure 7.9, 'Hybridogenetic species possess a hybrid genome: half is clonally inherited (hemiclonal reproduction) while the other half is obtained each generation by sexual reproduction with a parental species' (Semlitsch *et al.* 1996). Such is the case for the hybridogenetic taxon *Rana esculenta*. This species had (and continues to have) repeated origins through natural hybridization between *Rana lessonae* and *Rana ridibunda* (Hellriegel and Reyer 2000). However, the simple definition for hybridogenesis does not encompass the complexity, or reflect the dynamic nature, of this instance of genetic exchange. There are three types of breeding system contained within the category *R. esculenta*, as reflected by LE (*lessonae esculenta*), RE (*ridibunda esculenta*) and all-hybrid populations (Figure 7.18; Christiansen *et al.* 2005). The categories of *lessonae esculenta* and *ridibunda esculenta* indicated population types in which *lessonae* or *ridibunda* acted as the sexual donor, respectively. In contrast, the all-hybrid system reflected those populations in which neither of the progenitors (for the hybridization event that gave rise to *R. esculenta*) is present (Figure 7.18; Christiansen *et al.* 2005). Finally, there are a number of LE populations into which *R. ridibunda* have been introduced. In these cases, this latter species has replaced—due to asymmetric success of different genotypes—both the gynogen and *R. lessonae* (Vorburger and Reyer 2003).

Given the widely varying breeding systems present in *R. esculenta*, it is perhaps not surprising that a diverse array of additional evolutionary processes has also been detected. First, Semlitsch and others have produced a wealth of experimental data indicating environmentally mediated, fluctuating fitness estimates for *R. esculenta*, *R. ridibunda*, and *R. lessonae* (Semlitsch and Reyer 1992; Semlitsch 1993a, b; Semlitsch *et al.* 1997; Plénet *et al.* 2000, 2005). It is thus likely that *R. esculenta* exhibits lower or higher fitness in nature—relative to its

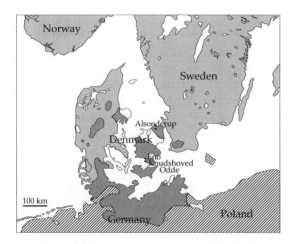

Figure 7.18   Geographic distribution of *R. esculenta*. The portion of the range map covered with dark gray shading indicates regions where this hybridogenetic species does not co-occur with either of its two progenitors (*R. ridibunda* and *R. lessonae*). The hatching indicates regions where *R. esculenta* occurs with its progenitors. The locations of three Danish all-hybrid populations (Alsønderup, Enø, and Knudshoved Odde) from which reproductive data were collected are also indicated (from Christiansen *et al.* 2005).

progenitors—depending on the ecological setting of its populations. In addition to the varying patterns of fitness for the hybrid and its parents, there is also evidence that the hybridogens acted as a source for (i) the formation of sexually produced, diploid or polyploid progeny and (ii) gene flow between the parental taxa (Hotz *et al.* 1992; Schmeller *et al.* 2005; Christiansen *et al.* 2005). For example, Christiansen *et al.* (2005) found that some all-hybrid populations of *R. esculenta* produced viable diploid and triploid progeny arrays similar in proportions to those found in the adult animals. Christiansen *et al.* (2005) inferred that this population was likely to be a long-term, stable, and independently reproducing unit. Like other 'asexual' lineages, the past and present evolutionary pathway of *R. esculenta* reflects a multi-branching route.

7.5.3   Evolution by homoploid hybrid lineage formation in animals—sexually reproducing taxa

Rhagoletis
The North American *Rhagoletis pomonella* species complex is a paradigm for studies of sympatric

diversification due to host shifts. The development of this tephritid fruitfly group as a model evolutionary system began in earnest with the work of Bush (1966). His landmark studies revealed patterns of morphological and genetic variation indicative of organismic divergence through the accumulation of genetic variation, thus allowing individuals to exploit a new habitat—in this case a new 'host' that served as food source and mating site. In Bush's (1966) words, the findings suggested '. . . that some members of certain groups of sibling species may have evolved sympatrically as a result of minor alterations in genes associated with host plant selection.'

Since the foundational work by Bush, numerous genetical, behavioral, and ecological analyses have confirmed his original hypothesis—that the *R. pomonella* group contained lineages that diverged in sympatry (e.g. Feder *et al.* 1997, 2003; Filchak *et al.* 1999, 2000; Linn *et al.* 2003, 2004; Dambroski *et al.* 2005). Specifically, the various *Rhagoletis* host races have apparently diverged due to the presence of genetic variation allowing selection for differential utilization of diverse host species (Filchak *et al.* 2000). In addition, recent findings by Schwarz *et al.* (2005) have documented an instance of apparent homoploid hybrid speciation associated with a shift to a novel host. Schwarz *et al.* (2005) discovered flies of the *R. pomonella* species complex on introduced, and invasive, honeysuckle (genus *Lonicera*). This plant complex consists of both parental species and introgressed forms introduced to North America from Asia. The intriguing finding made by these workers was that the *Rhagoletis* infesting this parental/hybrid complex were themselves hybrids (Figure 7.19). Thus the so-called *Lonicera* fly samples consisted of individuals possessing an admixture of mtDNA and nuclear (allozyme and DNA sequence variation) markers from *Rhagoletis mendax* and *Rhagoletis zephyria* (Schwarz *et al.* 2005). Although it was possible that the occurrence of these hybrid tephritid flies reflected a dynamic hybrid zone (i.e. with ongoing formation of new hybrids) between the two parental species, this hypothesis was discounted by the lack of any detectable F_1 *Lonicera* flies (Schwarz *et al.* 2005). Furthermore, these authors argued that the shift to the novel host by the hybrid tephritids would have

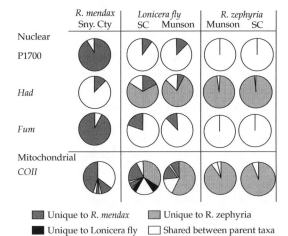

Figure 7.19 Frequencies detected at nuclear and mtDNA loci for the *Lonicera* fly (a hybrid *Rhagoletis* fly that occurs on the introduced—and hybridizing—honeysuckle, genus *Lonicera*) and its putative parents, *R. mendax* and *R. zephyria*. All collections came from central Pennsylvania populations (from Schwarz *et al.* 2005).

acted as a reproductive barrier between the new hybrid lineage and its progenitors. Schwarz *et al.* (2005) also concluded that such homoploid hybrid speciation is likely to be more common than previously thought in host–parasite systems. In particular, they suggested that the difficulty in detecting hybrids between highly morphologically similar lineages, coupled with the normal discounting of hybrid speciation as an important evolutionary process in animals, had contributed to the lack of identification of other examples like the *Lonicera* fly. To remedy this situation, Schwarz *et al.* (2005) offered the following conceptual- and data-based solution. First, 'Hybridization should be considered as a viable hypothesis for the origin of other host-specific animals . . .' Given this change in mindset, they predicted '. . . that future studies will discover more populations with a *Lonicera* fly-like evolutionary history' (Schwarz *et al.* 2005).

Macaca

The macaque primates are notable for having the widest geographical range (portions of northern Africa and over 20 Asian countries) of any non-human primate (discussed in Evans *et al.* 1999 and Tosi *et al.* 2003). They are also well known as a group that has been impacted by reticulate evolution. Thus a number of comparisons of molecular and

morphological data for species of *Macaca* have detected non-concordance. Furthermore, Evans *et al.* (1999) concluded that due to the likely effect of differential nuclear versus mtDNA introgression (i.e. resulting from male migration and female philopatry), the discordance between morphological and molecular phylogenies should not necessarily be used to restructure species designations. This conclusion reflected (i) the numerous instances of introgression within the genus *Macaca* and (ii) caution concerning the affect of this introgression on conservation-management decisions. In fact, findings from various studies have indicated the occurrence of introgression in at least 13 species of *Macaca* (Hayasaka *et al.* 1996; Bynum *et al.* 1997; Morales and Melnick 1998; Tosi *et al.* 2000, 2002, 2003; Evans *et al.* 2001, 2003; Bynum 2002).

From the above, it seems likely that genetic exchange has been of major evolutionary importance for species of macaques. However, for the present discussion it is significant that at least one example of homoploid hybrid lineage formation has also been hypothesized. This hypothesis was tested most rigorously in a study that reported patterns of molecular variation for loci inherited maternally (i.e. mtDNA), paternally (i.e. Y-chromosome sequences), or bi-parentally (i.e. two autosomal intron sequences) in *Macaca* species (Tosi *et al.* 2003). The sensitivity for detecting episodes of reticulate evolution, derived from using the three diverse marker systems, was not lost on Tosi *et al.* (2003). Indeed, they concluded, '. . . episodes of reticulate evolution often go undetected in analyses employing a single genetic system'. Comparing the sequence data from these differentially inherited loci, they detected non-concordant phylogenies and paraphyly for several species. In regard to the question of homoploid hybrid formation, they inferred the origin of the *Macaca arctoides* lineage from ancient hybridization between members of the *Macaca sinica* and *Macaca fascicularis* species complexes ('proto-*Macaca assamensis/thibetana*'; Tosi *et al.* 2003). They argued for a Pleistocene formation of this homoploid hybrid, during the time period when the *Macaca* species existed in forest refugia. A homoploid hybrid origin would explain not only the discordance between the different genetic markers in *M. arctoides*, but also might help to account for this species' unique sexual anatomy (Tosi *et al.* 2000).

Earlier in this chapter I argued that there was a need to recognize that definitional constraints can limit an appreciation for the number of organisms for which web-of-life processes have impacted evolutionary trajectories. Macaque lineages are an excellent example of possible, definitional constraints. Although one hybrid lineage has been given a taxonomic designation—thus falling within the conceptual framework of a homoploid hybrid species—other lineages identified as 'introgressed' have not. I am not arguing that all introgressed lineages be awarded taxonomic recognition, but I am suggesting that what is 'introgression' to one worker, or during one portion of a clade's evolutionary history, might be termed 'homoploid hybrid speciation' by another worker, or during another time period of a clade's existence. The importance is to thus recognize that reticulate evolution has been not only pervasive and creative in past and present-day populations, but will also presumably continue to be pervasive and creative in clades such as *Macaca*.

7.6 Viral recombination, lateral exchange, introgressive hybridization, and evolutionary diversification in microorganisms

As emphasized repeatedly, the modes of genetic exchange that result in the web of life are diverse and creative. However, if one was to rank the evolutionary importance of each mode, it is likely that, by many criteria—such as numbers of lineages affected, biomass produced, impact on other organisms, etc.—the processes of viral recombination and lateral exchange would far out-distance introgressive hybridization. Thus no treatment of genetic exchange-mediated evolutionary diversification would be complete without a discussion of the origin of novel lineages through these two classes of genetic transfer. Indeed, the diversity of examples of such exchange warrants a greater discussion than given here; that fact is the impetus for the additional treatments scattered throughout this book. However, genetic exchange-facilitated

lineage formation in microorganisms includes not only instances of viral recombination and lateral transfer, but introgressive hybridization as well. Thus in this section I will highlight three different organismal categories—viruses (i.e. bacteriophages), bacteria, and protozoa—to illustrate evolutionary diversification that is affected alternately by the three major classes of exchange. These groups do not reflect the entire breadth of organisms that diversify due, at least partially, to web processes. However, they do represent a variety of extremely divergent microorganisms that have been greatly impacted by the origin of recombinant/hybrid lineages.

7.6.1 Viral recombination, lateral exchange, and the evolution of bacteriophages

Viral elements known as bacteriophages make fundamentally important contributions to the evolution of bacterial diversity. They form the basis for bacteria to transduce the cells of their hosts, and act as intermediates for the bacterial cells to acquire new functions (Kwan *et al.* 2005). The potential for such large effects on their bacterial hosts is at least partially due to these organisms being the most abundant of life forms (Wommack and Colwell 2000; Kwan *et al.* 2005).

From the above, it is apparent that bacteriophages have a profound effect on the evolution of their bacterial hosts (as facilitators of gene-transfer events). However, viral recombination and horizontal transfer affect the evolution of bacteriophages as well. The dynamic nature of their evolution was first indicated by the extremely labile nature of the globally distributed bacteriophage lineages. For example, when considered as a single 'population', the replacement of the global bacteriophage pool was estimated to occur every few weeks (Wilhelm *et al.* 2002; Breitbart *et al.* 2004). However, deciphering the role of various processes, including viral recombination and lateral exchange, in the evolution of these organisms requires specific information on the genetic variability in bacteriophage lineages. To accomplish this, Kwan *et al.* (2005) carried out an analysis of the entire genomes of 27 bacteriophages that infect *Staphylococcus aureus*. By comparing the sequence information from these bacteriophages,

Kwan *et al.* (2005) were able to detect extensive genetic variation among the various lineages. Furthermore, comparisons of the 27 genomes with genomic sequences of other bacteriophages, with *S. aureus*, and with the entire prokaryotic sequence database, resolved the highly mosaic nature of these genomes (Figure 7.20). Approximately 49, 24, 3, and 0.1% of the proteins of the 27 bacteriophages showed homology to other *S. aureus* bacteriophages, to unrelated bacteriophages, to *S. aureus*, and to other bacterial hosts, respectively (Figure 7.20; Kwan *et al.* 2005). However, 24% of the sequences showed no significant match with any prokaryotic sequences (Figure 7.20; Kwan *et al.* 2005). These results reflect the extensive reshuffling of genetic material during the evolutionary history of these 27 bacteriophages.

Kwan *et al.* (2005) also illustrated the extreme mosaicism present among the *S. aureus* bacteriophage by comparing the DNA sequences of two phages (G1 and K) in the region of the DNA-replication module and by comparing phage 47 with other, highly similar *S. aureus* bacteriophages. These analyses revealed (i) insertion/deletion events

unique to either G1 or K, (ii) no homology of open reading frames unique to either G1 or K to any other bacterial or bacteriophage sequences, and (iii) a patchwork of sequence cassettes in phage 47 that show high similarities to different bacteriophage lineages (Kwan *et al.* 2005). In summarizing their findings, these authors concluded the following: 'This large amount of mosaicism found among phages supports the idea of large-scale genetic exchange in prokaryotic viruses' (Kwan *et al.* 2005). They also pointed out that this level of mosaicism, and thus genetic exchange, was consistent with previous findings from studies of genetic variation among bacteriophage lineages (Brüssow *et al.* 1998; Juhala *et al.* 2000; Pedulla *et al.* 2003).

7.6.2 Lateral exchange and the evolution of bacterial lineages

As with bacteriophages, prokaryotic taxa are characterized by extensive, evolutionarily effective, genetic exchange (see Zhaxybayeva *et al.* 2004; Beiko *et al.* 2005; Gogarten and Townsend 2005; Simonson *et al.* 2005; Sørensen *et al.* 2005 and Chapter 1 for discussions of this observation). It follows that lateral exchange—mediated by processes such as transduction, transformation, and conjugation—will have led to the formation of new evolutionary lineages. An example that illustrates this type of diversification involves the origin and evolution of *S. aureus* strains (the hosts for the bacteriophages discussed in the previous section). Holden *et al.* (2004a) presented a list of human diseases caused by *S. aureus*—a species now thought to be endemic in a large proportion of hospitals of the UK (Johnson *et al.* 2001). The human diseases recorded by Holden *et al.* (2004a) included, '. . . carbuncles and food poisoning, through more serious device and wound-related infections, to life threatening conditions, such as bacteremia, necrotizing pneumonia, and endocarditis.' Of critical importance for understanding evolutionary processes and for deciding upon treatment regimes, several new strains have evolved through the development of resistances to various antibiotics (Holden *et al.* 2004a).

To test for the factors that led to the evolution of antibiotic resistance, Holden *et al.* (2004a)

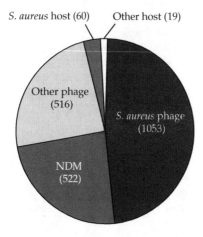

Figure 7.20 Frequencies of homologies between the *S. aureus* bacteriophage *proteome* to (i) other *S. aureus* bacteriophage (blue), (ii) non-*S. aureus* phage (yellow), (iii) the *S. aureus* bacterium (green), and (iv) other bacterial hosts (gray). The red portion of the diagram indicates those *S. aureus* bacteriophage proteins that had no detectable homologies to sequences found in the database. Numbers associated with the various portions of the diagram indicate the number of proteins in each category (from Kwan *et al.* 2005). See also Plate 5.

sequenced the genomes (approx. 2.8 Mbp in length each) of methicillin-resistant (clone MRSA252) and methicillin-sensitive (clone MSSA476) *S. aureus* strains. The resistant bacterial lineage was identified as the strain causing 50% of the methicillin-resistant *S. aureus* (MRSA) infections in the UK, and is also one of the strains of most concern in the USA. In addition to providing information for a strain that is sensitive to methicillin, MSSA476 was also selected because of its known potential to cause serious invasive diseases, and because it is also a major lineage of *S. aureus* within the UK (Holden *et al.* 2004a).

Comparison of the MRSA252 and MSSA476 genomes revealed several aspects relating to the evolution of these lineages (Holden *et al.* 2004a). First, these strains shared a highly conserved core sequence—a core shared with other *S. aureus* strains as well. Indeed, high levels of similarity, detected by multilocus sequence typing, were apparently due to the presence of this core sequence in the various *S. aureus* lineages (Feil *et al.* 2003; Holden *et al.* 2004a). Second, the resistant strain is genetically diverse relative to all other *S. aureus* strains. Homologs of approximately 6% of its sequences were undetectable in other *S. aureus* genomes (Holden *et al.* 2004a). Third, there were radically different distributions of genomic islands in the five sequenced genomes. This finding led Holden *et al.* (2004a) to conclude that genomic islands transfer frequently among *S. aureus* lineages via mobile element-mediated exchange. In further support of this conclusion these authors noted, '. . . all but one of the antibiotic-resistance determinants that account for the antibiotic-resistance profile of MRSA252 are encoded on mobile genetic elements.' Fourth, comparisons of the genomes of the methicillin-sensitive MSSA476 to another, highly similar, but methicillin-resistant, form known as MW2 revealed five distinct genomic acquisition/loss events (Holden *et al.* 2004a). Thus, lateral transfer-mediated evolution has been of profound importance in the origin of new *S. aureus* lineages containing a range of novel adaptations. Although reflective of higher fitness in *S. aureus*, these new lineages—founded on novel adaptations—lead to significant mortality in our own species.

7.6.3 Introgressive hybridization and the evolution of the protozoan genus *Trypanosoma*

Trypanosoma cruzi is the causative agent of Chagas' disease (or American trypanosomiasis; Gaunt *et al.* 2003). The Centers for Disease Control and Prevention (www.cdc.gov) estimate that 16–18 million people suffer from Chagas' disease, with approx. 50 000 deaths per year caused by these infections. Significantly, data from analyses of *T. cruzi* suggest the affect of genetic exchange on the origin and evolution of this human pathogen.

That *T. cruzi* is both genetically and phenotypically highly variable has been recognized for some time (Machado and Ayala 2001). First, allozyme analyses identified 43 isoenzyme classes (Tibayrenc *et al.* 1986; Tibayrenc and Ayala 1988). Second, studies of RAPD and rDNA variation (Souto and Zingales 1993; Tibayrenc *et al.* 1993) identified two evolutionary lineages (*T. cruzi* I and II; Machado and Ayala 2001). Third, variation at an additional set of RAPD loci recognized *T. cruzi* I and also supported the division of *T. cruzi* II into five lineages (Brisse *et al.* 2000).

Machado and Ayala (2001) tested the hypothesis that the genetic heterogeneity detected within *T. cruzi* was due at least partially to introgressive hybridization between the various lineages. Results from analyses of both nuclear and mtDNA sequence data supported this hypothesis. Machado and Ayala (2001) thus concluded, 'The results provide evidence of hybridization between strains from two divergent groups of *T. cruzi*, demonstrate mitochondrial introgression across distantly related lineages, and reveal genetic exchange among closely related strains'. Likewise, analyses of experimental *T. cruzi* hybrids supported the hypothesis that *T. cruzi* was capable of genetic exchange through natural hybridization. Specifically, Gaunt *et al.* (2003) produced hybrid clones that possessed striking genetic similarities to natural strains of this protozoan species. Each of these findings suggested that introgressive hybridization had played a fundamentally important role in the origin and evolution of the genetic and phenotypic variability (e.g. growth rate, pathogenicity, infectivity, and drug susceptibility;

Machado and Ayala 2001) within *T. cruzi*. It is significant to note that genetic exchange has also impacted the evolutionary trajectory of the related species, *Trypanosoma brucei* (the causative agent of African sleeping sickness). In the case of this latter species, lateral exchange has contributed to the variability of key genes and genetic regions (Berriman *et al.* 2005; El-Sayed *et al.* 2005).

7.7 Summary and conclusions

As illustrated in this chapter, genetic exchange is responsible for the origin of novel evolutionary lineages. The preceding examples reflect the taxonomic diversity of organisms that have originated from web processes. However, it is equally important to recognize that the diversity of processes underlying the multiplication of lineages is also great. From WGD, to host shifts, to the development of resistances to antibiotics, the number of identified classes of genetic exchange-mediated lineage multiplication continues to increase. It is also likely that as the number of in-depth analyses of the genetic variation in organisms accrue, additional mechanisms leading to web-like processes will be identified. However, regardless of whether or not additional processes contributing to reticulate evolution are detected, the pattern already identified in nature is indicative of a fundamentally important role for the well-defined processes that lead to genetic exchange. This is true for viral, prokaryotic, and eukaryotic clades.

CHAPTER 8

Implications for endangered taxa

In fact, the present population may be better off as a result of acquisition of new genes because of the multiple congenital difficulties that apparently emerged as a result of inbreeding . . .

(O'Brien and Mayr 1991)

Introduced species (or subspecies), however, can generate another kind of extinction, a genetic extinction by hybridization and introgression with native flora and fauna.

(Rhymer and Simberloff 1996)

. . . hybridization between rarer and more numerous taxa potentially results in a genetic enrichment of the endangered form. The rare form is aided by such interaction through elevated fitness, the addition of genetic variability that facilitates habitat expansion, and the hybrid population acting as a genetic reservoir for reconstituting the parental genotypes/phenotypes.

(Arnold 1997)

Here, the risk of extinction likely outweighs the risk or detriment of introgression, and this option would protect those species most in danger even though they may also be most likely to hybridize . . .

(McElroy *et al.* 1997)

This is a significant reduction in genetic differentiation, and represents a breakdown in species integrity most likely due to hybridization.

(Mank *et al.* 2004)

8.1 Introgressive hybridization and the conservation and restoration of endangered taxa

The effects of genetic exchange (specifically introgressive hybridization) on rare and endangered taxa have been presented almost exclusively as negative. This is understandable and logical given the viewpoint that subspecies, species, etc. are defined by levels of reproductive isolation, genetic/ecological cohesiveness, and/or strict reciprocal monophyly. Furthermore, it is impossible to divorce a discussion of genetic exchange and conservation from sociological/political realities. In this regard, it is presumably easier to explain (to those who allocate conservation funding) the need to protect a biological form if that form is seen to be highly distinct from all other related taxa. However, if the web-of-life metaphor accurately typifies evolutionary pattern and process, the assertion of a uniformly detrimental outcome from introgression on the genetic, ecological, and evolutionary trajectory of endangered taxa needs to be re-examined (Arnold 1997). Genetic exchange is part of the natural order of the biological world. Furthermore, given enough time all lineages will have been affected by lateral transfer, viral recombination, and/or introgressive hybridization. In some regards then, introgressive hybridization between rare and more common forms is merely another example of what is occurring continuously through evolutionary time. Yet with the whole-scale modification of the biosphere by humans, genetic exchange-mediated effects on taxa are likely to become more pronounced and, in concert with the loss of habitat, have affects on conservation efforts.

In this chapter I will focus on examples of introgressive hybridization in plant and animal clades as the mode of genetic exchange leading to conservation opportunities and risks for evolutionary lineages. However, microorganisms impacted by viral recombination and lateral transfer also fit within the conceptual framework of conservation biology. For example, when influenza, bacteriophage, or HIV lineages brought together in human (or human-associated) populations change through recombination with other strains, some forms may become extinct. In addition, the disappearance and origin of bacterial taxa through selection for recombinant progeny due to antibiotic application is conceptually and biologically identical to any other human-mediated perturbation that drives to extinction another component of nature.

Given the potential for both creative and destructive roles for introgression between rare and common taxa, topics and examples in this chapter will be discussed in the context of both positive and negative effects from genetic exchange on the fate of endangered flora and fauna. I must state that when I use results from studies to discuss possible gains for rare forms, I will almost always be drawing conclusions *not* arrived at by the authors of the particular references cited. For example, I will allude to the possibility that introgression may enrich otherwise genetically limited populations or, alternatively, cause the extinction of the rare form by genetic assimilation by a more numerous, related taxon. The latter conclusion is a common theme in the conservation literature. The former is not. Likewise, I will argue that if evolutionary diversification is indeed a web-like process we should not use the occurrence of genetic exchange alone to determine either (i) a taxonomic unit for the hybridizing forms or (ii) a value for their conservation.

8.2 Introgressive hybridization involving endangered plant taxa

8.2.1 *Eucalyptus*

Eucalyptus benthamii
Individuals of the Camden white gum, *Eucalyptus benthamii*, can be up to 40 m in height and live for over 150 years (NSW National Parks and Wildlife Service 2000). Though locally abundant, this species is restricted to the alluvial margins of the Nepean River (and tributaries of this river) southwest of Sydney, Australia (Figure 8.1; Butcher *et al.* 2005). Its distribution (i) within agriculturally favorable regions (fertile alluvial zones) and (ii) at low elevations in river bottoms has resulted in losses due to agricultural clearing and through flooding from dam projects (NSW National Parks and Wildlife Service 2000; Butcher *et al.* 2005).

There are two main populations of *E. benthamii*—the Bents Basin consisting of approximately 300 individuals and the Kedumba Valley consisting of

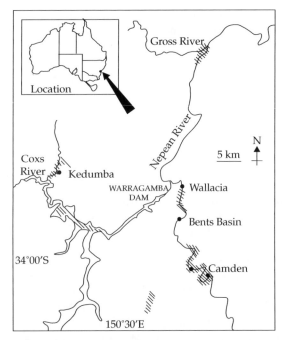

Figure 8.1 Distribution of sampling sites (Kedumba, Wallacia, Bents Basin, and Camden) and the geographical range (inset) of *E. benthamii*. The filled circles indicate the sampling localities, while the hatched areas reflect *E. benthamii's* distribution prior to land clearing for agricultural use (from Butcher *et al.* 2005).

approximately 6500 individuals (Figure 8.1; NSW National Parks and Wildlife Service 2000; Butcher *et al.* 2005). In addition, approx. 30 trees occur near Camden and nine individuals exist near Wallacia, New South Wales (Figure 8.1). Because of its restricted distribution, and the presence of various threats to its populations, this species has been named a vulnerable species as defined by the New South Wales Threatened Species Conservation Act 1995 (NSW National Parks and Wildlife Service 2000).

Butcher *et al.* (2005) concluded that limited regeneration—resulting from competition with non-native species and perturbed flooding and fire cycles—was the major concern for the long-term existence of *E. benthamii*. However, they also hypothesized that introgressive hybridization with related *Eucalyptus* species might negatively impact the ability of *E. benthamii* to have sustainable populations. In particular, Butcher *et al.* (2005) postulated that hybridization might be occurring with *Eucalyptus viminalis*, a species found in the Camden area and one known to hybridize with other species of *Eucalyptus* (Griffin *et al.* 1988). Data providing a test for the effects of hybridization in the *E. benthamii* populations came from microsatellite and morphological surveys of seedlings (Butcher *et al.* 2005). Both data sets indicated a large effect from interspecific gene flow (via pollen) on the production of progeny by *E. benthamii*. In particular, seedling morphology and microsatellite variation indicated that approximately 30 and 10% of the outcrossed progeny in Wallacia and Camden, respectively, were hybrids (Butcher *et al.* 2005). Cumulative estimates of the frequency of hybrid and selfed progeny (also not useable for regeneration) resulted in the following conclusion by Butcher *et al.* (2005): 'The high incidence of hybrid and selfed progeny means that only 25% of the seed collected from trees in the Wallacia population and 35% of the seed from Camden would be suitable for revegetation.' Thus, introgression (and inbreeding) is diminishing the chances for natural or artificial regeneration of *E. benthamii*.

Eucalyptus cordata

As with *E. benthamii*, the future of the southeastern Tasmanian rare endemic *Eucalyptus cordata* will likely include the disappearance of this species from parts of its limited range (McKinnon *et al.* 2004). This species shows widely varying growth forms depending upon the ecological setting; it possesses a stunted, mallee shrub-habit in dry sites, but demonstrates a tree phenotype at wetter locations (McKinnon *et al.* 2004). Both its fascinating evolutionary history and the endangered nature of this taxon are illustrated by the following quote: 'Its distribution, which falls within an area of 6000 km^2, coincides closely with a modeled former glacial refuge (Kirkpatrick and Fowler 1998) and appears to be relictual . . . About 37 scattered populations are known, most of which are sympatric with at least one other species from the same section . . .' (McKinnon *et al.* 2004). In contrast to its close genetic contact with other *Eucalyptus* species, most of the *E. cordata* populations are thought to be genetically isolated from one another (McKinnon *et al.* 2004).

The above description indicates that, again like the situation for *E. benthamii*, *E. cordata* may be in danger of genetic assimilation by a more common congener. However, unlike the research on *E. benthamii*, the hypotheses of McKinnon *et al.* (2004) reflected the potentially significant effects of introgression from the rare form into a more common gum tree species. These authors, using an extremely fine-scaled phylogeographic approach, tested the hypothesis that gene flow from *E. cordata* into the more widespread species *Eucalyptus globulus* had contributed significantly to the genetic constitution of this latter species. McKinnon *et al.* (2004) were also interested in testing the hypothesis that exchange from a rare form to a common taxon might affect conclusions drawn in phylogeographical analyses of the common form. In other words, they wished to test for the confounding effects from introgression from a rare species that had gone extinct through genetic assimilation with the more common lineage (McKinnon *et al.* 2004; Figure 8.2).

Both of the above hypotheses were suggested by findings from analyses of cpDNA variation (McKinnon *et al.* 2001) and phylogeographical structure (Jackson *et al.* 1999; Freeman *et al.* 2001). For example, *E. globulus* populations from eastern Tasmania possessed a high level of cpDNA variation, including a haplotype not found in its mainland

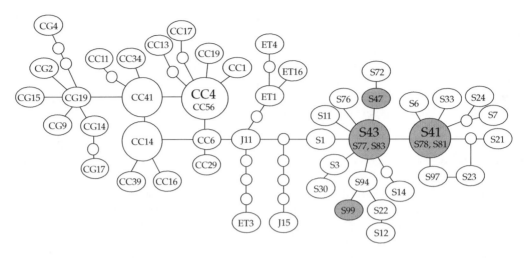

Figure 8.2   Chloroplast haplotype network for *E. globulus*. Connections between haplotypes indicate a single character-state change. Empty circles (nodes) indicate missing haplotypes. Larger nodes indicate haplotypes detected in >10 populations. Those haplotypes shared with the rare species, *E. cordata*, are shaded (from McKinnon *et al.* 2004).

Australian populations (Jackson *et al.* 1999). This haplotype was, however, found in high frequency in the rare endemics *E. cordata* and *Eucalyptus morrisbyi* (Jackson *et al.* 1999). In addition, haplotype sharing was detected among southeastern Tasmanian populations of other *Eucalyptus* species, but was absent in populations of the same species in central and northern portions of the state (McKinnon *et al.* 2001). From these data, McKinnon *et al.* (2004) concluded that introgression from the rare *E. cordata* into the more common *E. globulus* had occurred (Figure 8.2). They argued that such a scenario was likely due to the demographic and reproductive biology attributes associated with these two taxa. First, the less numerous *E. cordata* plants were more likely to act as the pollen recipients (maternal plants and thus cpDNA donors) for any F_1 hybrids due to the much larger *E. globulus* pollen pool. Second, McKinnon *et al.* (2004) reviewed data from experimental crosses between these species that demonstrated a trend for seed production to occur when *E. globulus* and *E. cordata* acted as the paternal and maternal parents, respectively.

The data gathered in prior analyses along with the findings of McKinnon *et al.* (2004) (Figure 8.2) also supported the hypothesis that phylogeographical inferences could be profoundly affected by past introgression between the species of interest and locally or globally extinct forms. Indeed, these

authors cautioned that the dynamic and interdependent nature of climate change, taxa distributions, and introgressive hybridization posed '. . . a major challenge for phylogeographic analysis' (McKinnon *et al.* 2004). Yet, this conclusion also reflects the evolutionarily creative aspect of introgressive hybridization resulting in the genetic enrichment of a widespread, and common, lineage by one containing only a few individuals (McKinnon *et al.* 2004).

8.2.2   *Carpobrotus*

Carpobrotus edulis and Carpobrotus chilensis

Increasing numbers of invasive plants threaten the benefits we acquire from wildland ecosystems.

(D'Antonio *et al.* 2004)

This statement, from a paper entitled 'Invasive plants in wildland ecosystems: merging the study of invasion processes with management needs', hints at the negative impacts—in terms of human utilization—from the introduction of non-native plant species. For the present discussion, it is significant that there are many potential examples of introgression-mediated evolution of invasive taxa. Indeed, Ellstrand and Schierenbeck (2000) and

Ellstrand (2003) discussed extensively the possibility that natural hybridization could be the causal step in the evolution of invasiveness. These authors did not claim to have detected the sole trigger for a shift to an invasive life-history, but they did argue that hybridization was '. . . clearly an underappreciated mechanism worthy of more consideration in explaining the evolution of invasiveness in plants' (Ellstrand and Schierenbeck 2000).

One possible example of the causal role for introgressive hybridization in the evolution of invasiveness involves the non-native species of iceplant, *Carpobrotus edulis*, and its native congener, *Carpobrotus chilensis* (Albert *et al.* 1997; Gallagher *et al.* 1997; Vilà and D'Antonio 1998a, b; Weber and D'Antonio 1999). The invasive species reflects an introduction from South Africa that is considered a threat to numerous California native plant species (see discussion by Albert *et al.* 1997). In contrast, *C. chilensis* is a less aggressive California native (or possibly a form naturalized to California several hundred years ago) not considered of concern for other native taxa (also discussed by Albert *et al.* 1997).

To test the hypothesis that introgression between native and introduced lineages has resulted in the evolution of invasiveness, it is first necessary to document the occurrence of genetic exchange. For the two *Carpobrotus* species, Albert *et al.* (1997) and Gallagher *et al.* (1997) collected morphological and allozyme data for a range of populations containing the parental forms and putative natural hybrids. Both data sets indicated the occurrence of introgressive hybridization. The morphological data indicated that the hybrids resembled the invasive *C. edulis* more than they did the less-aggressive growth form of *C. chilensis*. Likewise, allozyme variation indicated asymmetric introgression of *C. chilensis* alleles into *C. edulis*. From these findings it was concluded that introgression had given rise to a *Carpobrotus* hybrid swarm in California and that incorporation of *C. chilensis* alleles by *C. edulis* may have resulted in relatively fit hybrids capable of invading *C. chilensis* habitats (Albert *et al.* 1997; Gallagher *et al.* 1997). Vilà and D'Antonio (1998a, b) and Weber and D'Antonio (1999) tested the latter hypothesis by estimating environment-dependent fitness for both parental taxa and their hybrids. Overall, their findings supported the contention

that introgressive hybridization had facilitated the invasiveness of *C. edulis*. The introgressed *C. edulis* individuals demonstrated (i) similar responses (relative to both parental species) to salinity and (ii) higher fitness relative to *C. chilensis*, with regard to resistance to mammalian herbivores, vegetative growth, fruit preferences of native frugivores, and seed survival after passage through the fugivores' digestive systems (Vilà and D'Antonio 1998a, b; Weber and D'Antonio 1999). These results thus indicated the potential for introgressive hybridization between native and introduced forms to facilitate the evolution and spread of a highly invasive taxon.

Carpobrotus edulis and Carpobrotus acinaciformis

Invasive species have become a growing global concern . . . and are now regarded as the second-most important threat to the maintenance of biodiversity, after the fragmentation and/or destruction of habitats . . .

(Suehs *et al.* 2004a)

Like that of D'Antonio *et al.* (2004), the introduction of Suehs *et al.* (2004a) heralds the danger of invasive species to native flora (and fauna). Also, like the work of D'Antonio and her colleagues, research by Suehs *et al.* (2004a, b, 2005) involved analyses of *Carpobrotus* species. However, this latter research involved two species (*C. edulis* and *Carpobrotus acinaciformis*, also called *Carpobrotus affine acinaciformis*), both of which were invasive, found in the Mediterranean Basin. Suehs *et al.* (2004a, b, 2005) used a combination of morphological, allozyme, and reproductive biology analyses to determine the presence, genetic makeup, and relative fitness of various *Carpobrotus* genotypes/phenotypes in mainland and island (the Hyères archipelago) habitats in southeastern France.

The morphological and allozymic variation detected in the populations of *Carpobrotus* on the islands of the Hyères archipelago indicated the presence of both *C. edulis* and *C. acinaciformis* (Suehs *et al.* 2004a). In addition, a third genotypic class was also identified. Possessing an admixture of parental alleles and morphological characteristics, the third category was concluded to reflect introgression from *C. edulis* into *C. acinaciformis* (Suehs *et al.* 2004a). Due to the highly introgressed nature of *C. acinaciformis* plants located in southeastern France and nearby islands, Suehs *et al.* (2004b) named these individuals *C. affine acinaciformis*. Furthermore, Suehs *et al.* (2004a) postulated that the potential for a hybridization-mediated increase in invasiveness—particularly in the genetically enriched, introgressed *C. affine acinaciformis*—was great. Most significantly for the current topic, they argued that *C. affine acinaciformis* was of specific concern '. . . because of its strong clonality, high hybrid vigour, and potential for continued introgression from *C. edulis* genes' (Suehs *et al.* 2004b). Because of the risk of the continued invasion by introgressed *C. affine acinaciformis*, caused by its elevated fitness resulting from its acquisition of *C. edulis* genes, Suehs *et al.* (2004b) recommended a control strategy that included the minimizing of sympatric associations between *C. edulis* and *C. acinaciformis*.

8.2.3 *Taraxacum*

The final example of interactions between invasive and native plant congeners comes from the dandelion genus *Taraxacum*. In this case, it has been hypothesized that hybridization of the native alpine dandelion, *Taraxacum ceratophorum*, with the introduced, invasive species *Taraxacum officinale* might lead to genetic assimilation of the native form (Brock 2004; Brock and Galen 2005; Brock *et al.* 2005). While fossils of the native form have indicated its presence in North America for at least 100 000 years, *T. officinale* was introduced during the initial stages of European settlement of the New World (see discussion by Brock 2004).

That most species of *Taraxacum* reproduce asexually via agamospermy (Brock 2004) indicates a possible preadaptation for invasiveness; plants with uniparental reproduction may be more likely to colonize and spread than would individuals requiring a sexual partner. While agamospermy is associated with polyploid *Taraxacum* species, sexual reproduction (and self-incompatibility) characterizes the $2n = 16$ diploid lineages (Richards 1970; Brock 2004). However, experimental crosses among some apomictic (agamospermous), polyploid forms and

diploid individuals have resulted in hybrid progeny (Richards 1970).

Brock and his colleagues used a series of analyses of the mating system, ecophysiological responses, and phenotypic plasticity demonstrated by *T. ceratophorum*, *T. officinale*, and their hybrids to predict the likelihood of genetic swamping of the native species by the introduced lineage (Brock 2004; Brock and Galen 2005; Brock *et al.* 2005). The various data sets suggested different conclusions concerning the likelihood of such assimilation. First, the two species were found to be widely sympatric and, in some sympatric populations, were nearly identical in their flowering phenologies and pollen vectors (Brock 2004). Second, the diploid *T. ceratophorum* was determined to be an obligate outcrossing species and as the maternal parent in interspecific, experimental crosses demonstrated a much-reduced seed set (approx. 37%; Brock 2004). Third, although not more phenotypically plastic than *T. ceratophorum* (as predicted by some models for the evolution of invasiveness), *T. officinale* was found to possess characteristics that would increase its ability to colonize both open and vegetated habitats (Brock *et al.* 2005). Each of these observations indicated that the initial requirements for introgression-driven replacement of *T. ceratophorum* were met.

In contrast to the above results, several findings reflected at least partial barriers for the evolutionary sequence of hybridization→introgression→genetic assimilation. Although seed set from the interspecific crosses was approx. 37%, only one-third of the seeds were found to be of hybrid origin, with the remainder resulting from self-fertilization due to a disruption of the self-incompatibility system of *T. ceratophorum* (Brock 2004). Furthermore, studies of ecophysiological traits suggested a higher fitness for *T. ceratophorum* under drought conditions, relative to both *T. officinale* and their F_1 hybrids (Figure 8.3; Brock and Galen 2005). Specifically, individuals of the native species demonstrated higher water-use efficiency, carbon assimilation, and transpiration under arid conditions than did the invasive taxon or F_1 hybrids (Figure 8.3). This caused Brock and Galen (2005) to conclude the following: 'Arid habitats and occasional drought in mesic sites may provide native dandelions with refugia from negative interactions with invasives.'

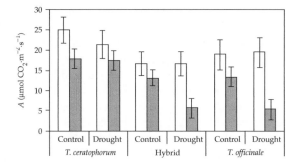

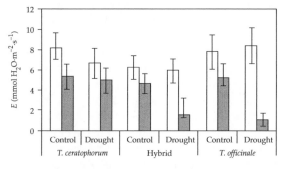

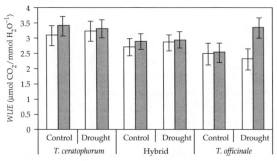

Figure 8.3 Estimates at the beginning (open bars) and end (shaded bars) of experiments for carbon assimilation (A), transpiration (E) and water-use efficiency (WUE) by *T. ceratophorum*, *T. officinale*, and their F_1 hybrids (from Brock and Galen 2005).

Such negative interactions would include genetic assimilation.

8.3   Introgressive hybridization involving endangered animal taxa

8.3.1   Felidae

Florida panther

In considering why some plant species were more widespread and numerous than others, Ledyard

Stebbins drew the following conclusion: '... most common and widespread species are genetically diverse, while rare and endemic ones contain relatively little genetic variability ... This homogeneity reduces the number of ecological niches in which rare species can compete successfully with other species ...' (Stebbins 1942). This led to the hypothesis that hybridization between formerly isolated populations of the same taxon, or between different taxa, could lead to the acquisition of genetic variability by rare and endangered taxa (Stebbins 1942). This infusion of genetic variability might increase the likelihood of the rare taxon spreading into novel habitats (Stebbins 1942).

In the case of one of the poster animals for conservation and recovery efforts, the Florida panther, O'Brien *et al.* (1990) and O'Brien and Mayr (1991) voiced a similar conclusion to that of Stebbins. These authors highlighted the irony that this taxon likely benefited from the '... genetic advantages of introducing some additional genetic material into a population suffering from inbreeding ...' and yet might be excluded from protection and restoration due to the so-called Hybrid Policy (O'Brien and Mayr 1991) of the Endangered Species Act 1973. Specifically, O'Brien *et al.* (1990) detected—from an examination of mtDNA variation among (i) Florida panther individuals, (ii) individuals from seven other North American subspecies, (iii) animals representing three South American subspecies, and (iv) animals from a captive breeding population—evidence of introgressive hybridization. The pattern of genetic variation indicated that the Florida panther populations included two divergent mtDNA lineages. One of the mtDNA haplotypes was more closely related to that of South American subspecies. This result was quite unexpected and suggested that the rare, endemic population of Florida panthers had recently undergone introgressive hybridization (O'Brien *et al.* 1990). The origin of the mtDNA introgression was concluded to have come from captive breeding individuals that had been released into the wild (O'Brien *et al.* 1990). Consistent with this hypothesis was the detection of an APRT-B allozyme allele that was in high frequency in the Everglades pumas, the captive population, and some South American samples, but not in other North American samples (O'Brien *et al.*

1990). Though problematic for the Florida panther's conservation—given that hybrids might not be afforded protection—the potential genetic advantages included (i) a reduction of genetic defects resulting from inbreeding and (ii) an increase in fitness of introgressants leading to species evolution (O'Brien *et al.* 1990). Thus human-mediated hybridization in the wild was viewed as potentially beneficial for this rare and endangered taxon. Indeed, an additional release of *Puma* individuals, this time involving animals from Texas, was carried out in 1995.

For the Florida panther, introgressive hybridization is viewed as a mechanism by which a rare form could be enriched genetically as a partial means for its recovery and evolution. However, it is also important to note that more recent genetic analyses (involving both mtDNA and nuclear sequences) would suggest that the Florida panther is not, relative to other North American *Puma* populations, an evolutionarily distinctive lineage (Culver *et al.* 2000). This latter observation does not, however, alter the conclusion that introgression could lead to more fit hybrid individuals acting as a biological bridge for the recovery of a rare form.

Canada lynx

Lynx rufus (bobcat) and *Lynx canadensis* (Canada lynx) belong to phylogenetically distinct lineages, with the bobcat lineage basal to the various lynx forms (Johnson and O'Brien 1997). These species are not normally found in the same habitats; however, they do overlap in some regions (see discussion by Schwartz *et al.* 2004). In addition, although a widely distributed species that is capable of long-distance dispersal (Schwartz *et al.* 2002), *L. canadensis* has been designated 'threatened' outside of its core distribution of Alaska and Canada. The periphery of the *L. canadensis* distribution is also an area in which it may overlap with *L. rufus*. Because of this, the peripheral regions have been identified as possible centers for hybridization between bobcat and lynx individuals, and therefore of conservation concern (Schwartz *et al.* 2004). In particular, Schwartz *et al.* (2004) stated, 'Hybridization between taxonomically similar species is an often-overlooked mechanism limiting the recovery of threatened and endangered species.'

One region of overlap for bobcat and lynx occurs in the state of Minnesota (Schwartz *et al.* 2004). Genetic data for animals collected in this area of sympatry allowed tests of the role of natural hybridization in the evolution of lynx and bobcat populations. In addition, these data facilitated the consideration of several conservation issues related to peripheral *L. canadensis* populations. Schwartz *et al.* (2004) reported that the first evidence for *L. rufus*×*L. canadensis* hybrids came from morphological analyses of three individuals tentatively assigned to *L. canadensis*. These three animals possessed a combination of phenotypic traits from both species. They had the large feet and mostly black tail band characteristic of *L. canadensis*, but the short ear tufts and compact bodies more reflective of *L. rufus*. An analysis of lynx- and bobcat-specific microsatellite and mtDNA markers for a total of 20 putative lynx samples confirmed the hybrid origin of these three animals (Schwartz *et al.* 2004). Although the authors of this study recognized that this was not necessarily a representative sample for the state of Minnesota, it seems significant that 15% of their samples were hybrids.

The observation of natural hybridization between *L. canadensis* and *L. rufus* was seen to have somewhat unique ramifications for the restoration/conservation of the rare taxon (Schwartz *et al.* 2004). The first conservation issue arises because it is legal to trap *L. rufus*, but not *L. canadensis*, in the contiguous states of the USA (the lower 48 states). Due to the lack of a clear governmental policy concerning the status of hybrids under the Endangered Species Act 1973 (Allendorf *et al.* 2001), Schwartz *et al.* (2004) concluded that it was unknown whether the bobcat–lynx hybrids were protected. If they were afforded protection, trapping of *L. rufus* in areas of overlap with *L. canadensis* would become of even greater concern due to the likelihood of inadvertent trapping of two (rather than one) categories of protected animals—both lynx and lynx–bobcat hybrids (Schwartz *et al.* 2004). The second conservation issue that arises from hybridization between Canadian lynx and bobcats concerns effects upon conservation policies. Schwartz *et al.* (2004) thus argued that 'Any factors that may favor bobcats in lynx habitat may lead to the production of hybrids . . .', thereby influencing policy recommendations by the US Fish and Wildlife Service concerning the conservation needs of *L. canadensis*.

8.3.2 African elephants

The African forest and savannah elephants (*Loxodonta cyclotis* and *Loxodonta africana*, respectively) are, based upon morphological and nuclear DNA characteristics, clearly-differentiated evolutionary lineages (Figure 8.4; Roca *et al.* 2001, 2005; Comstock *et al.* 2002). Nuclear DNA sequence differences and microsatellite variation support a divergence time for these two lineages of approx. 2.6 million YBP (Roca *et al.* 2001; Comstock *et al.* 2002). Arguments in favor of the taxonomic/evolutionary distinctiveness, and the need for the conservation management, of both the savannah and forest forms of *Loxodonta* were also based upon the observation of few hybrids as detected by

nuclear DNA data (Figure 8.4; Roca *et al.* 2001; Comstock *et al.* 2002). These data seemed even more compelling because hybrid genotypes were rare in the transitional zone between the major ecological types alternately inhabited by *L. cyclotis* and *L. africana* (Roca *et al.* 2001; Comstock *et al.* 2002). In contrast to the nuclear loci, mtDNA analyses detected high levels of variation in savannah elephants, and widespread haplotype sharing between the savannah and forest lineages (Eggert *et al.* 2002; Nyakaana *et al.* 2002; Debruyne 2005). In some studies, the levels and distribution of mtDNA variation led to conflicting phylogenetic hypotheses (Nyakaana and Arctander 1999; Eggert *et al.* 2002; Debruyne 2005).

Roca *et al.* (2005)—using sequence data from maternally, paternally, and bi-parentally transmitted loci—tested the hypothesis that introgressive hybridization was the cause of the

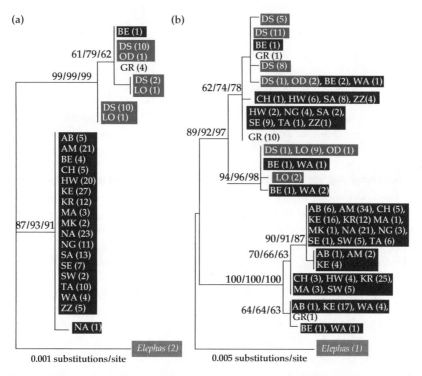

Figure 8.4 Phylogenetic hypotheses for Asian (*Elephas*; red), African forest (*Loxodonta cyclotis*; green), and African savanna (*Loxodonta africana*; blue) elephants using DNA sequence data and maximum likelihood procedures. Phylogenies were derived from sequence information for (a) the paternally transmitted, Y-chromosome AMELY gene and (b) the maternally transmitted mitochondrial ND5 gene. The numbers of individuals with identical sequences are indicated in parentheses. Individuals from Garamba—a transition zone between *L. cyclotis* and *L. africana*—are unshaded. Numbers to the left of clades reflect bootstrap support for (left to right) maximum likelihood, neighbor-joining, and maximum parsimony methods. All of the tree-construction methodologies resulted in phylogenies with the same topology (from Roca *et al.* 2005). See also Plate 6.

incongruent phylogenies for *L. cyclotis* and *L. africana*. They sequenced mtDNA (maternally inherited), Y-chromosome (paternally inherited), and X-chromosome (biparentally inherited) loci for populations used in the studies that detected conflicting patterns. Their results reflected high, but not perfect, concordance between the taxonomic/ecological origin of the samples and placement of the evolutionary relationships of the individuals based upon the nuclear sequences (Figure 8.4). Specifically, the placement of all but one of the Y-chromosome loci sequences agreed with the taxonomic designation of savannah or forest. The one aberrant placement involved a savannah male from a locale near the transition zone between *L. cyclotis* and *L. africana* that possessed the Y-chromosome haplotype of the forest species (Figure 8.4; Roca *et al.* 2005). This animal was also one of two individuals that possessed a mixture of species-specific, biparentally inherited nuclear loci (Roca *et al.* 2005). The mtDNA results of Roca *et al.* (2005) contrasted sharply with the sequence data from both the biparentally and paternally inherited nuclear loci (Figure 8.4). One indication of this non-concordance was the lack of the expected cytonuclear associations; diagnostic mtDNA of *L. cyclotis* was present in animals possessing only *L. africana* nuclear markers. Furthermore, Roca *et al.* (2005) found similar phylogeographic patterns for all the nuclear, but not mtDNA, loci.

A high frequency of elephants carrying savannah nuclear markers possessed forest mtDNA haplotypes (Figure 8.4). In contrast, forest elephants did not possess savannah mtDNA. These observations suggested the occurrence of (i) asymmetric introgressive hybridization of forest mtDNA, but not nuclear loci, into savannah populations and (ii) repeated backcrossing of hybrids with savannah individuals resulting in the loss of forest nuclear alleles (Roca *et al.* 2005). African elephants are typified by female philopatry and male-biased dispersal (Nyakaana *et al.* 2001). Furthermore, fully grown savannah males are approximately twice as large as forest males. These observations suggested a scenario that included preferential mating of savannah males with forest females as forest habitat was altered through climatic changes (Roca *et al.* 2005).

The above results suggest two conclusions concerning the interrelationship of introgressive hybridization and the conservation of *L. cyclotis* and *L. africana*. First, these species have likely been exchanging genetic material throughout their evolutionary history. The mtDNA variation suggested that this exchange had been extensive. Similarly, the nuclear loci also indicated a significant, albeit lower, level of introgression between the savannah and forest forms. Thus once again genetic exchange is seen to have played a significant, and likely creative, role in the evolution of a species complex. There are, however, legitimate concerns relating to the effect that human-mediated introgressive hybridization may have on the long-term survival of these two taxa. In this regard, Roca *et al.* (2005) concluded 'This pattern also implies that ongoing deforestation may foster genetic replacement of forest elephants by opening their habitat to reproductive competition and aggressive hybridization by larger savannah males in those regions where both species persist.'

8.3.3 Aves

Spotted owls
Possibly no other North American species better epitomizes the tension between human-mediated utilization and human-mediated conservation of the natural world than spotted owls, specifically the northern spotted owl, *Strix occidentalis caurina*. In particular, 'Its occurrence in old-growth forests has led to conflict with human activities such as timber harvesting, which has resulted in fragmentation and loss of spotted owl habitat' (Barrowclough *et al.* 2005). This quote indicates the major problem for all endangered taxa—loss of available habitat due to human activity. However, as with the other examples in this chapter, introgressive hybridization has been viewed as a possible impediment for conservation efforts as well (Haig *et al.* 2004a, b; Barrowclough *et al.* 2005).

Recent analyses of the population genetics and phylogeographic structure of owl populations in the western United States have allowed tests for the presence and pattern of hybridization between spotted owl subspecies (*S. o. caurina* and the California spotted owl, *Strix occidentalis occidentalis*) and between the northern spotted owl and the barred owl (*Strix varia*; Haig *et al.* 2004a, b; Barrowclough *et al.* 2005). Both Haig *et al.* (2004a) and Barrowclough *et al.* (2005) defined largely reciprocally monophyletic clades to which many of the northern and California spotted owls alternately belonged (Figure 8.5). However, data from both of these studies also indicated past, and ongoing, genetic exchange between northern and California spotted owls. For example, Haig *et al.*

(2004a) found that approx. 13% of geographically classified northern spotted owl individuals possessed mtDNA haplotypes characteristic of the California subspecies (Figure 8.5). In contrast, Haig *et al.* (2004a) and Barrowclough *et al.* (2005) concluded that a hybrid zone between *S. o. caurina* and *S. o. occidentalis* was both stable and narrow (relative to the dispersal distances possible for spotted owls). They suggested that the overlap region was a tension zone (Barton and Hewitt 1985) and thus reflective of a significant barrier for introgression.

Notwithstanding the conclusions of Haig *et al.* (2004a) and Barrowclough *et al.* (2005), it would seem that introgression is ongoing between the northern and California subspecies (Figure 8.5) and between barred owls and northern spotted owls as well (Haig *et al.* 2004b). The former may have resulted in as much as 13% of northern spotted owls being introgressed with California spotted owl mtDNA haplotypes. Furthermore, the 'stability' of the zone of overlap is dependent upon a lack of habitat change—which seems unlikely from either a long-term or short-term perspective. Thus it seems unwise to conclude that these lineages are deserving of restoration and protection on the basis of a lack of significant levels of introgression. Instead, conservation efforts for these lineages might be better served by emphasizing the retained uniqueness of the various spotted owls, in spite of ongoing introgression.

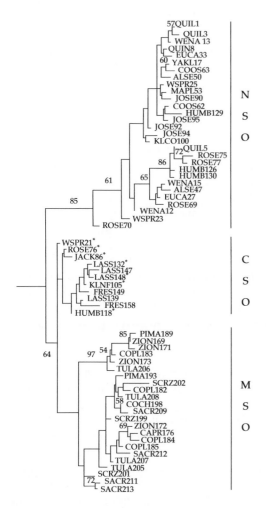

Figure 8.5 Phylogeny for spotted owls based upon DNA sequences from the mitochondrial control region. Numbers to the left of clades indicate bootstrap values. * indicate California spotted owl (CSO; *S. o. occidentalis*) haplotypes detected in the geographic range of northern spotted owls (NSO; *S. o. caurina*). The placement of haplotypes from Mexican spotted owls (MSO; *S. o. lucida*) is also indicated (from Haig *et al.* 2004a).

Black ducks

The taxonomic identity of *Anas rubripes* (the American black duck) has been of considerable interest from both evolutionary and conservation standpoints. In particular, the distinctiveness of this species relative to mallard ducks (*Anas platyrhynchos*) has been questioned. It is clear that *A. rubripes* belongs within the mallard species complex (McCracken *et al.* 2001). What has been debated is whether the black duck belongs to a separate evolutionary lineage relative to the mallard or, instead, is merely a color variant of *A. platyrhynchos* (Hepp *et al.* 1988). The debate stems largely from the lack of reciprocal monophyly detected by numerous genetic analyses of black duck and mallard populations (Figure 8.6; Ankney *et al.* 1986; Avise *et al.* 1990; McCracken *et al.* 2001).

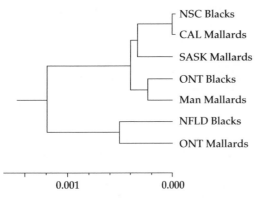

Figure 8.6 Genetic relatedness (allozyme variation was used to calculate Nei's *D* values; Nei 1978) of black duck (*A. rubripes*) and mallard (*A. platyrhynchos*) populations. NSC, Nova Scotia; CAL, California; SASK, Saskatchewan; ONT, Ontario; MAN, Manitoba; NFLD, Newfoundland (from Ankney *et al.* 1986).

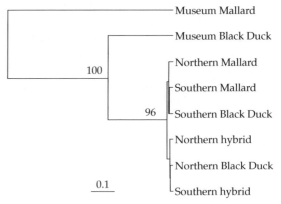

Figure 8.7 Genetic relatedness (based on Nei's genetic distance estimates; Dst = 0.1 is indicated by the scale bar; Nei 1978) of pre-1940 (Museum) and contemporary samples (Northern and Southern) for mallard (*A. platyrhynchos*), black duck (*A. rubripes*), and their natural hybrids. The two significant bootstrap values are indicated to the left of their respective clades (from Mank *et al.* 2004).

An alternate explanation for the absence of phylogenetic distinctiveness between *A. rubripes* and *A. platyrhynchos*, and the cause for conservation concerns for the rarer black duck, is human-induced introgressive hybridization. Thus, 'Until recently, black ducks were an isolated allopatric offshoot of the much larger mallard population. Habitat alteration that accompanied European settlement and game-farm mallard releases during the 20th century have enabled mallards to colonize territory east of the Appalachian mountains, where they had been only rare wanderers before . . .' (Mank *et al.* 2004). For example, in the state of Georgia, the total percentage of mallards was only 29% in the period 1900–1939, but increased to 68–78% in the period 1960–1964 (Johnsgard 1967). Concomitant with the frequency increase of mallards in the geographic range of black ducks was the increase in the opportunity for introgressive hybridization (Johnsgard 1967).

Mank *et al.* (2004) tested the alternative hypotheses that the lack of phylogenetic separation between *A. rubripes* and *A. platyrhynchos* indicated that black ducks were (i) a color variant within the mallard lineage or (ii) undergoing genetic assimilation through introgressive hybridization with their more numerous congener. They accomplished this by examining microsatellite variation in contemporary populations of black ducks and mallards, and in pre-1940 museum specimens of

both forms. Two measures of genetic relatedness illustrated clearly a pattern of increased genetic similarity after the human-mediated expansion of *A. platyrhynchos*. First, Gst values (a reflection of genetic differentiation) between mallards and black ducks decreased significantly from 0.146 to 0.008 in museum and modern samples, respectively (Mank *et al.* 2004). Second, Mank *et al.* (2004) detected reciprocal monophyly based on the DNA variation in the black duck and mallard samples from the museum specimens (Figure 8.7). In contrast, the microsatellite frequencies in the 1998 samples resulted in a mixed clade of both species and their natural hybrids (Figure 8.7).

The results from Mank *et al.* (2004) suggested that prior to the invasion of the *A. rubripes'* geographic range by *A. platyrhynchos*, these two forms belonged to well-differentiated evolutionary lineages. After the expansion, the differentiation eroded. Of all the examples discussed in this chapter, the black duck appears to be the best illustration of extinction through genetic assimilation. Mank *et al.* (2004) reflected this conclusion with the following statement: 'The implications of our findings for the conservation of the black duck are grim. Without preventing hybridization, conservation of pristine black duck habitat will be ineffective in preserving the species . . .' Yet, it is also likely that the phenotypic/genotypic convergence between black ducks and mallards is reflective of the action of selection that favors the hybrid/mallard phenotypes in a changing environment. As with the convergence of the Darwin's finch species, *G. scandens* and *G. fortis*, due to environmental fluctuations (see section 1.2.3), black ducks have apparently been drawn toward the mallard genotypic and phenotypic state by human-mediated environmental changes leading to a change in the selective sieve.

8.4 Summary and conclusions

As noted in section 8.2.2, Suehs *et al.* (2004a) have argued that the threat from invasive species, sometimes due to genetic assimilation, is great. Not to downplay the implications of genetic exchange for conservation efforts (see Beebee 2005; Streiff *et al.* 2005), I must nonetheless agree with the assessment of Detwiler *et al.* (2005) of the main threats to primate populations: 'Whatever the hazard posed by hybridization, it is minimal compared to the current threat of logging, hunting, and the displacement of . . . habitats by hostile, anthropogenic

environments . . .' Minimizing human-mediated modifications to ecosystems remains the best hope for the conservation of rare and endangered forms. However, Detwiler *et al.* (2005) also reflected the conceptual framework of this book and, more specifically, this chapter when they concluded that introgressive hybridization '. . . is best treated as one of many natural evolutionary processes that have played an important role in shaping the biodiversity that conservation aims to protect, and will continue to do so as long as a variety of population structures and zones of contact and overlap are allowed to exist' (Detwiler *et al.* 2005).

Genetic exchange may indeed lead to the assimilation of a rare and endangered lineage by a more numerous one. Alternatively, Stebbins (1942) may have been correct in his suggestion that the rare form could be enriched, and reinvigorated, genetically, ecologically, and evolutionarily through the introgression of loci. Unfortunately, like experimental tests of drugs that may or may not be curative, only time will reveal which trajectory occurs for specific rare and endangered lineages.

CHAPTER 9

Humans and associated lineages

Nicotiana tabacum (cultivated tobacco) is an extensively analyzed natural allotetraploid ($2n = 4x = 48$).

(Volkov *et al.* 1999)

When challenged about the significance of the shape of the cigars he smoked, Sigmund Freud purportedly replied, 'Sometimes, a cigar is only a cigar.'

These authors argued that such hybridization was not surprising given the accidental or intentional release of pigs from Asian and European stocks by Polynesian and European explorers and settlers, respectively. The presence of hybrid genotypes in the Cook Islands would, however, indicate that natural hybridization in the feral populations was the final step in a domestic isolate→feral population→natural hybridization→domestic isolate cycle...

(Arnold 2004a)

Since horizontally acquired genes are often connected to pathogenic or resistance-mediating traits, the genome was searched for clusters whose genes show atypical codon usage. At least ten regions could be identified using the 'alien' gene cluster prediction program SIGI...

(Brüggemann *et al.* 2004)

9.1 The role of genetic exchange in the evolutionary history of humans and their food, drugs, clothing, and diseases

Human-mediated genetic modification of lineages is a highly visible and socially, philosophically, culturally, and politically contentious issue. However, if any message should be clear from this book, it is that *Homo sapiens* was neither the first, nor will it ever be the major, contributor to the process of genetic exchange-mediated organismic evolution. Ironically, this fact is nowhere better exemplified than by an examination of our own evolutionary history, and that of organisms with which we interact—either for our benefit or detriment (Arnold 2004a; Arnold and Meyer 2006).

In the following sections I will illustrate that our own genome and the genomes of organisms that affect our survival are no less mosaic than organisms that do not affect the fitness of humans. Given the wealth of information already presented, this conclusion—that humans and their associated taxa have also been impacted by genetic exchange—should come as no surprise. Yet, there is still much debate on whether our own species has been impacted by the processes of genetic exchange, and in particular that of introgressive hybridization with related lineages. As with the rejection of introgressive hybridization as an important component in the evolution of animal taxa in general, the resistance to considering our own species from the web-of-life paradigm may reflect as much a sociological/philosophical argument as a scientific one. The following discussion will reflect the data that illustrate the two sides of this controversy; however, the weight of evidence supports the hypothesis that the primate lineages leading to our own species, as well as *H. sapiens* itself, participated in reproductive interactions with related taxa (e.g. Templeton 2002, 2005; Osada and Wu 2005).

In addition to our own evolutionary history, the evolution of our food sources, drugs, parasites, diseases, and even our clothing, indicates the involvement of lateral transfer and natural hybridization events. I have chosen in this discussion not to include cases in which food sources etc. originated solely through artificial crosses by humans. Instead, I focus attention on examples where genetic exchange has occurred at least partially through processes outside of human intervention. I should reiterate—as I have in previous sections—that I am aware that some of the cases presented below can be used to exemplify multiple categories. For example, bovid and deer species are placed into a section containing examples of the evolution of human clothing. Obviously, these species have been a major food source for various human populations, yet they exemplify a key material for the production of clothing as well. Thus the main topic of this chapter, also emphasized in portions of previous

chapters, is that genetic exchange has played a significant role in the evolution of the taxa that we utilize or, alternatively, attempt to avoid, resist, or kill to increase our fitness.

9.2 Introgressive hybridization and the evolution of *Homo sapiens*

9.2.1 Hominoids

Gorillas, chimpanzees, humans
The phylogenetic relationships among chimpanzees, gorillas, and humans have been a focus of scientific enquiry and debate for decades (e.g. Figure 9.1; Sarich and Wilson 1967; Ruvolo *et al.* 1991). However, over the past 10–15 years data have accumulated indicating that humans and chimpanzees are sister species (Figure 9.2; Ruvolo *et al.* 1991). Yet, even as we enter the twenty-first century, findings still lead to conclusions such as the following: 'The consensus approach identifies the chimpanzee as the nearest living relative of humans, but the evidence supporting this conclusion is not overwhelming...Inconsistency in the

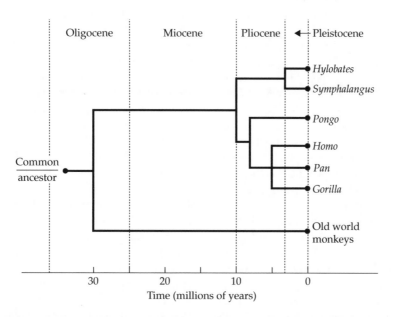

Figure 9.1 The phylogenetic relationships and divergence times resolved for selected hominoid taxa, based on quantitative microcomplement fixation analyses of the albumin proteins from these taxa. The divergence of the Old World Monkeys and the hominoids was assumed to have occurred *c*.30 million YBP. The unresolved trichotomy for the clade that includes humans indicates the highly similar structure of the albumin molecules of these species (from Sarich and Wilson 1967).

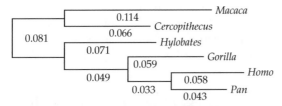

Figure 9.2 Evolutionary relationships between various primate taxa inferred from maximum-likelihood analysis of the DNA sequence variation of the mtDNA COII genes from these species. Numbers reflect the degree of sequence change along the various branches (from Ruvolo *et al.* 1991).

inferred patterns of shared-derived substitutions... is apparent both between and within loci of the three species comprising the trichotomy' (O'hUigin *et al.* 2002).

As illustrated in previous chapters, genetic exchange has been inferred from the type of phylogenetic inconsistencies detected by O'hUigin *et al.* (2002). Thus the lack of concordance between different data sets, in terms of phylogenetic patterns resolved, has been a major tool used to detect reticulate events (see Chapter 3). Data sets leading to an inference of genetic exchange among the lineages leading to *Gorilla*, *Pan*, and *Homo* have been collected from a variety of studies. For example, Richard *et al.* (2000) used chromosome rearrangement differences to resolve an evolutionary tree for hominoids including the gorilla, chimpanzee, and human clade. In spite of their inference of a dichotomously branching scheme for this primate group, they concluded that a 'tree' was an oversimplification of the evolutionary events that had occurred. Instead, they argued that the collection of additional chromosomal data would likely support their inference of 'a network' (i.e. a reticulating phylogeny; Richard *et al.* 2000).

In addition to the cytogenetic evidence, DNA sequence variation also suggested that introgression affected the genomic composition and evolution of the sister taxa *Homo* and *Pan*. Data sets that supported the inference of genetic exchange between the human and chimpanzee (and in some cases, gorilla) lineages included those for (i) *Alu* sequences (Salem *et al.* 2003), (ii) 57 DNA segments derived from 51 loci (O'hUigin *et al.* 2002), (iii) 115 autosomal genes (Navarro and Barton 2003), and

(iv) 345 coding and 143 intergenic sequences (Osada and Wu 2005). For both (i) and (ii) the presence of variation not fitting the model of a dichotomously branching tree was concluded to reflect incomplete lineage sorting of ancestral polymorphisms (O'hUigin *et al.* 2002; Salem *et al.* 2003). However, the recently completed sequencing of chromosome 22 in chimpanzees and the orthologous chromosome (i.e. 21) in humans (Watanabe *et al.* 2004) allowed the construction of a phylogenetic hypothesis for *Alu* sequences inserted into these chromosomes. Of particular interest was the detection of interspecific admixtures that apparently occurred subsequent to the divergence of the chimpanzee and human lineages from their common ancestor (Watanabe *et al.* 2004). Like the *Alu* sequence data, the DNA sequence variation at the 51 loci analyzed by O'hUigin *et al.* (2002) demonstrated a high frequency of non-concordance for the placement of the *Homo/Pan/Gorilla* lineages.

The findings from the above analyses are consistent with introgressive hybridization giving rise to a reticulate, rather than a simply dichotomous, relationship among humans, chimpanzees, and gorillas. Navarro and Barton (2003) and Osada and Wu (2005) drew this same conclusion from analyses of DNA sequence evolution in both genic and nongenic portions of the human and chimpanzee genomes. The main goal of the study by Navarro and Barton (2003) was to test whether differentially selected mutations giving rise to *Pan* and *Homo* accumulated more frequently in rearranged portions of the genome, where genes underlying new adaptations would be protected from recombination. Navarro and Barton's (2003) results also allowed a test for introgression during the divergence of *Homo* and *Pan*. A review of Navarro and Barton's study summarized well the findings from this test: 'The present paper... suggests that the most famous speciation event of all represents a kind of speciation with gene flow...' (Rieseberg and Livingstone 2003). Interestingly, results from a study by Prager and Wilson (1975) predicted this finding nearly 30 years before Navarro and Barton's analysis. Prager and Wilson (1975) concluded that mammalian taxa were unable to produce hybrids 2–3 million years after divergence from a common ancestor (see also Fitzpatrick 2004).

Given that the human and chimpanzee lineages diverged *c.*4–6 million YBP, introgression was possible for half of the time since their derivation. Indeed, Osada and Wu (2005) detected sequence variation at coding and non-coding loci indicative of a '. . . prolonged period of genetic exchange during the formation of these two species.' Thus introgression appears to have affected the origin and early evolutionary trajectory of these sister taxa.

Australopithecines

The data to test for introgression between taxa belonging to the genus *Australopithecus* must necessarily come from the fossil record. Arnold and Meyer (2006) argued that drawing such inferences for *Australopithecus*, or for that matter any other extinct lineage, was analogous to previous work on the non-primate taxon *Cerion*. For the snail genus *Cerion*, analyses of extinct and extant species located on the island of Great Inagua (Goodfriend and Gould 1996) detected the origin and evolution of introgressed populations. One population was inferred to have originated from introgressive hybridization between a fossil species that went extinct *c.*13 000 YBP and an extant species found near the ancient hybrid zone. In a second case, a hybrid zone persisted for thousands of years subsequent to the extinction of one of the parental forms (Goodfriend and Gould 1996).

As with the *Cerion* studies, data from *Australopithecus* have also been argued to reflect the presence of ancient, introgressive hybridization. Indeed, Holliday (2003) argued that 'Several possible examples of interspecific reticulation in human evolution exist. A relatively non-controversial one involves East African *Australopithecus* ('*Paranthropus*') *boisei* and South African *Australopithecus robustus*. Interbreeding between these taxa could have occurred much in the way that it does in the contact zone between parapatric populations of East African yellow baboons and South African chacma baboons . . .' (Holliday 2003). Furthermore, Holliday (2003) also invoked hybrid speciation as a possible explanation for some of the morphological variation present in the *Australopithecus* clade. In particular, he hypothesized that the presence of the derived morphological characteristics resembling *Australopithecus africanus*

in *A. robustus* reflected a hybrid speciation event involving *A. africanus* and *A. boisei* (Holliday 2003). If this hypothesis is correct, *A. robustus* reflects a stable hybrid derivative of the type found in extant animal and plant species complexes (see Chapter 7).

Homo erectus, Homo neanderthalensis, Homo sapiens

Two competing evolutionary models have been constructed to explain the evolution of the genus *Homo*. Termed the replacement and multiregional hypotheses, they differ largely on their assumption of whether or not—as anatomically modern *H. sapiens* expanded its geographical range—gene flow occurred between the modern and archaic taxa. Wolpoff *et al.* (2001) defined these models in the following manner: 'Two conflicting evolutionary models of modern human origins have emerged . . . complete replacement, in which modern humans are a new species that replaced all archaic populations, and multiregional evolution, in which modern humans are the present manifestation of an older worldwide species with populations connected by gene flow. . .'

The evidence compiled in this book and elsewhere (e.g. see Arnold and Meyer 2006 for a discussion of other cases of reticulate evolution in primates) leads to the hypothesis that the synchronic and spatial overlap between archaic and modern humans (Finlayson 2005) would most likely have resulted in some level of gene flow. However, this topic has generated enormous debate. This is somewhat surprising based on the evidence that quite distinct, contemporary primate species are involved in introgressive hybridization (e.g. Jolly 2001; Arnold and Meyer 2006). However, conclusions concerning the role of introgressive hybridization between our own species and other *Homo* taxa have been anything but unanimous. This lack of unanimity is partially the result of conflicting data, but it is also likely to be due somewhat to an entrenched philosophical/cultural view that results in the granting of an 'integrity' to species (Mayr 1963; Coyne and Orr 2004). Such a paradigm disavows the possibility of gene flow and for our own species, possibly even more so because we view ourselves as somehow divorced from behavior that would lead to reproduction with dissimilar phenotypes.

Numerous studies have been undertaken to test the predictions of the introgression (i.e. multiregional) versus no introgression (i.e. replacement) models of evolution within the genus *Homo* (e.g. Krings *et al.* 1997; Ovchinnikov *et al.* 2000; Hawks and Wolpoff 2001; Wolpoff *et al.* 2001; Templeton 2002, 2005; Caramelli *et al.* 2003; Garrigan *et al.* 2005). The findings that best support the replacement model for the evolution of *H. sapiens* are those that have defined mtDNA variation in anatomically modern human and Neanderthal samples (Krings *et al.* 1997; Ovchinnikov *et al.* 2000; Caramelli *et al.* 2003). The pattern of mtDNA variation detected in each of these analyses was consistent with the prediction that the gene pools of *H. sapiens* and *H. neanderthalensis* remained separate. For example, Caramelli *et al.* (2003) examined mtDNA sequence variation from (i) extant humans, (ii) 24 000-year-old Cro-Magnon samples (i.e. anatomically modern humans), and (iii) *H. neanderthalensis* individuals. The ancient mtDNA sequences from the Cro-Magnon individuals were significantly less divergent from extant *H. sapiens* samples than they were from the Neanderthals (Figure 9.3). This finding was even more striking when the ages of the

samples were taken into consideration; some of the Cro-Magnon and Neanderthal samples were temporally closer than were the Cro-Magnon and extant humans (Figure 9.3; Caramelli *et al.* 2003). Pääbo (2003) reflected well the underlying assumption of the replacement model, and the conclusions from several of the above analyses, when he stated, 'Thus, it seems likely that modern humans replaced archaic humans without extensive interbreeding . . .'.

Given the above, it is important to revisit the concept of the mosaic genome; that is, different portions of a genome will be more or less likely to introgress due to patterns of recombination, selection, etc. The use of a single genetic marker and especially one that is inherited uni-parentally, and largely without recombination, is not a rigorous test for introgression. As already indicated by the studies cited in this and the preceding chapters, the strongest test for gene flow between *H. sapiens* and archaic *Homo* taxa would necessarily come from the analysis of multiple loci.

Templeton (2002, 2005) concluded that mtDNA and nuclear sequence variation in extant human populations actually supported the multiregional rather than the replacement hypothesis. His analysis of mtDNA, Y-chromosome, X-chromosome, and autosomal sequence variation resulted in estimates of coalescent times for many, unlinked markers. These estimates led to the inference of extensive gene exchange throughout the history of the *Homo* lineages (see Figure 3.17). This conclusion has recently been supported by an analysis of the X-chromosome pseudogene *RRM2P4* by Garrigan *et al.* (2005). The geographic pattern of variation and the coalescence time estimate for this locus suggested to Garrigan *et al.* (2005) that '. . . this ancient lineage is a remnant of introgressive hybridization between expanding anatomically modern humans emerging from Africa and archaic populations in Eurasia.' Templeton's (2002, 2005) model (multiple expansions from Africa with genetic exchange between *Homo* lineages) also suggested an explanation for findings from an earlier study by Takahata *et al.* (2001). These latter workers detected 10% of the lineages with coalescence patterns contradicting an out-of-Africa origin. In considering all of the available data sets,

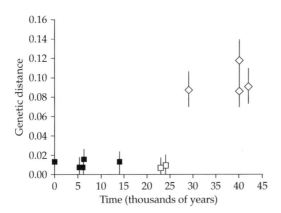

Figure 9.3 Genetic distances between ancient and extant mtDNA sequences from *Homo* individuals. The square positioned at time point 0 reflects the pairwise differences between 2566 sequences of modern Europeans. The age of the samples is indicated by the *x* axis (in thousands of years). Vertical lines represent two standard deviations above and below the mean for the given class of sample (■, anatomically modern humans; □, mtDNA sequence variation for Cro-Magnon samples; ◇, Neanderthal samples; from Caramelli *et al.* 2003).

it appears that our own genome, and that of our sister taxa, have been affected by introgressive hybridization. Not surprisingly, we too carry a mosaic genome.

9.3 Introgressive hybridization, hybrid speciation, and the evolution of human food sources

9.3.1 Animals

Domestic dogs

... we demonstrate that the numerous MHC *DRB* alleles that are present in modern domestic mammals implies that substantial backcrossing with wild ancestors, either accidental or intentional, has been important in shaping the genetic diversity of our domesticates...

(Vilà *et al.* 2005).

One of the taxa referred to by Vilà *et al.* (2005) was the domestic dog, *Canis familiaris*. Though not considered such by most recent immigrants to North America, *C. familiaris* has not only been a beast of burden and a companion, but also a dietary item for many people groups in the Old and New Worlds (e.g. see http://coombs.anu.edu.au/ ~vern/ wild-trade/eats/eats.html). As an example, some Native American groups utilized domestic dogs as a frequent source of protein. For example, at the so-called Laramie Council of 1851, 'Father De Smet, the famous Catholic missionary, who was there, declared that "no epoch in Indian annals probably shows a greater massacre of the canine race" ' (Ambrose 1996, p. 54).

As reflected by the quotes from Vilà *et al.* (2005), *C. familiaris* is a paradigm for demonstrating the role of introgressive hybridization between introduced, domesticated food sources and wild relatives. In particular, genetic evidence suggests that domestic dogs have repeatedly hybridized with their sister species, *Canis lupus* (i.e. the gray wolf; Figure 9.4; Vilà *et al.* 1997, 2003, 2005). Such hybridization is expected, given the occurrence of introgression between a variety of *Canis* species, including the close relatives of *C. familiaris*, the gray wolf, and the coyote. For example, introgressive

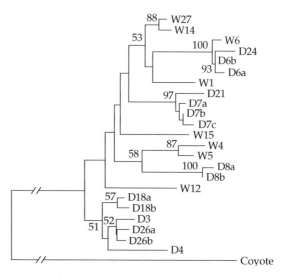

Figure 9.4 Evolutionary tree of eight wolf (W) and 15 domestic dog (D) genotypes based on a 1030-bp sequence of mtDNA. a, b, and c assigned to the haplotypes indicates sequences that were identical for a 261-bp mtDNA sequence, but differed for the 1030-bp region. Bootstrap values showing more than 50% support are indicated (from Vilà *et al.* 1997).

hybridization between coyotes and wolves—likely promoted by human-mediated habitat modification in North America—has resulted in an extensive hybrid zone (Lehman *et al.* 1991). Indeed, this case is illustrative for understanding the genetic exchange between the domestic and wild forms in North America that contributed to the evolution of the food source for Native Americans. Like the deforestation that provided areas of sympatry for coyotes and wolves, humans also modified the gray wolf habitat when they migrated into North America some 12 000–14 000 YBP accompanied by *C. familiaris* (Leonard *et al.* 2002; Savolainen *et al.* 2002). Like many other invaders, the domestic dog was capable of crossing with related, native species. The crosses between this species and gray wolves are not, however, restricted to North America, and indeed introgression among dogs and *C. lupus* has been suggested as the source for the genetic variation that forms the basis for the extreme phenotypic variability found in present-day breeds of *C. familiaris* (Vilà *et al.* 1997). That introgressive hybridization has contributed to the genetic structure of worldwide populations of the domestic dog indicates the role played by genetic exchange in the evolutionary trajectory of a major human food source.

Honeybees

Like dogs, honeybees reflect the role of hybridization between human-introduced and native forms. Thus, 'Before they were disseminated around the world by humans, populations of the western honeybee, *Apis mellifera*, were naturally distributed in Africa, Europe, and western Asia' (Franck *et al.* 1998). Since their distribution, *A. mellifera* has contributed substantially to human food production both directly (in the formation of honey) and as a major pollen vector in agricultural systems. This species is also a paradigm for the role of reticulate events in the evolution of a clade, specifically a clade that benefits the human population. Two examples can illustrate this conclusion. First, Franck *et al.* (2000) used nuclear and mtDNA markers to test for the genomic constitution of Italian populations of the two European subspecies, *A. mellifera ligustica* and *A. mellifera sicula*. This analysis resolved the hybrid derivation of both lineages. The importance for the current topic is that *A. m. ligustica* queens are the most frequently exported for use in beekeeping (Franck *et al.* 2000). Thus, the basis of much of the beekeeping industry is a hybrid lineage possibly formed through the secondary contact of lineages expanding from different glacial refugia (Franck *et al.* 2000).

The recent, whole-scale genetic perturbation of New World feral and managed honeybee populations has resulted from the introduction of the tropical-African subspecies, *A. mellifera scutellata* (e.g. Clarke *et al.* 2002; Pinto *et al.* 2005). Termed, 'africanization,' the process reflects the replacement of eastern and western European-derived honeybees with colonies consisting largely of introgressed individuals. The African lineage was imported into Brazil in 1956 for crosses with European bees that were less tolerant of tropical conditions (see Clarke *et al.* 2002 for a discussion). This introduction resulted in the formation of hybrid swarms throughout South and Central America that have now extended into North America. Pinto *et al.* (2004, 2005; Figure 9.5) have documented the genetic transition of one region in

southern Texas using nuclear and mtDNA markers. Both the mtDNA and nuclear markers indicated the occurrence of extensive introgressive hybridization beginning in *c.*1991 (Figure 9.5; Pinto *et al.* 2004, 2005). In fact, the pattern of nuclear and mtDNA variation was indicative of the evolution of a hybrid swarm. The origin and evolution of this population genetic structure involved bidirectional mtDNA and nuclear introgression between the European and African lineages, resulting in the replacement of a panmictic population of European extraction with one consisting of African/European admixtures (Figure 9.5; Pinto *et al.* 2005). Thus, as with the European lineages that formed the original basis of beekeeping in the New World, it appears that a new bout of introgressive hybridization is playing a significant role in the evolution of the biological material for this important food-producing industry.

Hares

The utilization of wild hares and rabbits as a meat protein is common. For example, one website (http://fooddownunder.com) lists over 60 recipes originating from around the world, including a number reflective of the tradition of widespread hunting for the various species. Furthermore, like the hybrid Italian honeybees, some European hare lineages reflect the role of introgression affected by glacial refugia. Thus one clade of Old World hares, genus *Lepus*, reflects the reticulate evolutionary pattern described for the species complexes discussed throughout this book. The first example of web-like processes within Old World *Lepus* species involves the genetic interactions between the native Swedish species *Lepus timidus* (the mountain hare) and the introduced brown hare, *Lepus europaeus*. As the name implies, the latter species originated from Europe, being introduced into Sweden as a game animal in the 1800s (see discussion in Thulin *et al.* 1997). Initial mtDNA analyses of animals from areas of sympatry and allopatry found that six of 18 hunter-collected *L. europaeus* possessed introgressed *L. timidus* mtDNA; all six hybrids were from sympatric populations (Thulin *et al.* 1997). A subsequent analysis of mtDNA variation in 522 brown hares and 149 mountain hares confirmed introgressive hybridization between the two

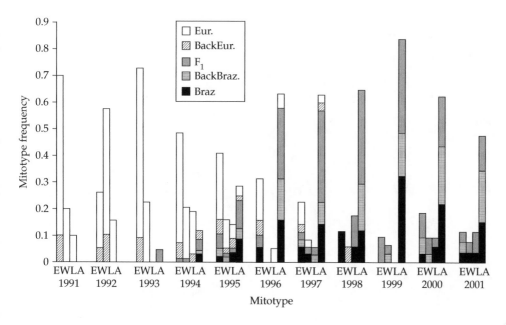

Figure 9.5   Patterns of associations across years for nuclear and mitochondrial (i.e. E, W, L = European mtDNA and A = African mtDNA) markers characteristic of African (Braz.) and European (Eur.) lineages of the honeybee *A. mellifera* in a south Texas population. Honeybees were assigned to the parental lineages, the F₁ hybrid between the parentals, or the backcrosses toward one of the parental taxa (BackEur. or BackBraz.) on the basis of their nuclear and mtDNA genotypes (from Pinto *et al.* 2005).

species (Thulin and Tegelström 2002). Furthermore, the asymmetric nature of the mtDNA introgression (i.e. *L. timidus*→*L. europaeus*) detected initially was also confirmed. Specifically, approx. 10% of the animals assigned to *L. europaeus* possessed mtDNA from *L. timidus*, but no mountain hare individuals were found with the mtDNA haplotype characteristic of the brown hare (Thulin and Tegelström 2002).

The second instance of reticulate evolution in *Lepus* also involves introgression of *L. timidus* mtDNA into related species (Figure 9.6). Specifically, Alves *et al.* (2003) and Melo-Ferreira *et al.* (2005) defined mtDNA variation in *Lepus* from the Iberian peninsula. Populations of *L. europaeus*, *Lepus granatensis* and the Cantabrian endemic, *Lepus castroviejoi* possessed mtDNA haplotypes characteristic of *L. timidus* (Figure 9.6). The remarkable nature of this finding was that *L. timidus* was thought to have disappeared from the peninsula at the end of the last Ice Age (discussed by Melo-Ferreira *et al.* 2005). To explain the presence of the mountain hare mtDNA haplotypes in 23

of 37 populations of the three species, and indeed their predominance in the northern Iberian populations (Figure 9.6), Melo-Ferreira *et al.* (2005) invoked 'Multiple hybridizations and, potentially, a selective advantage for the *L. timidus* . . .' mtDNA variants.

9.3.2   Plants

Wheat

The evolutionary history of the tribe Triticeae reflects the extensive influence of genetic exchange. In particular, introgressive hybridization and allopolyploidy have been of prime importance in the origin and evolution of many lineages. Among the forms demonstrating reticulate patterns are both the diploid ancestors and their polyploid derivative, bread wheat (*Triticum aestivum* L.). Mason-Gamer (2005) reviewed the numerous phylogenetic hypotheses derived from various molecular markers, as well as presenting findings from the β-amylase gene. The general conclusion drawn was 'Although the phylogenetic

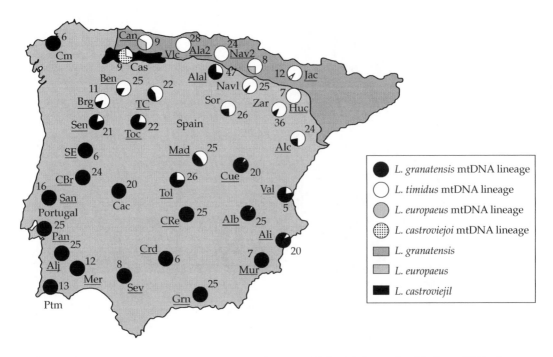

Figure 9.6 Species ranges of *L. granatensis, L. europaeus,* and *L. castroviejoi* and the frequencies of the four mtDNA haplotypes characteristic of these three species as well as *L. timidus* in the Iberian peninsula. Sample sizes and population codes are given adjacent to the pie diagrams (from Melo-Ferreira *et al.* 2005).

relationships among the Triticeae genera are not consistent with a single bifurcating tree, it should be possible to clarify reticulate patterns with the use of multiple gene trees, based on molecular markers from throughout the genome' (Mason-Gamer 2005). It was thus argued that instead of using the various data sets to discuss only where phylogenies were concordantly bifurcating, analyses should also be used to demonstrate where consistent discordance was found between phylogenies. These latter results would indicate probable evolutionary events that included genetic exchange (Kellogg *et al.* 1996; Mason-Gamer 2005). For example, like some of the previous molecular phylogenies (Kellogg *et al.* 1996), the β-amylase phylogenetic results indicated that the wild wheat species (i.e. belonging to *Triticum* and *Aegilops*) did not form a monophyletic assemblage, but rather were interspersed among taxa from other genera (Mason-Gamer 2005).

Because of the phylogenetic discrepancies among the various wheat species, the taxonomic legitimacy of the genera *Triticum* and *Aegilops* has been questioned. The paraphyletic groupings of members of these genera could reflect the sharing of common ancestors. However, the paraphyly might instead be the footprint of extensive genetic exchange throughout the evolutionary history of not only these two genera, but also numerous sister genera. For example, Yamane and Kawahara (2005) examined cpDNA variation within and among diploid wheat species. One conclusion drawn from this analysis was that taxa belonging to *Aegilops* should be placed into the genus *Triticum*. However, like previous studies, the cpDNA analysis detected several inconsistencies that may reflect the action of introgressive hybridization. The taxa that demonstrated non-concordant placements—either demonstrated by the cpDNA data alone, or in comparisons of the cpDNA results with findings from previous studies—included *Aegilops mutica, Aegilops speltoides, Aegilops caudate,* and two species of *Secale*. The latter result did not lead to the suggestion that the genus *Secale* should also be combined into *Triticum*. Instead, Yamane and Kawahara (2005) concluded that this seemingly aberrant placement might '. . . reflect early introgression of organelle DNA between *Triticum-Aegilops* and *Secale*.' It would seem more consistent to also conclude that the discordance in the placement of *Triticum* and *Aegilops* species, based as it is on the same data set (Yamane and Kawahara 2005), might be due to the action of reticulate evolution, rather than the divergence from common ancestor.

In addition to the impact of introgressive hybridization on the evolution of diploid Triticeae species, allopolyploid reticulate events have led to the derivation of the final products that we now utilize as agricultural crops. For example, cultivated wheat is a hexaploid produced from the combination of three separate genomes (the A, B, and D genomes; see Figure 6.5). The allopolyploidization leading to the origin of *T. aestivum* was thus multistaged and involved three separate progenitors. The diploid sources for the A, B, and D genomes have been assigned (for B somewhat tentatively) to *Triticum urartu*, an ancestor of *Aegilops speltoides*, and *Aegilops tauschii*, respectively (see Hegde and Waines 2004 for a discussion). Genetic exchange has apparently affected the evolutionary trajectory of both the progenitors and

their domesticated derivatives that contribute a large proportion of the plant protein consumed by *H. sapiens*.

Breadfruit

The genus *Artocarpus* includes the widely distributed, tropical cultivar *Artocarpus altilis* (breadfruit). Scores of cultivars of *A. altilis*, developed over thousands of years, are now used as a starch source across Melanesia, Micronesia, and Polynesia (Zerega *et al.* 2004). Some regions of cultivation utilize seeded varieties, and other regions seedless varieties, with the latter necessarily being propagated via vegetative means. Of the approx. 60 wild species of *Artocarpus*, the phylogenetically closest relatives to the cultivar have been inferred to be *Artocarpus camansi* and *Artocarpus mariannensis* (Zerega *et al.* 2004, 2005). However, the geographic point of origin and the mode of derivation of *A. altilis* have both been the focus of much debate (see Zerega *et al.* 2004 for a review of the relevant literature).

With regard to genetic exchange-mediated evolution of human foodstuffs, Fosberg (1960) hypothesized that the *A. altilis* variants found in the Philippines may have derived from introgression involving *Artocarpus blancoi*. He also postulated further introgressive hybridization between the cultivar and *A. mariannensis* in Micronesia (Fosberg

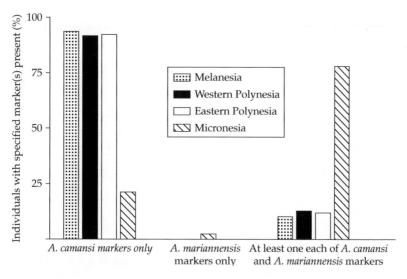

Figure 9.7 Geographic distributions of *A. camansi*. and *A. mariannensis* AFLP markers in samples of Melanesian, Polynesian, and Micronesian breadfruit (*A. altilis*) cultivars (from Zerega *et al.* 2004).

1960). Zerega *et al.* (2004, 2005) used nuclear AFLP data to test the various hypotheses proposed to explain the dispersal and genetic makeup of wild and cultivated species of *Artocarpus*. From their data, they were able to infer a role for both geographic spread through vegetative propagation and the evolution of wild and cultivated taxa involving introgressive hybridization. First they detected genetic patterns consistent with the hypothesis that the majority of Melanesian and Polynesian *A. altilis* had originated through asexual propagation and artificial selection of *A. camansi* variants (Zerega *et al.* 2004). Second the AFLP variation detected in Micronesian breadfruit indicated their origin through introgressive hybridization between the introduced, *A. camansi* derivative and the native *A. mariannensis* (Zerega *et al.* 2004). Specifically, '... Polynesian cultivars have only *A. camansi*-specific markers present, while most Micronesian cultivars have both *A. camansi*- and *A. mariannensis*-specific markers present within individual cultivars' (Figure 9.7; Zerega *et al.* 2004). As with domestic dogs, human-mediated introductions of previously domesticated breadfruit led to introgressive hybridization with a wild taxon, followed by the incorporation of the introgressed lineages into the populations utilized as a source of food.

9.4 Introgressive hybridization, hybrid speciation, and the evolution of human drugs

9.4.1 Coffee

Two species from the genus *Coffea*, *Coffea arabica* and *Coffea canephora*, are a major economic component of numerous Old World and New World countries. Raina *et al.* (1998) reviewed data indicating that coffee plants were cultivated in more than 50 countries, accounting for a trade of approx. US$18 billion. *C. arabica* accounts for approx. 70% and *C. canephora* approx. 30% of this trade (Moncada and McCouch 2004). The greater proportion of coffee cultivation involving *C. arabica* reflects the higher-quality beverage (i.e. low content of caffeine and pleasant aroma) produced by this species (Raina *et al.* 1998). *C. canephora*'s lower quality product is utilized in the manufacturing of instant coffee (Steiger *et al.* 2002).

Of the >80 *Coffea* species, *C. arabica* is the only tetraploid and self-fertile form ($2n = 4x = 44$; Raina *et al.* 1998; Steiger *et al.* 2002). The remainder of the species, including *C. canephora*, are self-sterile and possess $2n = 22$ chromosomes (Raina *et al.* 1998; Steiger *et al.* 2002). A native of the southwestern highlands of Ethiopia, *C. arabica* has spread worldwide through human-mediated movement (Figure 9.8) beginning with an initial introduction to Yemen possibly as early as 575 AD, followed by transfers to Java and Reunion Island and then to the Amsterdam Botanical gardens and finally to the Americas (Anthony *et al.* 2002). In the process of its cultivation and spread, two varieties arose: Typica and Bourbon (Figure 9.8; Anthony *et al.* 2002).

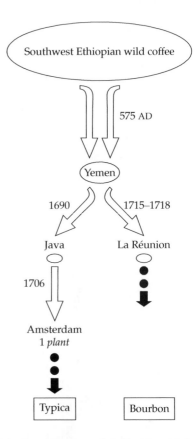

Figure 9.8 The origin and spread of coffee (*C. arabica*) cultivation. *C. arabica* was dispersed via human-mediated transport from its native range in Ethiopia through Yemen, Java, Reunion Island (La Réunion), and Amsterdam. The two varieties, Typica and Bourbon, were derived during the various stages of cultivation and now act as the source of *C. arabica* plants grown worldwide (from Anthony *et al.* 2002).

As mentioned above, *C. arabica* is the only known polyploid in the genus, having arisen through a natural allopolyploidy event (see discussion by Ruas *et al.* 2003). Though a well-defined allopolyploid, the identity of the diploid progenitors for this species have been debated. For example, Raina *et al.* (1998) used genomic and fluorescent *in situ* hybridization to test the genomic constitution of *C. arabica*. Their results suggested that the two diploid genomes possessed by this allotetraploid were derived from *Coffea eugenioides* and *Coffea congensis*. This was indicated by the strong hybridization of 22 *C. arabica* chromosomes each by the DNA from these two species (Raina *et al.* 1998). Lashermes *et al.* (1999), using both genomic fluorescent *in situ* hybridization and RFLP analyses of nuclear loci, also concluded that *C. eugenioides* had contributed one of the genomes present in the cultivar. However, in contrast to the findings of Raina *et al.* (1998), this latter study suggested that the diploid, cultivated species, *C. canephora*, rather than *C. congensis*, was the other progenitor lineage (Lashermes *et al.* 1999). Furthermore, the degree of genomic divergence, as indicated by the RFLP data, led Lashermes *et al.* (1999) to conclude that '. . . the speciation of *C. arabica* took place in relatively recent times (i.e. from historical time to 1 million years ago).'

Ruas *et al.* (2003) utilized inter-simple sequence repeat markers to test the alternate hypotheses outlined above concerning the derivation of *C. arabica*. Consistent with the findings of Lashermes *et al.* (1999), Ruas *et al.*'s (2003) data indicated that, of the eight diploid species analyzed, *C. eugenioides* and *C. canephora* shared the highest similarity with *C. arabica*. Significantly, Ruas *et al.* (2003) also pointed out that the only wild, diploid species that has been shown to possess 'fine aroma and flavor' was *C. eugenioides*. They argued that the participation of this species as a parent of the allotetraploid *C. arabica*, might have resulted in the presence of these characteristics in the cultivar. However, the reticulation that led to the derivation of *C. arabica* resulted in a species that is adapted to environmental conditions different from its two progenitors—this is reflected by its occupation of a native range outside the area of distribution of the diploid species (Lashermes *et al.* 1999). Thus it appears

likely that the genetic exchange leading to the formation of *C. arabica* resulted in a uniquely adapted lineage that now forms the basis of a widely applied human drug.

9.4.2 Chocolate

The genus *Theobroma* is confined to tropical America and comprises 22 species . . . divided into six sections. *Theobroma cacao* is alone in one section of this classification.

(Kennedy 1995)

Like *C. arabica*, the plant from which humans isolate chocolate, *T. cacao*, represents a major crop plant of the humid tropics—particularly regions of West Africa, Brazil, and Malaysia (Kennedy 1995; Motamayor *et al.* 2003). Also like *C. arabica*, major consumers of the *T. cacao* product are located in Europe and North America (Kennedy 1995; Wrigley 1995). However, unlike the *Coffea* cultivar, the plants that produce cacao beans are diploid-level derivatives ($2n = 20$) of wild species. Three major cultivars of *T. cacao* have historically been recognized and utilized for the production of chocolate—Criollo, Forastero, and Trinitario (Kennedy 1995). As stated by Motamayor *et al.* (2003), 'Today, 70% of cacao production is still derived from these traditional cultivars . . . much of which is composed of Trinitario.' Though the three cultivar lineages have been defined on the basis of properties associated with the beans they produce (Kennedy 1995), they were not found to be reciprocally monophyletic on the basis of RFLP and microsatellite data (Motamayor *et al.* 2002).

In the current context, the significance of Trinitario's predominant position as the major cultivar for the production of chocolate relates to its origin through hybridization between Criollo and Forastero. Specifically, the Trinitario lineage was formed due to introgressive hybridization following the introduction of Forastero trees from the Lower Amazon into the native Trinidadian habitat of the Criollo form (see Motamayor *et al.* 2003 for a discussion). It is important to note that the bias toward the adoption of Trinitario as the major cultivar is a recent phenomenon, but it is most important to understand why the replacement took place. For example, though cacao was cultivated by Central American cultures from pre-Colombian times, Trinitario did not begin to replace other forms in Central and South America until the 1800s (Motamayor *et al.* 2003). There are two oft-cited reasons for why this introgressive hybrid lineage became the major crop plant for cacao production: (i) Trinitario plants were found to be more disease resistant relative to Criollo and Forastero individuals; and (ii) Trinitario individuals were found to combine the higher productivity of the Forastero cultivar and the desirable flavor characteristics of the Criollo variant (Trinitario is referred to as 'fine cocoa'; Kennedy 1995; Motamayor *et al.* 2003). Thus the natural hybridization event on the island of Trinidad—once again caused by a human-mediated introduction—was the basis for the evolution of this drug of choice for many in the Old and New Worlds.

9.4.3 Tobacco

Though it is impossible to draw appropriate comparisons in terms of i) the artistry needed for the construction of or ii) the quality of fine, hand-rolled, long-leaf cigars, with a packet of cigarettes or a can of snuff, the tobacco used in each product comes from the same evolutionary lineage, *Nicotiana tabacum*. Yet historically this species was not the sole source for the intoxicating effects of nicotine. Thus *N. tabacum* was utilized by the indigenous peoples of Central and South America, *Nicotiana bigelovii*, *Nicotiana attenuata*, and *Nicotiana trigonophylla* by the inhabitants of western North America, *Nicotiana rustica* by people groups in eastern North America, northern Mexico, and the West Indies, and *Nicotiana benthamiana* by native Australians (Gerstel and Sisson 1995). The origin of tobacco farming in the New World (in Virginia) initially involved the species used by the local Native Americans, *N. rustica* (Gerstel and Sisson 1995). However, this species was quickly replaced by the better-flavored *N. tabacum* (Gerstel and Sisson 1995).

For tobacco, like chocolate and coffee, the basis of the product is a plant that reflects genetic exchange-mediated evolution. In particular, *N. tabacum*, like *C. arabica*, demonstrates an allopolyploid genomic constitution. Recent molecular analyses have determined that (i) the maternal and paternal progenitors of the tobacco cultivar were ancestors of the diploid species *Nicotiana sylvestris* and *Nicotiana tomentosiformis*, respectively (e.g. see Figure 9.9; Lim *et al.* 2000; Kitamura *et al.* 2001; Ren and Timko 2001; Fulnecek *et al.* 2002) and (ii) the derivation of *N. tabacum* occurred approx. 5–6 million YBP (see discussion by Fulnecek *et al.* 2002). As with many of the examples discussed throughout this book, the combination of the parental *Nicotiana* genomes gave rise to a lineage with novel adaptations relative to its progenitors. Gerstel and Sisson (1995) highlighted this fact when they stated, '. . . the wild parents of *N. tabacum* . . . possess a dominant "converter" gene which . . . demethylates nicotine into undesirable nornicotine in the leaves. For this reason these species may have been useless. . .' Thus, like the previous two examples, the web of life appears to have provided the biological starting material for a drug-based, agricultural industry.

9.5 Introgressive hybridization, hybrid speciation, and the evolution of pathogens of plants utilized by humans

In Arnold (2004a) I posed the following question: 'Has natural hybridization led to the origin of pathogens and pests that attack agricultural or natural plant populations that are important to humans?' In the following discussion I will give several examples for which findings suggest such an evolutionary scenario. Though diseases affecting animal species utilized by humans also may form via reticulate evolution (e.g. influenza viruses and trypanosomes), I will focus this discussion on plant pathogens to illustrate the role genetic exchange plays in the evolutionary history of organisms that impact us secondarily; that is, through their interaction with another lineage. Furthermore, I will use two examples from tree species (alders and elms) that are not food sources, but rather have been utilized for construction and fuel for heating.

9.5.1 Phytophthora

In a review of the evolution of introduced plant pathogens, Brasier (2001) observed that epidemics

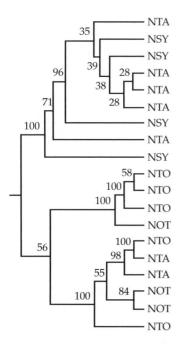

Figure 9.9 Phylogenetic hypothesis for samples of the allotetraploid cultivar *N. tabacum* (NTA) and its putative dipoid progenitors *N. sylvestris* (NSY) and *N. tomentosiformis* (NTO). The phylogeny was dervived using sequence data from spacer regions of the 5 S rDNA genes. Numbers to the left of nodes reflect bootstrap values. NOT = *N. otophora* (from Fulnecek *et al.* 2002).

resulting from the translocation of fungal pathogens to new geographic regions were well known. He (and Brasier *et al.* 1999) also concluded that the bringing together of related, introduced and native pathogens presented ' . . . an opportunity for rapid emergence of new or modified pathogens via interspecific gene flow . . .' Brasier *et al.* (1999) argued for the importance of testing for the action of natural hybridization in the evolution of plant pathogens because the outcome of genetic exchange events might include the origin of novel adaptations allowing their spread to new hosts. Indeed, these authors defined pathogenic variants from the fungal complex belonging to the genus *Phytophthora* that were shown to be (i) hybrid derivatives but (ii) genetically very similar to the non-pathogenic species *Phytophthora cambivora* (Brasier *et al.* 1999). The evolution of this novel pathogen was thus consistent with the hypothesis that 'Interspecific hybrids are more likely to survive if

they have a fitness advantage over the parent species, such as increased aggressiveness or the ability to exploit a new host . . .' (Brasier *et al.* 1999). The alder trees attacked by this new pathogen were not a human food source; however, they were a significant component of an important ecosystem and thus of critical concern for conservation efforts.

9.5.2 Dutch elm disease

Similar to the alder tree pathogen, the spread of Dutch elm disease has been marked by introgressive hybridization and hybrid lineage formation. Specifically, Brasier and Kirk (2001) defined two subspecies of the organism recognized as the cause of the present-day Dutch elm disease pandemic, the ascomycete fungus *Ophiostoma novo-ulmi*. The two subspecies, *O. novo-ulmi novo-ulmi* and *O. novo-ulmi americana* were originally designated as Eurasian and North American variants, respectively. Brasier and Kirk (2001) hypothesized that these two subspecies might have arisen through two separate hybrid subspeciation events. In particular, they posited that after its origin in Eurasia, *O. novo-ulmi* hybridized with Eurasian *Ophiostoma ulmi* and that *O. novo-ulmi* introduced into North America in the 1940s might have hybridized with the North American *O. ulmi*, giving rise to *O. novo-ulmi americana*. Consistent with this hypothesis was the observation of diagnostic differences between the two subspecies shared with the respective (i.e. Eurasian or North American) forms of *O. ulmi* (Brasier and Kirk 2001).

The hypothesized hybridization events leading to the formation of the two Dutch elm disease subspecies were not the only genetic exchange affecting the evolutionary trajectory of this disease organism. Subsequent to their formation, *O. novo-ulmi novo-ulmi* and *O. novo-ulmi americana* have come into contact in Europe (Figure 9.10). This sympatry has resulted in introgressive hybridization between these two forms (Brasier 2001; Brasier and Kirk 2001). The recombinant forms were shown to have intermediate phenotypes between the parental subspecies, but were determined to be no less pathogenic for their hosts (Brasier 2001; Brasier and Kirk 2001). The geographically extensive area of sympatry and hybridization

(Figure 9.10) resulted in Brasier's (2001) hypothesis that 'From such a mélange of recombinants, natural selection may in the future favor...a new race or subspecies of the pathogen.' Indeed, the introgression of the *MAT-1* and *vic* loci was subsequently hypothesized to reflect '...rapid adaptation of invasive organisms to a new environment' (Paoletti *et al.* 2006). In this case, the diversification of these loci allowed sexual reproduction, which for this fungal pathogen provided a means of escaping viral infections that were otherwise perpetuated through clonal lineages (Paoletti *et al.* 2006).

9.5.3   Erwinia carotovora

The bacterial family Enterobacteriaceae is probably best known for human pathogens such as *Salmonella* and *Yersinia*. Indeed, I have used both of these human disease-causing organisms in previous chapters to exemplify the web-of-life paradigm. However, the family Enterobacteriaceae also includes organisms that act as significant phytopathogens of human food sources. For example, the genus *Erwinia* contains a number of lineages that cause soft rot in potatoes and other crops, both in temperate and tropical regions (Pérombelon 2002). *Erwinia carotovora* ssp. *carotovora*, *Erwinia carotovora* ssp. *atroseptica*, and *Erwinia chrysanthemi* are the primary causes of the decay of stored potatoes (all

three taxa) and blackleg in agricultural fields (only *E. carotovora* ssp. *atroseptica* and *E. chrysanthemi*; reviewed by Pérombelon 2002). That these species are widespread in the majority of commercial seed-grade stocks of potatoes (Pérombelon 2002) makes them a major concern for this key food source.

An analysis of the genome sequence of *E. carotovora* ssp. *atroseptica* by Bell *et al.* (2004) included a test for genes that may have been laterally transferred and also for the effect of these putatively transferred loci on the pathogenicity of this species. The results from their analyses indicated that approx. 33% of *E. carotovora* ssp. *atroseptica*'s genes were not shared with other sequenced genomes from members (i.e. human pathogens) of this bacterial family. A number of these loci were characterized by atypical G+C ratios, proximity to tRNA genes and the presence of phage/plasmid genes (Bell *et al.* 2004), all traits suggestive of horizontal transfer events. Furthermore, among these loci were some associated with pathogenicity as well as unexpected metabolic traits (Bell *et al.* 2004). Significantly, when mutants for the pathogenicity loci were constructed, they demonstrated virulence levels significantly reduced relative to the original *E. carotovora* ssp. *atroseptica* genotypes. The similarity of these genes to those found in other plant pathogens and the reduction of the disease-causing potential of mutants, led Bell *et al.* (2004) to

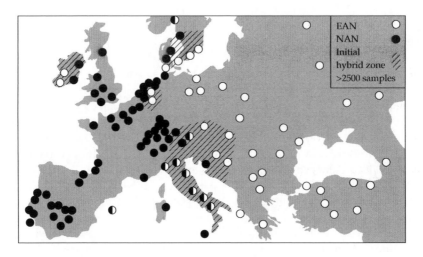

Figure 9.10   European geographic distribution of the Dutch elm disease pathogens *O. novo-ulmi novo-ulmi* (EAN) and *O. novo-ulmi americana* (NAN). Known sympatric regions occurred in Ireland, the Netherlands, Scandinavia, Germany, and Italy. Hatching indicates demonstrated or postulated regions of introgression between the two subspecies (from Brasier 2001).

conclude that lateral exchange of genomic islands containing pathogenicity loci had contributed to the evolution of *E. carotovora* ssp. *atroseptica*. Thus, as with the fungal species that attack alder and elm trees, *E. carotovora* ssp. *atroseptica* was seen to have evolved due to genetic exchange. However, for this latter species horizontal exchange, rather than introgressive hybridization, had been the mechanism of genetic transfer.

9.6 Introgressive hybridization, hybrid speciation, and the evolution of human clothing materials

9.6.1 Leather

. . . for the species where domestic and wild lineages have co-existed over extensive areas, such as cattle, dogs, and pigs, occasional backcrosses might have increased the diversity of the domestic lines. . . Our results, together with the archaeological evidence, suggest that the separation between wild and domestic might not have always been as well defined in mammals as suggested by previous genetic studies. . .

(Vilà *et al.* 2005)

As discussed above, the lack of complete reproductive barriers and thus introgression did indeed occur between the domestic dog lineage and its wild relative, the gray wolf. However, as indicated by Vilà *et al.* (2005), genetic enrichments through

introgression between domesticated forms and their wild sister taxa were not limited to Canids. Furthermore, Vilà *et al.* (2005) hypothesized that the pattern of variation found at mitochondrial, compared with nuclear, loci suggested a specific scenario for gene exchange between wild and domesticated forms. In particular, low levels of mtDNA haplotype diversity, relative to the allelic diversity present at nuclear loci, was consistent with asymmetric crossing involving wild males and domestic females. This mode of crossing would likely result in the association of nuclear loci from the wild lineage with the mtDNA (i.e. maternal) haplotype of the domesticated form (Vilà *et al.* 2005).

Götherström *et al.* (2005) reviewed the following findings concerning the domestication process of one animal used in the manufacturing of clothing, cattle used as a source of leather: (i) cattle were derived from aurochs (wild ox) *c.*10 000 years ago; (ii) the domestication process probably occurred in both the Near East and Asia; (iii) cattle have been of major economic significance since *c.*5500 BC; (iv) cattle were likely brought to Europe from the Near East center of origin; and (v) the wild ox found in Europe (and elsewhere) went extinct in the early seventeenth century. In regard to the evolution of domesticated organisms—and particularly those used for clothing—bovid lineages are illustrative of the non-concordant mtDNA and nuclear genetic variation predicted by the above hypothesis. Thus, the mtDNA

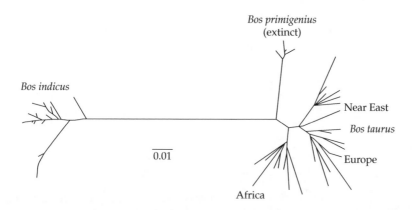

Figure 9.11   Phylogenetic relationships inferred using mtDNA sequence data from contemporary populations of the two forms of domesticated cattle, *B. indicus* and *B. taurus*, and the extinct aurochs or wild ox (*B. primigenius*). A scale bar of genetic divergence of 0.01 is also shown (from Troy *et al.* 2001).

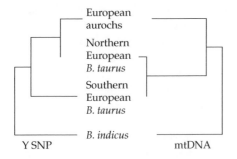

Figure 9.12 Phylogenetic trees based on interpopulation divergence estimates among extinct European aurochs, present-day northern and southern European cattle (*B. taurus*), and African samples of the domesticated, *B. indicus*. The left- and right-hand phylogenies were derived from Y-chromosome single-nucleotide-polymorphism (Y SNP) and mtDNA loci, respectively (from Götherström *et al.* 2005).

variation found in present-day European cattle populations is found in significant frequency in Near East populations, but is distinct from mtDNA haplotypes detected in samples from the extinct European aurochs (Figure 9.11 Troy *et al.* 2001). In contrast, Götherström *et al.* (2005) found that Y-chromosome haplotypes from contemporary, northern European cattle were more similar to those present in the extinct wild ox than to either present-day, southern European or Near East populations of cattle (Figure 9.12). Furthermore, these workers found a significant north–south cline in the frequency of Y-chromosome haplotypes and they also detected the northern European variant in 20 of 21 aurochs or early domesticated cattle samples from 9500 to 1000 BC (Götherström *et al.* 2005). Taken together, these results suggest that the foundation of the northern European cattle (and thus leather) industry (and presumably that of North America and elsewhere) were breeds developed from hybrids between male wild aurochs and female domesticated cattle (Götherström *et al.* 2005).

9.6.2 Deer skin

The use of deerskins for clothing by North American indigenous peoples was widespread, as reflected by rituals such as dances that commemorated the killing of the animals (Hibben 1992). Two species, *Odocoileus virginianus* (the white-tailed deer)

and *Odocoileus hemionus* (mule deer) were commonly utilized in western North America. Furthermore, there was a transfer of this clothing technology to explorers of European extraction as recorded in the writings of these later travelers. For example, the May 13, 1805 journals from the Lewis and Clark expedition recorded, 'Captain Clark who was on shore the greater part of the day killed a mule and a common [i.e. white-tailed] deer, the party killed several deer and some elk principally for the benefit of their skins which are necessary to them for clothing . . .'(www.lewisclarkeandbeyond.com/journals/). It is thus apparent that from their initial migration until at least the nineteenth century one of the key ingredients for the survival of *H. sapiens* in North America was the protection afforded by the articles of clothing made from the skins of the white-tailed and mule deer.

The significant morphological differentiation between *O. virginianus* and *O. hemionus* (Roosevelt, 1996, pp. 127–128) is also reflected by behavioral differences leading to different habitat preferences (Avey *et al.* 2003). Theodore Roosevelt, in his *Hunting Trips of a Ranchman* (1996, p. 128), described the habitat differences of white-tailed and mule deer in the following manner: 'One of the noticeable things in western plains hunting is the different zones or bands of territory inhabited by different kinds of game. Along the alluvial land of

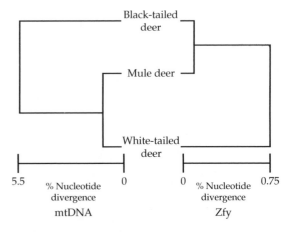

Figure 9.13 Phylogenetic placements of mule (*O. hemionus*), black-tailed (also *O. hemionus*), and white-tailed (*O. virginianus*) deer species based on mitochondrial (mtDNA) or Y-chromosome (Zfy) sequences (from Cathey *et al.* 1998).

the rivers and large creeks is found the white-tail. Back of these alluvial lands generally comes a broad tract of broken, hilly country, scantily clad with brush in some places; this is the abode of the black-tailed [i.e. mule] deer.'

Notwithstanding the significant phenotypic and behavioral differences between *O. virginianus* and *O. hemionus*, recent analyses have found non-concordant phylogenetic patterns based on sequence variation at mtDNA and nuclear loci (Carr *et al.* 1986; Cathey *et al.* 1998; Bradley *et al.* 2003). Figure 9.13 illustrates the different phylogenetic hypotheses derived from either maternally inherited mtDNA sequences or paternally inherited Y-chromosome sequences (i.e. the zinc-finger gene, *Zfy*; Cathey *et al.* 1998). The *Zfy* sequence data resolved a phylogenetic tree consistent with morphological and behavioral data sets. Thus the *conspecific* black-tailed and mule deer individuals were resolved as sister-lineages relative to the white-tailed deer samples (Figure 9.13; Cathey *et al.* 1998). In contrast, mtDNA sequences resulted in the close association of mule deer and white-tailed deer in a clade separate from the black-tailed deer lineage (Figure 9.13). This phylogenetic discordance is consistent with either incomplete lineage sorting or genetic exchange. However, the closely associated habitats of the two species (Roosevelt 1996; Avey *et al.* 2003) and the detection of contemporary hybrid zones (Carr *et al.* 1986; Bradley *et al.* 2003), supported a hypothesis of mtDNA introgression leading to phenotypically defined mule deer individuals carrying white-tailed deer mtDNA haplotypes. Over the western North American range of these species, hybrid individuals (especially given that all mule deer possess haplotypes captured from the white-tailed deer lineage) would have thus provided the raw material for a large proportion of the clothing for the indigenous peoples and later explorers.

9.6.3 Cotton

The genus *Gossypium* is unique in the annals of plant domestication in that a minimum of four similar crop plants emerged independently on opposite sides of the world…These species collectively provide the world's most important

textile fibre and its second most valuable oil and meal seed.

(Wendel 1995)

Of these four cotton lineages, *Gossypium hirsutum* ('upland' cotton) is the source of the vast majority (>90%) of the raw material used in clothing (Wendel 1995), and thus accounts for much of the multi-billion dollar commerce (Jiang *et al.* 1998) in cotton fiber. Like many other cultivars, *G. hirsutum* reflects the evolution of a new domestic species through allopolyploidy. As discussed in Chapter 6, this species—and *Gossypium* in general—is a model for defining genomic changes associated with genome duplication. However, *G. hirsutum* is also a paradigm for the role of genetic exchange in the origin of a new evolutionary lineage that is of significant importance for humans. The allopolyploid derivation of *G. hirsutum* (and other New World *Gossypium* species) involved hybridization between diploid forms whose contemporary derivatives are now allopatric (Wendel 1989). Specifically, the *G. hirsutum* genome consists of the so-called A and D genomes of Asian–African and New World diploids, respectively (Wendel 1989, 1995). Furthermore, the derivation of *G. hirsutum* and the other New World allotetraploids involved an Old World species acting as the maternal parent. This is suggested by *G. hirsutum* possession of cpDNA haplotypes with little sequence divergence from those of Old World taxa (Wendel 1989).

Of particular significance for the current discussion, the formation of the allotetraploid, *G. hirsutum* *c.*1–2 million YBP (Wendel 1989) resulted in genetic combinations that gave rise to unique phenotypes (in terms of fiber quality) relative to the diploid progenitors (Jiang *et al.* 1998). In particular, the combination of the diploid genomes from the lineages resembling *Gossypium herbaceum* (the A genome donor) and possibly *Gossypium raimondii* (the D genome donor) resulted in gene interactions that contributed to the high yield and easily spinnable fibers of upland cotton. Surprising, the majority of the QTLs contributing to these characteristics derived from the D genome—a diploid ancestor that does not possess spinnable fibers (Jiang *et al.* 1998). In contrast, the A genome contained fewer loci that contributed to the high quality of *G. hirsutum* even though the donor of this genome produces fiber. These discoveries once again reflect the power of the web-of-life processes in the creation of novel evolutionary trajectories. Jiang *et al.* (1998) voiced this same conclusion in the following manner: 'The joining in a common nucleus of A and D genomes, with very different evolutionary histories, appears to have created unique avenues for response to selection in AD-tetraploid cottons.'

9.7 Introgressive hybridization, hybrid speciation, and the evolution of human disease vectors

9.7.1 Anopheles funestus

Though the *An. gambiae* species complex has been extremely well defined genetically and evolutionarily (e.g. see section 4.3.4), it is only one of the two significant African malarial vectors. The other taxon that is a primary contributor to human malarial mortality in Africa is *Anopheles funestus* (Cohuet *et al.* 2005; Michel *et al.* 2005). Cytogenetic analyses of *An. funestus* have detected genetic structuring in certain geographical regions, but not in others. For example, in West Africa Lochouarn *et al.* (1998) found evidence for three chromosomally differentiated races of this species; the races were defined by different inversions. Furthermore, in areas of sympatry between two of the karyotypic forms, Lochouarn *et al.* (1998) detected a heterozygote deficit consistent with restricted introgression. In contrast to these findings, Sharakhov *et al.* (2001) found no significant departures from Hardy–Weinberg expectations for Kenyan populations that also demonstrated inversion polymorphisms. It is thus apparent that factors—likely to be at least partially reflective of ecological selection—affecting introgression between differentiated lineages within the *An. funestus* complex vary across space.

Two recent analyses by Cohuet *et al.* (2005) and Michel *et al.* (2005) have resulted in a greater resolution of the genetic structuring within and among *An. funestus* populations. These studies have also contributed to a definition of some of the evolutionary processes affecting the genetic patterns. The analysis by Cohuet *et al.* (2005) involved the collection of both cytogenetic and molecular marker (i.e. microsatellite) data for Cameroon populations. As with previous analyses of cytogenetic markers in West Africa, Cohuet *et al.* (2005) found significant structuring indicative of limits to genetic exchange involving the inverted chromosomal regions. The genetic structure revealed by the inversion data included the presence of clinal variation and a hybrid zone with heterozygote deficiencies (Cohuet *et al.* 2005). In contrast to the spatial differentiation indicated by the chromosomal markers, the microsatellite frequencies were consistent with high levels of gene flow between chromosomal forms (Cohuet *et al.* 2005).

To determine the population genetic structure of *An. funestus* throughout its range, Michel *et al.* (2005) analyzed microsatellite alleles (from each autosomal chromosome arm and the X-chromosome) and mtDNA haplotypes for populations from 11 African countries. The microsatellite data led to the definition of three lineages, termed eastern, western, and central on the basis of the geographic location of the respective populations (Michel *et al.* 2005). Though the mtDNA data were largely consistent with the microsatellite findings, the distribution of mtDNA haplotype variation did allow the definition of additional subdivisions within the *An. funestus* clade. In particular, two lineages were defined that possessed approx. 2% sequence divergence. This level of differentiation suggested a divergence time of *c.*1 million YBP. The observation of the high level of divergence, coupled with the fact that one mtDNA lineage was distributed throughout the range of the species, while the second lineage was found only in the samples from Mozambique and Madagascar, resulted in an inference of mtDNA introgression either between *An. funestus* lineages or between *An. funestus* and another *Anopheles* species (Michel *et al.* 2005).

The findings from the earlier cytogenetic analyses of *An. funestus*, combined with those from later cytogenetic and molecular studies, indicate that this species complex reflects the evolutionary importance of genetic exchange. Furthermore, like the other major African malarial vector, *An. gambiae*, genetic exchange has contributed to the evolution of an organism that is highly detrimental to *H. sapiens*.

9.7.2 Culex pipiens

Unlike similar outbreaks of West Nile virus in European urban settings (which do not reoccur and are geographically localized), the North American occurrences of this disease have persisted across years and have spread geographically (Hayes 2001). The nature of the North American outbreaks was surprising because of the biology of the disease and its vector. Thus the spread of West Nile virus in humans requires what is called a 'bridge vector' '... because humans and other mammals do not usually generate high enough viremia to infect biting mosquitoes ...' (Fonseca *et al.* 2004).

Though the common vectors responsible for spreading West Nile virus are mosquitoes of the *Culex pipiens* clade, European and North American populations of this vector demonstrate behavioral differences. In Europe two genetically differentiated forms prefer either birds or mammals (including humans) as sources for their blood meals (Fonseca *et al.* 2004). In contrast, certain lineages of the North American *Cx. pipiens* species complex have been shown to feed from both humans and birds (Spielman 2001) and may thus explain the unique ability of West Nile virus to spread and persist in a North American setting.

The most significant finding related to the North American clade of *Culex* was that it included a hybrid lineage between the two divergent European forms that fed differentially from mammals and birds. Indeed, introgression between these taxa (named *Cx. pipiens* and *Culex quinquefasciatus*) was found to have produced a genotypically diverse and geographically widespread form (Fonseca *et al.* 2004; Smith and Fonseca 2004). Taken together, the observations suggested that, '... hybrids between human-biters and bird-biters may be the bridge vectors contributing to the unprecedented severity and range of the West Nile virus epidemic in North America' (Fonseca *et al.* 2004).

9.8 Lateral transfer and the evolution of human diseases

9.8.1 Propionibacterium acnes

Acne is the most common skin disease and affects approx. 80% of adolescents in the United States

(Brüggemann *et al.* 2004). Surprisingly, the role played by the human skin-commensal bacterium, *Propionibacterium acnes*, in the development of acne is still debated. Some of the processes that may contribute to this skin disease—and in which *P. acnes* may be involved—include damage to host tissues due to bacterial lipases (Miskin *et al.* 1997) or inflammation caused by heat-shock proteins (Farrar *et al.* 2000). Notwithstanding the lack of clarity concerning its mode of action, this bacterial species likely plays a significant role in the development of acne as well as a plethora of other, more serious, human diseases (e.g. Yamada *et al.* 2002).

Brüggemann *et al.* (2004) reported the entire genome sequence of *P. acnes*. In addition to their analysis of the genomic structure of this human commensal bacterium, these authors also discussed the interaction of genetic-exchange processes in the development of its putative, pathogenic characteristics. Specifically, Brüggemann *et al.* (2004) detected a number of regions that may have been added through lateral gene transfers. These regions demonstrated numerous traits known to be characteristic for horizontally transferred sequences, including being flanked by tRNA genes and possessing an unusual G+C content. Most significantly, numerous key functions were identified for the transferred islands. Some of the regions corresponded to (i) cryptic prophages, (ii) genes for a conjugation system, (iii) substrate uptake and utilization, (iv) putative pathogenicity loci, and (v) genes possibly associated with DNA repair and recombination (Brüggemann *et al.* 2004). *P. acnes* is yet another likely example of both a mosaic genome and an effective pathogen that has evolved at least partially via the incorporation of loci from a wide array of evolutionary lineages.

9.8.2 Burkholderia pseudomallei

Melioidosis is the disease caused by the Gram-negative bacterial species *Burkholderia pseudomallei* (formerly *Pseudomonas pseudomallei*; Dance 1991; White 2003). This bacterial infection has been identified as a significant cause of sepsis in both East Asia and northern Australia. For example, *B. pseudomallei* infections cause 20% of the community-acquired septicemias in Thailand, with a 40% mortality rate in those treated for the disease. The illness was identified in Burmese populations early in the twentieth century, being described as a 'glanders-like' malady (i.e. an abscess-forming disease of horses, mules, and donkeys caused by the related species *Burkholderia mallei*; see Nierman *et al.* 2004 for a discussion; White 2003). Melioidosis, like glanders, results in abscesses, particularly in both internal organs and muscle tissue (White 2003). However, unlike the glanders-causing bacterium, which is an obligate parasite of its equine hosts, *B. pseudomallei* is a soil saprophyte that is widespread throughout tropical regions (Dance 1991).

Because of its pathogenicity and also its potential as a tool for bioterrorism, Holden *et al.* (2004b) sequenced the genome of *B. pseudomallei*. Holden *et al.* (2004b) detected evidence for the pervasive effect of horizontal acquisitions of genomic islands in the evolution of *B. pseudomallei*. In particular, they noted that this bacterium possessed a large genome relative to other prokaryotes and had numerous genomic islands with characteristics reflective of recent lateral exchange events (e.g. G+C anomalies and the presence of gene sequences related to mobile elements). Furthermore, Holden *et al.* (2004b)

noted that a major contributor to the larger genome size of *B. pseudomallei* relative to *B. mallei* were the genomic islands. Consistent with their role in genome amplification was the demonstration of a high frequency of transmission and turnover of these regions. Figure 9.14 illustrates the presence/absence of the different islands among diverse isolates of *B. pseudomallei* (Holden *et al.* 2004b). The highly variable occurrence of the sequences among the clinical and natural strains was indicative of their frequent transmission through horizontal exchange.

In summarizing the significance of their findings, Holden *et al.* (2004b) emphasized the same concept emphasized throughout this book: genetic exchange provides opportunities for evolutionary transitions. Their conclusions also reflect the danger of this creative potential for human populations when the exchanges involve organisms that may become pathogenic. Furthermore, in the context of the final topic of this chapter—human-mediated genetic exchange—the following statement causes serious reflection: 'Coupled with the status of *B. pseudomallei* as a biothreat agent, there is therefore a pressing need to gain a better understanding of the role that horizontal gene

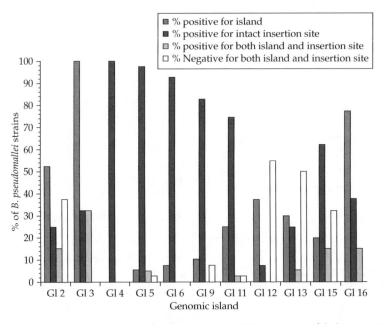

9.14 Frequency of occurrence of putatively horizontally-transferred genomic islands (GI) among strains of the bacterium *B. pseudomallei*. The histogram indicates the presence or absence of the insertion sites and/or the genomic islands in the various isolates (*n* = 40) as detected by multiplex PCR (from Holden *et al.* 2004b).

transfer plays in the genomic plasticity of this potent pathogen' (Holden *et al.* 2004b). Sadly, terrorists may be able to inadvertently capitalize on the amenability of certain pathogens to evolution through genetic exchange to modify the organisms for their own use.

9.9   Human-mediated genetic exchange

Evidence involving various species, then, suggests that evolution is best depicted by closely or distantly related strands of a web that diverge, converge, and intersect. Yet humans are not mere passive actors caught in this web. For millennia, through crop and animal breeding, we have actively contributed to the process of hybridizing closely related species. Most recently, with the advent of genetic engineering and biotechnology, we have begun to contribute as well to the web process of horizontal gene transfer.

(Arnold and Larson 2004)

This quote by Arnold and Larson (2004) reflects two main conclusions concerning human-engineered gene-transfer events. First, web processes have not only affected humans, humans have also participated in, if you will forgive the pun, web design. Second, *H. sapiens* has been in the business of making such transfers since the dawn of agriculture in the form of introgressive hybridization. Third, the most recent methodologies employed by our species to engineer plants, microbes, and animals have involved lateral gene-transfer-like events.

In regard to point three, Arnold and Larson (2004) continue, 'Whether transferring "delayed ripening" genes from a disease of bacteria (i.e. a bacteriophage) into cantaloupe to prevent our breakfast from going mushy too quickly, or splicing a gene for pesticide resistance from bacteria into corn to keep insects from feeding on the plants in a field, or implanting the gene for human interferon into the DNA of chickens so that their eggs contain the protein used to battle hepatitis C, or introducing the gene for a red fluorescent protein from a sea anemone into zebra fish so that they look more attractive to us in an aquarium, biotech researchers now move genes between species so unrelated to

one another that it's difficult to imagine the natural web of life ever accomplishing the same task. Yet these researchers' work perfectly represents how evolution has proceeded through the ages to produce the diversity of life on earth. Different evolutionary strands have been brought into association … and the result has been mosaic genomes.'

As alluded to at the end of section 9.8.2, the implications of *H. sapiens*' contribution to the web of life go beyond the possible benefits derived from improved crop and animal yields to the horrendous consequences of fanatics using the technology to produce biological agents that will kill and maim other humans. Yet the most widely discussed and debated questions concerning genetic modification remain whether the potential benefits from 'improving' substrates for human use outweigh the potential dangers engendered by introducing the modified organisms into the environment.

There are numerous reviews reflecting the plethora of possible benefits and costs from human-mediated genetic exchange (e.g. Ellstrand and Hoffman 1990; Ellstrand 2000; Hails and Morley 2005; Chapman and Burke 2006). However, I will use a single example to illustrate the balance between potentially beneficial and detrimental results from genetic modification. In particular, I will discuss the likelihood that the introgression of transgenes from crop plants into related, native lineages might occur and thus change the evolutionary trajectory of the native forms.

As discussed in previous chapters, research on the annual sunflower genus *Helianthus* has provided wonderful examples of the evolutionary significance of introgressive hybridization and hybrid lineage formation. However, this plant complex has also been used as a model for testing the outcome of introgression between genetically modified crops and their wild relatives. In particular, the domesticated sunflower, *H. annuus*, has been investigated with regard to the likelihood of transgene 'escape' (i.e. the introgression of genes introduced into crop plants into wild relatives). In an initial study, Whitton *et al.* (1997) documented the long-lasting genetic effects initiated by a single generation of hybridization between cultivated and wild *H. annuus*. By monitoring the frequencies of crop-specific alleles in neighboring wild populations, these workers documented the maintenance of a high level of the foreign markers in the native plants. In fact, Whitton *et al.* (1997) found no significant decrease from the initial crop-specific allele frequencies—even though hybrid plants likely demonstrated lower fitness relative to non-hybrid genotypes, at least for some fitness components (Cummings *et al.* 2002). Whitton *et al.*'s (1997) findings suggested that transient hybridization could lead to the long-term maintenance of introgressed crop loci in native populations. Consistent with cultivar-native hybridization being of potentially great importance for *H. annuus*, Burke *et al.* (2002) found evidence for geographically widespread genetic exchange. Specifically, these authors found that two-thirds of cultivated sunflower fields were spatially associated, and had overlapping flowering times, with the native *H. annuus*. In addition, Burke *et al.* (2002) detected morphological indications of hybridization in up to one-third of the populations.

The above studies indicate the potential for introgression between crop plants and their native relatives. However, the key issue for the current discussion is whether genes introduced into crop plants through genetic modification will introgress into wild populations thus causing evolutionary changes that lead to deleterious effects, such as the evolution of invasiveness (e.g. see Ellstrand 2000). Two analyses addressed directly the hypothesis that the introgression of trangenes from crop to wild plants would generate genotypes possessing higher fitness relative to their parental lineages. First, Snow *et al.* (2003) examined the fitness effects from the introgression of a herbivore resistance transgene (*Bt*) from cultivated to native *H. annuus*. Though the action of some herbivore castes were unaffected by the presence of the *Bt* gene in introgressed, native plants, lepidopteran damage was significantly reduced. Thus the introgression of the *Bt* gene increased the fitness of wild, introgressed plants under natural conditions (Snow *et al.* 2003).

In contrast to the results from Snow *et al.* (2003), Burke and Rieseberg (2003) found no evidence for a change in fitness from the presence of a transgene that protected plants against white mold infections. Specifically, they detected offsetting effects from the variation in the likelihood and severity of infection.

The cumulative effect was a tradeoff resulting in no fitness change between genotypes with or without the transgene, leading to the conclusion that the '. . . transgene will do little more than diffuse neutrally after its escape' from the crop plants into native individuals (Burke and Rieseberg 2003). As a whole, the data from the *H. annuus* system indicate the likelihood of extensive crop→wild introgression. However, the degree to which effects from such genetic exchange are positive or negative will likely depend greatly upon the transgenes involved and the ecological setting in which the introgression occurs.

9.10 Summary and conclusions

Genetic exchange—in the form of lateral transfer events, introgressive hybridization, and viral recombination—has played a significant role in the evolutionary histories of our own lineage and those lineages with which we interact on a daily basis. Our own genome is a mosaic of elements inherited from genomes of ancestors on divergent evolutionary lineages of the primate assemblage. Furthermore, introgression appears to have continued into the clade that contains the genus *Homo*. Thus, the human genome is only 'human' when considered as a point-in-time product. Like other genomes, it is better perceived as an ongoing construction project, with additions and deletions having occurred in the past, continuing to occur at present (e.g. retroviral insertions), and almost certainly continuing into the future.

Likewise, organisms that feed us, clothe us, make us sick, and even kill us have as diverse an array of ancestors as we. This recognition helps us to design strategies for crop and livestock 'improvement'. It also helps us understand why it will be so difficult to escape from those pathogens and parasites that utilize *H. sapiens* as a host. Hopefully, this recognition will also act as a cautionary tale as we design arrays of genetically modified organisms. Thus genetic exchange is, and will continue to be, a pervasive and powerful process in nature. If we choose to contribute to this process, we may produce incredible benefits for our existence. We may create a bane to our existence as well.

CHAPTER 10

Emergent properties

Hybridization between populations having very different genetic systems of adaptation may lead to several different results.

(Anderson and Stebbins 1954)

The introduction of genes from another species can serve as the raw material for an adaptive evolutionary advance.

(Lewontin and Birch 1966)

Vertical transmission of heritable material, a cornerstone of the Darwinian theory of evolution, is inadequate to describe the evolution of eukaryotes, particularly microbial eukaryotes. This is because eukaryotic cells and eukaryotic genomes are chimeric, having evolved through a combination of vertical (parent-to-offspring) and lateral (trans-species) transmission.

(Katz 2002)

. . . for the diverse prokaryotes in our sample, we find a pervasive coherent vertical genetic signal with significant modulation by [lateral gene transfer], particularly among thermophiles, pathogens, and cyanobacteria.

(Beiko *et al.* 2005)

Although each character exhibits intermediates ranging from one extreme to the other, interactions among different characters can create unique morphologies in other characters. In addition, in this case, distinct geographic variation arises not by isolation but by hybridization and introgression, revealing the importance of hybridization as a source of evolutionary novelty.

(Chiba 2005)

10.1 Genetic exchange is pervasive

Studies of natural hybridization in plants and animals indicate that this process cannot be ignored as a type of evolutionary noise. Rather, examples of introgression and reticulate evolution continue to be reported for an increasing number of plant and animal taxa. These reports take on added significance because fitness estimates for some hybrid genotypes are equal to or greater than those of their parents. However, this leads to the following, frequently asked question: If hybrids are relatively fit, why don't natural populations consist of hybrids rather than well-defined species?

(Arnold 1997, p. 182)

In answering my own question, I emphasized two points. First, the frequency of introgressive hybridization, though unequally distributed across taxonomic groups, had been grossly underestimated. Second, species were not 'well defined' as required by descriptors such as the biological

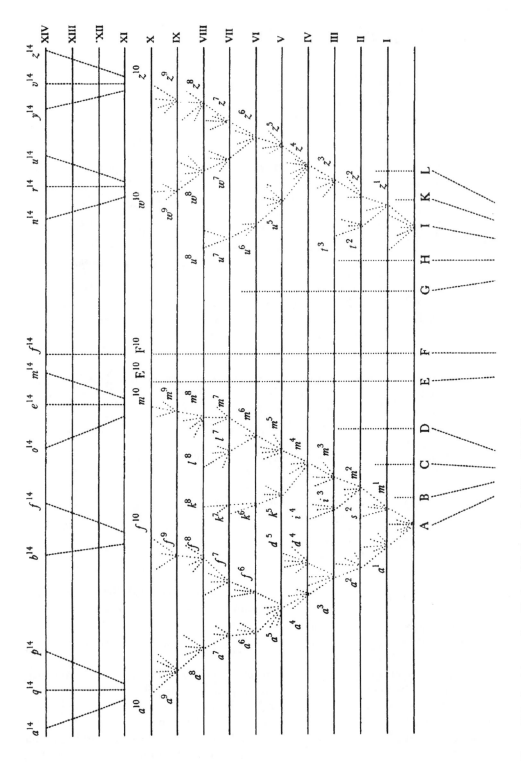

Figure 10.1 Darwin's perception of the tree of life (from Darwin 1859, pp. 514–515).

species concept, but instead were often the products of recent or ancient reticulate events.

From the data presented in this book, and the enormous amount of additional evidence not reviewed here, I would have to correct my 1997 conclusions in the following manner. Exchange is much more pervasive than I appreciated. This is a consequence of the variety of modes that have acted as bridges for the transfer of genetic material between closely or distantly related organisms. Notwithstanding my fondness for Darwin's illustration of the evolutionary process (Figure 10.1) as an ever-diversifying pattern of branching, as well as my desire for the simplicity and clarity that this paradigm provides, it would be inaccurate to continue to promote such a model given so much contradictory data. Instead, the relationships among organisms are much more complex, and I would argue wonderful, than we have been taught or have, for many of us, taught to others. Because of the pervasive nature of this process, a metaphor such as the web of life (Figure 10.2) necessarily reflects evolutionary patterns and processes better than does the tree of life. Given my penchant for neat, clean answers, such a viewpoint leads me to the uncomfortable position of accepting nature as even more messy than I formerly believed it to be.

Some will inevitably argue that the above, and indeed this entire book, is hyperbole. This argument often, but not always, comes from those who work on a very limited set of organisms—usually termed

'higher'. I find it ironic that these organisms—such as *Drosophila*, birds, and *Homo*—used so often to justify a very limited view of evolutionary pattern and process now provide us with some of our very best examples of evolution via web processes. On my worst days, when I hear such comments as, 'this does not apply to animals', I want to counter with 'Fine, let's assume you are correct and you keep those organisms for your own research agendas and the rest of us will keep the vast majority of organismic diversity for our own studies.' However, this would be an inaccurate and disingenuous answer since we now know that many of the "exceptions" are also products of genetic exchanges and thus contain mosaic genomes.

Some might complain that this treatment lacks the necessary tables and graphs that quantify how many organisms actually reflect the paradigm that is presented. I make no apologies for not including such tables because they would be misleading in the context of what has been presented. In other words, the major point addressed is that every genome examined thus far has the signatures from one or more of the avenues of genetic exchange.

10.2 Genetic exchange: research directions

The weight of evidence leads to the conclusion that the representation of evolutionary pattern and process by simple bifurcating diagrams (with all of the conceptualization that accompanies this model, such as divergence mainly in allopatry) is like a chicken whose head has recently been removed: it is dead, but the realization has yet to set in. Thus the realignment of the way we evolutionary biologists think about and study our various systems will change, but it will likely take some time.

The above conclusion begs the question of what types of studies might encourage this realignment. I would argue that as genomic information increases the balance will tip more and more to an appreciation of the frequency and diversity of genetic exchange events. The breadth of evidence will assuredly also continue to increase as phylogenetic treatments, both for 'model' and 'non-model' organisms, multiply. I argued (and by no means was I the first to do this) in *Natural Hybridization and*

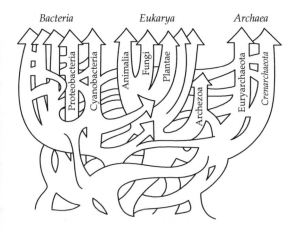

Figure 10.2   A representation of biological diversification reflecting the web-of-life paradigm (from Doolittle 1999).

Evolution that phylogenetic analyses provided some of the most powerful tests for reticulate evolution. This remains true, but these types of analysis remain largely qualitative in nature. Most inferences concerning reticulate events result from discordances detected by the manual comparison of phylogenies derived from different data sets. Thus there is a need for analytical methodologies that would allow the quantitative testing of multiple phylogenies, as well as the testing of data distributions in single phylogenies, for patterns indicative of genetic exchange among lineages (e.g. Vriesendorp and Bakker 2005; Huson and Bryant 2006).

The genomic and phylogenetic data are mainly limited to describing the pattern of evolutionary diversification, and not necessarily the evolutionary processes that underlie the web of life. In particular, though we now have many good estimates of the fitness of recombinants/hybrids under a variety of experimental conditions, we still lack large amounts of similar information from natural settings (but see examples in Chapters 1 and 5). In particular, this leaves us in the uncomfortable position of recognizing numerous cases of, for example, lateral exchange in bacteria that may or may not reflect adaptive gains. The same can be concluded for instances of introgressive hybridization: most

examples lack explicit tests for whether the transfers lead to differentially adaptive genotypes or phenotypes. Major gains in our understanding of the role of genetic exchange in adaptive evolution—though such a role is consistent with many data sets already in existence—will depend on the multiplication of this class of studies.

I concluded *Natural Hybridization and Evolution* (Arnold 1997, p. 185) with a quote from a later edition of *On the Origin of Species*. I believe the slightly different passage from the first edition of this work captures as well the essence of what I have tried to express in this book:

It is interesting to contemplate an entangled bank, clothed with many plants of many kinds, with birds singing on the bushes, with various insects flitting about, and with worms crawling through the damp earth, and to reflect that these elaborately constructed forms, so different from each other, and dependent on each other in so complex a manner, have all been produced by laws acting around us . . .

(Darwin 1859, p. 489)

It is now apparent that genetic exchange is one of those laws.

Glossary

adaptive radiation The evolution of ecological and phenotypic diversity within a rapidly multiplying lineage (Seehausen 2004).

adaptive trait introgression *See* adaptive trait transfer.

adaptive trait transfer The transfer of genes and thus the phenotype of an adaptive trait through viral recombination, lateral gene transfer, or introgressive hybridization.

agamic species complex This type of complex results '. . . from the combination of *agamospermy* and hybridization . . .' (King 1993).

agamospermy Seeds develop on the maternal plant without being fertilized (Grant 1981, p. 5).

allopolyploid species A species formed through hybridization between members of evolutionary lineages with 'strongly differentiated genomes' followed by chromosomal doubling, trebling, etc. (Stebbins 1947).

apogamous taxa Those taxa that reproduce by the development of an embryo '. . . from some cell or nucleus other than the egg in the embryo sac . . .' (Grant 1981, p. 10).

autopolyploidy A type of polyploidy in which the derivatives originate from crosses within a single evolutionary lineage (Stebbins 1947).

biological species concept 'Species are groups of actually or potentially interbreeding natural populations, which are reproductively isolated from other such groups' (Mayr 1942).

bounded hybrid superiority model 'When a secondary contact occurs, it would usually involve whole ecological communities expanding from refugia. Hybrid zones . . . would form in the ecotone and remain there because hybrids would be maladapted in the parental communities but relatively well adapted in the intermediate or mixed ecology' (Moore and Buchanan 1985). Also known as the hybrid superiority model (Moore 1977).

clonal microspecies Microspecies are '. . . plant populations which reproduce mainly if not exclusively by uniparental methods, are morphologically uniform, occupy a definite geographical or microgeographical

area, are differentiated morphologically—often slightly—from related species and microspecies, and frequently possess a hybrid constitution . . . Clonal microspecies are those that reproduce by various means of vegetative propagation' (Grant and Grant 1971).

cohesion species concept '. . . an evolutionary lineage or set of lineages with genetic exchangeability and/or ecological interchangeability' (Templeton 2001).

concerted evolution Molecular processes that facilitate changes in a single genetic element to be incorporated into all genes belonging to a multi-gene family (Zimmer *et al.* 1980; Hartwell *et al.* 2004).

conjugation The transfer of DNA through physical contact between donor and recipient cells. This process can mediate the transfer of genetic material between such divergent evolutionary lineages as bacteria and plants (Ochman *et al.* 2000).

cyclic parthenogenesis Occurs when '. . . unisexual propagation via eggs is periodically interrupted by a bisexual phase' (Lynch 1984). For example, in the cladoceran *Daphnia*, 'Cyclic parthenogens produce sexual resting eggs; that is, the eggs are haploid and require fertilization by sperm to develop' (Crease *et al.* 1989).

donor An organism that acts as the source for DNA sequences transferred through lateral gene transfer.

epigenetic 'A state of gene functionality that is not encoded within the DNA sequence but that is still inheritable from one generation to the next. It can be accomplished and maintained through a chemical modification of DNA such as methylation' (Hartwell *et al.* 2004).

floral preference '. . . preference refers to the net overvisitation of one flower type by pollinators, when more than one flower type is present . . .' (Meléndez-Ackerman *et al.* 1997, from the concept of Cock 1978).

gamete competition The finding that contaxon gametes are superior to heterotaxon gametes in fathering progeny, when both gamete types are present.

gene conversion The transfer of the gene sequence from one DNA element to another related, but non-allelic, element resulting in the 'conversion' of the recipient into the DNA sequence of the donor (Hartwell *et al.* 2004).

genome downsizing The loss of DNA following polyploidization. A concept derived from the observation that the amount of DNA in polyploids does not increase in direct proportion to the ploidy level (Leitch and Bennett 2004).

gynogenesis Mode of asexual reproduction in which the presence of sperm from a related bisexual species is required to stimulate egg development in an asexually reproducing female. In concert with the stimulation, '. . . the female's nuclear genome is transmitted intact to the egg, which then develops into an offspring genetically identical to the mother' (Avise *et al.* 1992).

Haldane's Rule 'When in the F_1 offspring of two different animal races one sex is absent, rare, or sterile, that sex is the heterozygous sex' (Haldane 1922), '. . . that is to say, the sex which produces two sorts of gametes, namely the male in most animal groups . . .' (Haldane 1990, p 42). Also known as Haldane's Law.

homeobox gene 'Homeobox genes encode DNA-binding proteins characterized by the presence of the 60 or 63 amino acid homeodomain' (Castro and Holland 2003).

horizontal gene transfer *See* lateral gene transfer.

hybrid index A quantitative assessment of species-specific characters, in which the characters from the different species are assigned different numerical values (e.g. 1 for every character from species A and 0 for every character from species B). The index values are often displayed as a bar graph distribution for each population or species.

hybrid speciation Process in which natural hybridization results in the production of an evolutionary lineage that is at least partially reproductively isolated from both parental lineages, and which demonstrates a distinct evolutionary and ecological trajectory (Arnold and Burke 2006).

hybrid species At least partially reproductively isolated lineages arising as a result of natural hybridization. These lineages demonstrate distinct evolutionary and ecological trajectories as defined by distinguishable (and heritable) morphological, ecological and/or reproductive differences relative to their progenitors (Arnold and Burke 2006).

hybrid superiority model *See* bounded hybrid superiority model.

hybrid zone Geographical region in which natural hybridization occurs (Arnold 1997 as adapted from Harrison 1990).

hybridogenesis Mode of asexual reproduction in which '. . . an ancestral genome from the maternal line is trans-mitted to the egg without recombination, while paternally-derived chromosomes are discarded only to be replaced in each generation through fertilization by sperm from a related sexual species' (Avise *et al.* 1992).

introgressive hybridization The transfer of DNA between individuals from two populations, or groups of populations, that are distinguishable on the basis of one or more heritable characters via hybridization followed by repeated backcrossing between hybrid and parental individuals (Anderson and Hubricht 1938). Also known as introgression.

lateral gene transfer The transfer of genetic material between individuals from two populations, or groups of populations, that are distinguishable on the basis of one or more heritable characters through the processes of transformation, transduction, conjugation, or vector-mediated transfer.

Linneon '. . . a group of individuals which resemble one another more than they do any other individuals' (Lotsy 1916, p. 27).

methylation The selective addition of methyl groups (-CH_3) to DNA by enzymes known as methylases (Hartwell *et al.* 2004).

mosaic hybrid zone A hybrid zone demonstrating a complex patchwork of parental and hybrid populations. The different populations are associated with different habitats, indicating the importance of environment-dependent selection (Howard 1982, 1986; Harrison 1986, 1990).

natural hybrid An offspring resulting from a cross in nature between individuals from two populations, or groups of populations, that are distinguishable on the basis of one or more heritable characters (Arnold 1997 as adapted from Harrison 1990).

natural hybridization Successful matings in nature between individuals from two populations, or groups of populations, that are distinguishable on the basis of one or more heritable characters (Arnold 1997 as adapted from Harrison 1990).

neo-functionalization 'When one of two duplicate genes acquires a mutation in coding or regulatory sequences that allows the gene to take on a new and useful function' (Prince and Pickett 2002).

nested clade analysis Phylogenetic construction of nested clades of genotypes/phenotypes into a single evolutionary tree that is then used to test for associations between clades and geographical location (Templeton 1993; Templeton *et al.* 1995).

non-functionalization (silencing) 'When one of two duplicate genes acquires a mutation in coding or regulatory sequences that ultimately renders the gene non-functional' (Prince and Pickett 2002).

nucleolar dominance '. . . epigenetic phenomenon that describes the formation of nucleoli around rRNA genes inherited from only one parent in the progeny of an interspecific hybrid' (Chen and Pikaard 1997).

obligate parthenogenesis When the only reproduction possible for an organism is parthenogenetic. For example, in the cladoceran *Daphnia*, 'Obligate parthenogens produce their resting eggs ameiotically; the eggs are diploid and genetically identical to the mother' (Crease *et al.* 1989).

paralogous genes Gene copies originating through duplications within the same species (e.g. via autopolyploidy) and often within the same linkage group (Hartwell *et al.* 2004). Also known as paralogs.

parthenogenesis Defined as 'reproduction in which offspring are produced by an unfertilized female' (Hartwell *et al.* 2004) and 'asexual reproduction via eggs' (Futuyma 1998).

phylogenetic species concept Two concepts can be grouped under this general heading: character-based and history-based (Baum and Donoghue 1995). The character-based phylogenetic species is '. . . the smallest aggregation of populations (sexual) or lineages (asexual) diagnosable by a unique combination of character states in comparable individuals . . .' (Nixon and Wheeler 1990). The history-based phylogenetic species is '. . . an exclusive group of organisms whose members are more closely related to each other than to any organisms outside the group' (Olmstead 1995).

prokaryotic species concept Two concepts can also be placed under this general descriptor. A phylo-phenetic species is '. . . a monophyletic and genomically coherent cluster of individual organisms that show a high degree of overall similarity in many independent characteristics, and is diagnosable by a discriminative phenotypic property' (Rosselló-Mora and Amann 2001). A phylogenetic and ecological divergence species is '. . . a group of organisms whose divergence is capped by a force of cohesion; divergence between different species is irreversible; and different species are ecologically distinct' (Cohan 2002).

proteome 'The complete set of proteins encoded by a genome' (Hartwell *et al.* 2004).

recipient An organism that receives the DNA from a donor through lateral gene transfer.

reticulate evolution Web-like phylogenetic relationships reflecting genetic exchange (through lateral transfer, viral recombination, introgressive hybridization, etc.) between diverging lineages.

Robertsonian rearrangement The fusion (translocation) or fission of chromosomes leading to metacentric or acrocentric chromosomes, respectively. Such changes result in a decrease (fusions) or increase (fissions) in chromosome number, but the number of chromosome 'arms' remains the same (Robertson 1916, as discussed by Mayr 1963).

semi-permeable genome The observation that portions of the genomes of hybridizing taxa introgress, because they are neutral or positively selected, whereas others are selected against and thus do not introgress (Key 1968; Harrison 1986).

serotype The genotype of a unicellular organism as defined by antisera against antigenic determinants expressed on the surface (www.biology-online.org).

serovar A subdivision of a species or subspecies distinguishable from other strains on the basis of antigenic character. Synonymous with *serotype*.

silencing *See* non-functionalization.

species flock A taxonomic group '. . . in which several to many species have diversified from a single common ancestor in a geographically restricted area, often over an evolutionarily short period of time' (Sullivan *et al.* 2002).

sub-functionalization '. . . after duplication, the two gene copies acquire complementary loss-of-function mutations in independent sub-functions, such that both genes are required to produce the full complement of functions of the single ancestral gene' (Prince and Pickett 2002).

syngameon '. . . an habitually interbreeding community . . .' (Lotsy 1931) or '. . . the most inclusive unit of interbreeding in a hybridizing species group . . .' (Grant 1981).

tension zone A hybrid zone maintained by a balance between dispersal of parental forms into the region of overlap and selection against recombinant (i.e. hybrid) genotypes (Barton and Hewitt 1985).

transduction The transfer of DNA through a bacteriophage intermediate (Ochman *et al.* 2000).

transformation The uptake of naked DNA from the environment (Ochman *et al.* 2000).

vector-mediated transfer The transfer of DNA from a donor to a recipient by a vector intermediate (e.g. between insect species through a shared parasite).

virion A complete virus particle, as it exists outside the cell, with—in the case of HIV—a core made up of two RNA strands and a protein coat (Hutchinson 2001).

References

Abbott, R.J. (1992) Plant invasions, interspecific hybridization and the evolution of new plant taxa. *Trends in Ecology and Evolution* **7**, 401–405.

Abbott, R.J. and Forbes, D.G . (2002) Extinction of the Edinburgh lineage of the allopolyploid neospecies, *Senecio cambrensis* Rosser (Asteraceae). *Heredity* **88**, 267–269.

Abbott, R.J. and Lowe, A.J. (2004) Origins, establishment and evolution of new polyploid species: *Senecio cambrensis* and *S. eboracensis* in the British Isles. *Biological Journal of the Linnean Society* **82**, 467–474.

Abbott, R.J., Ashton, P.A., and Forbes, D.G. (1992) Introgressive origin of the radiate groundsel, *Senecio vulgaris* L var *hibernicus* Syme: *Aat-3* evidence. *Heredity* **68**, 425–435.

Abbott, R.J., James, J.K., Milne, R.I., and Gillies, A.C.M. (2003) Plant introductions, hybridization and gene flow. *Philosophical Transactions of the Royal Society of London Series B* **358**, 1123–1132.

Abila, R., Barluenga, M., Engelken, J., *et al.* (2004) Population-structure and genetic diversity in a haplochromine fish cichlid of a satellite lake of Lake Victoria. *Molecular Ecology* **13**, 2589–2602.

Abrahamsen, M.S., Templeton, T.J., Enomoto, S., *et al.* (2004) Complete genome sequence of the apicomplexan, *Cryptosporidium parvum*. *Science* **304**, 441–445.

Achtman, M., Zurth, K., Morelli, G., *et al.* (1999) *Yersinia pestis*, the cause of plague, is a recently emerged clone of *Yersinia pseudotuberculosis*. *Proceedings of the National Academy of Sciences USA* **96**, 14043–14048.

Achtman, M., Morelli, G., Zhu, P., *et al.* (2004) Microevolution and history of the plague bacillus, *Yersinia pestis*. *Proceedings of the National Academy of Sciences USA* **101**, 17837–17842.

Adamowicz, S.J., Gregory, T.R., Marinone, M.C., and Hebert, P.D.N. (2002) New insights into the distribution of polyploid *Daphnia*: The Holarctic revisited and Argentina explored. *Molecular Ecology* **11**, 1209–1217.

Adamowicz, S.J., Hebert, P.D.N., and Marinone, M.C. (2004) Species diversity and endemism in the *Daphnia* of Argentina: A genetic investigation. *Zoological Journal of the Linnean Society* **140**, 171–205.

Adams, K.L. and Wendel, J.F. (2004) Exploring the genomic mysteries of polyploid cotton. *Biological Journal of the Linnean Society* **82**, 573–581.

Adams, K.L. and Wendel, J.F. (2005) Novel patterns of gene expression in polyploid plants. *Trends in Genetics* **21**, 539–543.

Adams, K.L., Cronn, R., Percifield, R., and Wendel, J.F. (2003) Genes duplicated by polyploidy show unequal contributions to the transcriptome and organ-specific reciprocal silencing. *Proceedings of the National Academy of Sciences USA* **100**, 4649–4654.

Adams, M., Foster, R., Hutchinson, M.N., *et al.* (2003) The Australian scincid lizard *Menetia greyii*: a new instance of widespread vertebrate parthenogenesis. *Evolution* **57**, 2619–2627.

Ainouche, M.L., Baumel, A., Salmon, A., and Yannic, G. (2003) Hybridization, polyploidy and speciation in *Spartina* (Poaceae). *New Phytologist* **161**, 165–172.

Ainouche, M.L., Baumel, A., and Salmon, A. (2004) *Spartina anglica* C.E. Hubbard: a natural model system for analyzing early evolutionary changes that affect allopolyploid genomes. *Biological Journal of the Linnean Society* **82**, 475–484.

Alarcón, R. and Campbell, D.R. (2000) Absence of conspecific pollen advantage in the dynamics of an *Ipomopsis* (Polemoniaceae) hybrid zone. *American Journal of Botany* **87**, 819–824.

Albert, M.E., D'Antonio, C.M., and Schierenbeck, K.A. (1997) Hybridization and introgression in *Carpobrotus* spp. (Aizoaceae) in California. I. Morphological evidence. *American Journal of Botany* **84**, 896–904.

Alberts, S.C. and Altmann, J. (2001) Immigration and hybridization patterns of yellow and anubis baboons in and around Amboseli, Kenya. *American Journal of Primatology* **53**, 139–154.

Albertson, R.C. and Kocher, T.D. (2005) Genetic architecture sets limits on transgressive segregation in hybrid cichlid fishes. *Evolution* **59**, 686–690.

Albertson, R.C., Markert, J.A., Danley, P.D., and Kocher, T.D. (1999) Phylogeny of a rapidly evolving clade: the cichlid fishes of Lake Malawi, East Africa. *Proceedings of the National Academy of Sciences USA* **96**, 5107–5110.

Albertson, R.C., Streelman, J.T., and Kocher, T.D. (2003) Directional selection has shaped the oral jaws of Lake Malawi cichlid fishes. *Proceedings of the National Academy of Sciences USA* **100**, 5252–5257.

Albertson, R.C., Streelman, J.T., Kocher, T.D., and Yelick, P.C. (2005) Integration and evolution of the cichlid mandible: the molecular basis of alternate feeding strategies. *Proceedings of the National Academy of Sciences USA* **102**, 16287–16292.

Aldridge, G. (2005) Variation in frequency of hybrids and spatial structure among *Ipomopsis* (Polemoniaceae) contact sites. *New Phytologist* **167**, 279–288.

Allender, C.J., Seehausen, O., Knight, M.E., *et al.* (2003) Divergent selection during speciation of Lake Malawi cichlid fishes inferred from parallel radiations in nuptial coloration. *Proceedings of the National Academy of Sciences USA* **100**, 14074–14079.

Allendorf, F.W., Leary, R.F., Spruell, P., and Wenburg, J.K. (2001) The problems with hybrids: setting conservation guidelines. *Trends in Ecology and Evolution* **16**, 613–622.

Alves, P.C., Ferrand, N., Suchentrunk, F., and Harris, D.J. (2003) Ancient introgression of *Lepus timidus* mtDNA into *L. granatensis* and *L. europaeus* in the Iberian Peninsula. *Molecular Phylogenetics and Evolution* **27**, 70–80.

Ambrose, S.E. (1996) *Crazy Horse and Custer—The Parallel Lives of Two American Warriors*. Random House, New York.

Anderson, E. (1948) Hybridization of the habitat. *Evolution* **2**, 1–9.

Anderson, E. (1949) *Introgressive Hybridization*. John Wiley and Sons, New York.

Anderson, E. (1951) Concordant versus discordant variation in relation to introgression. *Evolution*, **5**, 133–141.

Anderson, E. and Hubricht, L. (1938) Hybridization in *Tradescantia*. III. The evidence for introgressive hybridization. *American Journal of Botany* **25**, 396–402.

Anderson, E. and Stebbins, G.L., Jr (1954) Hybridization as an evolutionary stimulus. *Evolution* **8**, 378–388.

Ankney, C.D., Dennis, D.G., Wishard, L.N., and Seeb, J.E. (1986) Low genic variation between black ducks and mallards. *Auk* **103**, 701–709.

Anthony, F., Combes, M.C., Astorga, C., *et al.* (2002) The origin of cultivated *Coffea arabica* L. varieties revealed by AFLP and SSR markers. *Theoretical and Applied Genetics* **104**, 894–900.

Arnegard, M.E., Bogdanowicz, S.M., and Hopkins, C.D. (2005) Multiple cases of striking genetic similarity between alternate electric fish signal morphs in sympatry. *Evolution* **59**, 324–343.

Arnold, M.L. (1986) The heterochromatin of grasshoppers from the *Caledia captiva* species complex III. Cytological organization and sequence evolution in a dispersed highly repeated DNA family. *Chromosoma* **94**, 183–188.

Arnold, M.L. (1992) Natural hybridization as an evolutionary process. *Annual Review of Ecology and Systematics* **23**, 237–261.

Arnold, M.L. (1993) *Iris nelsonii*: origin and genetic composition of a homoploid hybrid species. *American Journal of Botany* **80**, 577–583.

Arnold, M.L. (1994) Natural hybridization and Louisiana Irises. *BioScience* **44**, 141–147.

Arnold, M.L. (1997) *Natural Hybridization and Evolution*. Oxford University Press, Oxford.

Arnold, M.L. (2000) Anderson's paradigm: Louisiana Irises and the study of evolutionary phenomena. *Molecular Ecology* **9**, 1687–1698.

Arnold, M.L. (2004a) Natural hybridization and the evolution of domesticated, pest, and disease organisms. *Molecular Ecology* **13**, 997–1007.

Arnold, M.L. (2004b) Transfer and origin of adaptations through natural hybridization: were Anderson and Stebbins right? *The Plant Cell* **16**, 1–9.

Arnold, M.L. and Jackson, R.C. (1978) Biochemical, cytogenetic and morphological relationships of a new species of *Machaeranthera* section *Arida* (Compositae) *Systematic Botany* **3**, 208–217.

Arnold, M.L. and Shaw, D.D. (1985) The heterochromatin of grasshoppers from the *Caledia captiva* species complex II. Cytological organization of tandemly repeated DNA sequences. *Chromosoma* **93**, 183–190.

Arnold, M.L. and Bennett, B.D. (1993) Natural hybridization in Louisiana irises: genetic variation and ecological determinants. In R.G. Harrison, ed. *Hybrid Zones and the Evolutionary Process*, pp. 115–139. Oxford University Press, Oxford.

Arnold, M.L. and Hodges, S.A. (1995) Are natural hybrids fit or unfit relative to their parents? *Trends in Ecology and Evolution* **10**, 67–71.

Arnold, M.L. and Larson, E.J. (2004) Evolution's new look. *The Wilson Quarterly*, Autumn, 60–72.

Arnold, M.L. and Burke, J.M. (2006) Natural hybridization. In C.W. Fox and J.B. Wolf, eds. *Evolutionary Genetics: Concepts and Case Studies*, pp. 399–413. Oxford University Press, Oxford.

Arnold, M.L. and Meyer, A. (2006) Natural hybridization in primates: one evolutionary mechanism. *Zoology*, in press.

Arnold, M.L., Appels, R., and Shaw, D.D. (1986) The heterochromatin of grasshoppers from the *Caledia captiva* species complex I. Sequence evolution and conservation in a highly repeated DNA family. *Molecular Biology and Evolution* **3**, 29–43.

Arnold, M.L., Shaw, D.D., and Contreras, N. (1987a) Ribosomal RNA encoding DNA introgression across a narrow hybrid zone between two subspecies of grasshopper. *Proceedings of the National Academy of Sciences USA* **84**, 3946–3950.

Arnold, M.L., Wilkinson, P., Shaw, D.D., *et al.* (1987b) Highly repeated DNA and allozyme variation between sibling species: evidence for introgression. *Genome* **29**, 272–279.

Arnold, M.L., Contreras, N., and Shaw, D.D. (1988) Biased gene conversion and asymmetrical introgression between subspecies. *Chromosoma* **96**, 368–371.

Arnold, M.L., Bennett, B.D., and Zimmer, E.A. (1990a) Natural hybridization between *Iris fulva* and *I. hexagona*: pattern of ribosomal DNA variation. *Evolution* **44**, 1512–1521.

Arnold, M.L., Hamrick, J.L., and Bennett, B.D. (1990b) Allozyme variation in Louisiana Irises: a test for introgression and hybrid speciation. *Heredity* **65**, 297–306.

Arnold, M.L., Buckner, C.M., and Robinson, J.J. (1991) Pollen mediated introgression and hybrid speciation in Louisiana irises. *Proceedings of the National Academy of Sciences USA* **88**, 1398–1402.

Arnold, M.L., Robinson, J.J., Buckner, C.M., and Bennett, B.D. (1992) Pollen dispersal and interspecific gene flow in Louisiana irises. *Heredity* **68**, 399–404.

Arnold, M.L., Hamrick, J.L., and Bennett, B.D. (1993) Interspecific pollen competition and reproductive isolation in *Iris*. *Journal of Heredity* **84**, 13–16.

Arnold, M.L., Bulger, M.R., Burke, J.M., *et al.* (1999) Natural hybridization—How low can you go? (and still be important) *Ecology* **80**, 371–381.

Arnold, M.L., Bouck, A.C., and Cornman, R.S. (2004) Verne Grant and Louisiana Irises: is there anything new under the sun? *New Phytologist* **161**, 143–149.

Arntzen, J.W. (1978) Some hypotheses on the postglacial migrations of the fire-bellied toad *Bombina bombina* (L.) and the yellow-bellied toad, *Bombina variegata*. *Journal of Biogeography* **5**, 339–345.

Ashton, P.A., and Abbott, R.J. (1992) Multiple origins and genetic diversity in the newly arisen allopolyploid species, *Senecio cambrensis* Rosser (Compositae) *Heredity* **58**, 25–32.

Aubert, J., and Solignac, M. (1990) Experimental evidence for mitochondrial DNA introgression between *Drosophila* species. *Evolution* **44**, 1272–1282.

Avey, J.T., Ballard, W.B., Wallace, M.C., *et al.* (2003) Habitat relationships between sympatric mule deer and white-tailed deer in Texas. *Southwestern Naturalist* **48**, 644–653.

Avise, J.C. (2000a) Cladists in wonderland. *Evolution* **54**, 1828–1832.

Avise, J.C. (2000b) *Phylogeography*. Harvard University Press, Cambridge, MA.

Avise, J.C., Giblin-Davidson, C., Laerm, J., *et al.* (1979a) Mitochondrial DNA clones and matriarchal phylogeny within and among geographic populations of the pocket gopher, *Geomys pinetus*. *Proceedings of the National Academy of Sciences USA* **76**, 6694–6698.

Avise, J.C., Lansman, R.A., and Shade, R.O. (1979b) The use of restriction endonucleases to measure mitochondrial DNA sequence relatedness in natural populations. I. Population structure and evolution in the genus *Peromyscus*. *Genetics* **92**, 279–295.

Avise, J.C., Ankney, C.D., and Nelson, W.S. (1990) Mitochondrial gene trees and the evolutionary relationship of mallard and black ducks. *Evolution* **44**, 1109–1119.

Avise, J.C., Trexler, J.C., Travis, J., and Nelson, W.S. (1991) *Poecilia mexicana* is the recent female parent of the unisexual fish *P. formosa*. *Evolution* **45**, 1530–1533.

Avise, J.C., Quattro, J.M., and Vrijenhoek, R.C. (1992) Molecular clones within organismal clones: mitochondrial DNA phylogenies and the evolutionary histories of unisexual vertebrates. *Evolutionary Biology* **26**, 225–246.

Bacilieri, R., Ducousso, A., Petit, R.J., and Kremer, A. (1996) Mating system and asymmetric hybridization in a mixed stand of European oaks. *Evolution* **50**, 900–908.

Bailey, J.K., Schweitzer, J.A., Rehill, B.J., *et al.* (2004) Beavers as molecular geneticists: a genetic basis to the foraging of an ecosystem engineer. *Ecology* **85**, 603–608.

Baldwin, B.G. and Sanderson, M.J. (1998) Age and rate of diversification of the Hawaiian Silversword alliance (Compositae). *Proceedings of the National Academy of Sciences USA* **95**, 9402–9406.

Baldwin, B.G., Kyhos, D.W., and Dvorak, J. (1990) Chloroplast DNA evolution and adaptive radiation in the Hawaiian Silversword alliance (Asteraceae-Madiinae). *Annals of the Missouri Botanical Garden* **77**, 96–109.

Baldwin, B.G., Kyhos, D.W., Dvorak, J., and Carr, G.D. (1991) Chloroplast DNA evidence for a North American origin of the Hawaiian Silversword alliance (Asteraceae). *Proceedings of the National Academy of Sciences USA* **88**, 1840–1843.

Bangert, R.K., Turek, R.J., Martinsen, G.D., *et al.* (2005) Benefits of conservation of plant genetic diversity to arthropod diversity. *Conservation Biology* **19**, 379–390.

Baric, S., Salzburger, W., and Sturmbauer, C. (2003) Phylogeography and evolution of the Tanganyikan cichlid genus *Tropheus* based upon mitochondrial DNA sequences. *Journal of Molecular Evolution* **56**, 54–68.

Barluenga, M. and Meyer, A. (2004) The Midas cichlid species complex: incipient sympatric speciation in Nicaraguan cichlid fishes? *Molecular Ecology* **13**, 2061–2076.

Barluenga, M., Stölting, K.N., Salzburger, W., *et al.* (2006) Sympatric speciation in Nicaraguan crater lake cichlids. *Nature* **439**, 719–723.

Barrier, M., Baldwin, B.G., Robichaux, R.H., and Purugganan, M.D. (1999) Interspecific hybrid ancestry of a plant adaptive radiation: allopolyploidy of the Hawaiian Silversword alliance (Asteraceae) inferred from floral homeotic gene duplications. *Molecular Biology and Evolution* **16**, 1105–1113.

Barrier, M., Robichaux, R.H., and Purugganan, M.D. (2001) Accelerated regulatory gene evolution in an adaptive radiation. *Proceedings of the National Academy of Sciences USA* **98**, 10208–10213.

Barrowclough, G.F., Groth, J.G., Mertz, L.A., and Gutiérrez, R.J. (2005) Genetic structure, introgression, and a narrow hybrid zone between northern and California spotted owls (*Strix occidentalis*). *Molecular Ecology* **14**, 1109–1120.

Barry, C.E., III, Lee, R.E., Mdluli, K., *et al.* (1998) Mycolic acids: structure, biosynthesis and physiological functions. *Progress in Lipids Research* **37**, 143–179.

Barton, N.H. and Hewitt, G.M. (1985) Analysis of hybrid zones. *Annual Review of Ecology and Systematics* **16**, 113–148.

Batut, J., Andersson, S.G.E., and O'Callaghan, D. (2005) The evolution of chronic infection strategies in the a-proteobacteria. *Nature Reviews Microbiology* **2**, 933–945.

Baum, D.A. and Donoghue, M.J. (1995) Choosing among alternative 'phylogenetic' species concepts. *Systematic Botany* **20**, 560–573.

Baumel, A., Ainouche, M.L., and Levasseur, J.E. (2001) Molecular investigations in populations of *Spartina anglica* CE Hubbard (Poaceae) invading coastal Brittany (France). *Molecular Ecology* **10**, 1689–1701.

Baumel, A., Ainouche, M.L., Bayer, R.J., *et al.* (2002) Molecular phylogeny of hybridizing species from the genus *Spartina* Schreb (Poaceae). *Molecular Phylogenetics and Evolution* **22**, 303–314.

Baumel, A., Ainouche, M.L., Misset, M.T., *et al.* (2003) Genetic evidence for hybridization between the native *Spartina maritima* and the introduced *Spartina alterniflora* (Poaceae) in south-west France: *Spartina* x *neyrautii* re-examined. *Plant Systematics and Evolution* **237**, 87–97.

Beaton, M.J. and Hebert, P.D.N. (1988) Geographical parthenogenesis and polyploidy in *Daphnia pulex*. *American Naturalist* **132**, 837–845.

Beebee, T.J.C. (2005) Conservation genetics of amphibians. *Heredity* **95**, 423–427.

Beetle, A.A. (1971) An ecological contribution to the taxonomy of *Artemisia*. *Madroño* **20**, 385–386.

Beiko, R.G., Harlow, T.J., and Ragan, M.A. (2005) Highways of gene sharing in prokaryotes. *Proceedings of the National Academy of Sciences USA* **102**, 14332–14337.

Belahbib, N., Pemonge, M.-H., Ouassou, A., *et al.* (2001) Frequent cytoplasmic exchanges between oak species that are not closely related: *Quercus suber* and *Q. ilex* in Morocco. *Molecular Ecology* **10**, 2003–2012.

Bell, K.S., Sebaihia, M., Pritchard, L., *et al.* (2004) Genome sequence of the enterobacterial phytopathogen *Erwinia carotovora* subsp. *atroseptica* and characterization of virulence factors. *Proceedings of the National Academy of Sciences USA* **101**, 11105–11110.

Benedix, J.H., Jr and Howard, D.J. (1991) Calling song displacement in a zone of overlap and hybridization. *Evolution* **45**, 1751–1759.

Bennett, B.D. and Grace, J.B. (1990) Shade tolerance and its effect on the segregation of two species of Louisiana iris and their hybrids. *American Journal of Botany* **77**, 100–107.

Bergman, T.J. and Beehner, J.C. (2004) Social system of a hybrid baboon group (*Papio anubis* x *P. hamadryas*). *International Journal of Primatology* **25**, 1313–1330.

Bergthorsson, U., Adams, K.L., Thomason, B., and Palmer, J.D. (2003) Widespread horizontal transfer of mitochondrial genes in flowering plants. *Nature* **424**, 197–201.

Bergthorsson, U., Richardson, A.O., Young, G.J., *et al.* (2004) Massive horizontal transfer of mitochondrial genes from diverse land plant donors to the basal angiosperm *Amborella*. *Proceedings of the National Academy of Sciences USA* **101**, 17747–17752.

Bernatchez, L. (2001) The evolutionary history of brown trout (*Salmo trutta* L.) inferred from phylogeographic, nested clade, and mismatch analyses of mitochondrial DNA variation. *Evolution* **55**, 351–379.

Bernatchez, L., Vuorinen, J.A., Bodaly, R.A., and Dodson, J.J. (1996) Genetic evidence for reproductive isolation and multiple origins of sympatric trophic ecotypes of whitefish (*Coregonus*). *Evolution* **50**, 624–635.

Berriman, M., Ghedin, E., Hertz-Fowler, C., *et al.* (2005) The genome of the African trypanosome *Trypanosoma brucei*. *Science* **309**, 416–422.

Besansky, N.J., Lehmann, T., Fahey, G.T., *et al.* (1997) Patterns of mitochondrial variation within and between African malaria vectors, *Anopheles gambiae* and *An. arabiensis*, suggest extensive gene flow. *Genetics* **147**, 1817–1828.

Besansky, N.J., Krzywinski, J., Lehmann, T., *et al.* (2003) Semipermeable species boundaries between *Anopheles gambiae* and *Anopheles arabiensis*: evidence from multilocus DNA sequence variation. *Proceedings of the National Academy of Sciences USA* **100**, 10818–10823.

Bhaya, D., Dufresne, A., Vaulot, D., and Grossman, A. (2002) Analysis of the *hli* gene family in marine and freshwater cyanobacteria. *FEMS Microbiology Letters* **215**, 209–219.

Bímová, B., Karn, R.C., and Piálek, J. (2005) The role of salivary androgen-binding protein in reproductive isolation between two subspecies of house mouse: *Mus musculus musculus* and *Mus musculus domesticus*. *Biological Journal of the Linnean Society* **84**, 349–361.

Blair, W.F. (1955) Mating call and stage of speciation in the *Microhyla olivacea-M. carolinensis* complex. *Evolution* **9**, 469–480.

Bohlen, J. and Ráb, P. (2001) Species and hybrid richness in spined loaches of the genus *Cobitis* (Teleostei: Cobitidae), with a checklist of European forms and suggestions for conservation. *Journal of Fish Biology* **59** (suppl A), 75–89.

Boisselier-Dubayle, M.-C. and Bischler, H. (1999) Genetic relationships between haploid and triploid *Targionia* (Targioniaceae, Hepaticae). *International Journal of Plant Science* **160**, 1163–1169.

Bolnick, D.I. and Near, T.J. (2005) Tempo of hybrid inviability in centrarchid fishes (Teleostei: Centrarchidae). *Evolution* **59**, 1754–1767.

Bonhoeffer, S., Chappey, C., Parkin, N.T., *et al.* (2004) Evidence for positive epistasis in HIV-1. *Science* **306**, 1547–1550.

Bordenstein, S.R. and Reznikoff, W.S. (2005) Mobile DNA in obligate intracellular bacteria. *Nature Reviews Microbiology* **3**, 688–699.

Borge, T., Lindroos, K., Nádvorník, P., *et al.* (2005) Amount of introgression in flycatcher hybrid zones reflects regional differences in pre and post-zygotic barriers to gene exchange. *Journal of Evolutionary Biology* **18**, 1416–1424.

Boucher, Y., Douady, C.J., Papke, R.T., *et al.* (2003) Lateral gene transfer and the origins of prokaryotic groups. *Annual Review of Genetics* **37**, 283–328.

Bouck, A.C. (2004) *The Genetic Architecture of Reproductive Isolation in Louisiana Irises.* PhD thesis, University of Georgia, Athens, GA.

Bouck, A.C., Peeler, R., Arnold, M.L., and Wessler, S.R. (2005) Genetic mapping of species boundaries in Louisiana Irises using *IRRE* retrotransposon display markers. *Genetics* **171**, 1289–1303.

Boursot, P., Auffray, J.-C., Britton-Davidian, J., and Bonhomme, F. (1993) The evolution of house mice. *Annual Review of Ecology and Systematics* **24**, 119–152.

Bowers, J.E., Chapman, B.A., Rong, J., and Paterson, A.H. (2003) Unravelling angiosperm genome evolution by phylogenetic analysis of chromosomal duplication events. *Nature* **422**, 433–438.

Bozíková, E., Munclinger, P., Teeter, K.C., *et al.* (2005) Mitochondrial DNA in the hybrid zone between *Mus musculus musculus* and *Mus musculus domesticus*: a comparison of two transects. *Biological Journal of the Linnean Society* **84**, 363–378.

Bradley, R.D., Bryant, F.C., Bradley, L.C., *et al.* (2003) Implications of hybridization between white-tailed deer and mule deer. *Southwestern Naturalist* **48**, 654–660.

Brasier, C.M. (2001) Rapid evolution of introduced plant pathogens via interspecific hybridization. *BioScience* **51**, 123–133.

Brasier, C.M. and Kirk, S.A. (2001) Designation of the EAN and NAN races of *Ophiostoma novo-ulmi* as subspecies. *Mycological Research* **105**, 547–554.

Brasier, C.M., Cooke, D.E.L., and Duncan, J.M. (1999) Origin of a new *Phytophthora* pathogen through interspecific hybridization. *Proceedings of the National Academy of Sciences USA* **96**, 5878–5883.

Brassinga, A.K.C., Hiltz, M.F., Sisson, G.R., *et al.* (2003) A 65-kilobase pathogenicity island is unique to Philadelphia-1 strains of *Legionella pneumophila*. *Journal of Bacteriology* **185**, 4630–4637.

Breitbart, M., Wegley, L., Leeds, S., *et al.* (2004) Phage community dynamics in hot springs. *Applied and Environmental Microbiology* **70**, 1633–1640.

Brenner, D.J., Steigerwalt, A.G., and McDade, J.E. (1979) Classification of the Legionnaires' disease bacterium: *Legionella pneumophila*, genus novum, species nova, of the family Legionellaceae, familia nova. *Annals of Internal Medicine* **90**, 656–658.

Brisse, S., Dujardin, J.-C., and Tibayrenc, M. (2000) Identification of six *Trypanosoma cruzi* lineages by sequence-characterised amplified region markers. *Molecular and Biochemical Parasitology* **111**, 95–105.

Britch, S.C., Cain, M.L., and Howard, D.J. (2001) Spatio-temporal dynamics of the *Allonemobius fasciatus*—*A. socius* mosaic hybrid zone: a 14-year perspective. *Molecular Ecology* **10**, 627–638.

Britton-Davidian, J., Fel-Clair, F., Lopez, J., *et al.* (2005) Postzygotic isolation between the two European subspecies of the house mouse: estimates from fertility patterns in wild and laboratory-bred hybrids. *Biological Journal of the Linnean Society* **84**, 379–393.

Brochmann, C. (1992) Pollen and seed morphology of Nordic *Draba* (Brassicaceae): phylogenetic and ecological implications. *Nordic Journal of Botany* **12**, 657–673.

Brochmann, C. (1993) Reproductive strategies of diploid and polyploid populations of arctic *Draba* (Brassicaceae). *Plant Systematics and Evolution* **185**, 55–83.

Brochmann, C., Soltis, D.E., and Soltis, P.S. (1992a) Electrophoretic relationships and phylogeny of Nordic polyploids in *Draba* (Brassicaceae). *Plant Systematics and Evolution* **182**, 35–70.

Brochmann, C., Soltis, P.S., and Soltis, D.E. (1992b) Multiple origins of the octoploid Scandinavian endemic *Draba cacuminum*: electrophoretic and morphological evidence. *Nordic Journal of Botany* **12**, 257–272.

Brochmann, C., Soltis, P.S., and Soltis, D.E. (1992c) Recurrent formation and polyphyly of Nordic poly-ploids in *Draba* (Brassicaceae). *American Journal of Botany* **79**, 673–688.

Brochmann, C., Stedje, B., and Borgen, L. (1992d) Gene flow across ploidal levels in *Draba* (Brassicaceae). *Evolutionary Trends in Plants* **6**, 125–134.

Brochmann, C., Borgen, L., and Stedje, B. (1993) Crossing relationships and chromosome numbers of Nordic pop-ulations of *Draba* (Brassicaceae), with emphasis on the *D. alpina* complex. *Nordic Journal of Botany* **13**, 121–147.

Brochmann, C., Brysting, A.K., Alsos, I.G., *et al.* (2004) Polyploidy in arctic plants. *Biological Journal of the Linnean Society* **82**, 521–536.

Brock, M.T. (2004) The potential for genetic assimilation of a native dandelion species, *Taraxacum ceratophorum* (Asteraceae), by the exotic congener *T. officinale*. *American Journal of Botany* **91**, 656–663.

Brock, M.T. and Galen, C. (2005) Drought tolerance in the alpine dandelion, *Taraxacum ceratophorum* (Asteraceae), its exotic congener *T. officinale*, and interspecific hybrids under natural and experimental conditions. *American Journal of Botany* **92**, 1311–1321.

Brock, M.T., Weinig, C., and Galen, C. (2005) A compari-son of phenotypic plasticity in the native dandelion *Taraxacum ceratophorum* and its invasive congener *T. officinale*. *New Phytologist* **166**, 173–183.

Brosch, R., Gordon, S.V., Marmiesse, M., *et al.* (2002) A new evolutionary scenario for the *Mycobacterium tuber-culosis* complex. *Proceedings of the National Academy of Sciences USA* **99**, 3684–3689.

Brown, E.W., Mammel, M.K., LeClerc, J.E., and Cebula, T.A. (2003) Limited boundaries for extensive horizontal gene transfer among *Salmonella* pathogens. *Proceedings of the National Academy of Sciences USA* **100**, 1567–1581.

Brown, W.M. and Wright, J.W. (1979) Mitochondrial DNA analyses and the origin and relative age of partheno-genetic lizards (genus *Cnemidophorus*). *Science* **203**, 1247–1249.

Brüggemann, H., Henne, A., Hoster, F., *et al.* (2004) The complete genome sequence of *Propionibacterium acnes*, a commensal of human skin. *Science* **305**, 671–673.

Brumfield, R.T., Jernigan, R.W., McDonald, D.B., and Braun, M.J. (2001) Evolutionary implications of diver-gent clines in an avian (*Manacus*: Aves) hybrid zone. *Evolution* **55**, 2070–2087.

Brüssow, H., Bruttin, A., Desiere, F., *et al.* (1998) Molecular ecology and evolution of *Streptococcus thermophilus* bacteriophages—a review. *Virus Genes* **16**, 95–109.

Burke, J.M. and Rieseberg, L.H. (2003) Fitness effects of transgenic disease resistance in sunflowers. *Science* **300**, 1250.

Burke, J.M., Carney, S.E., and Arnold, M.L. (1998a) Hybrid fitness in the Louisiana Irises: analysis of parental and F_1 performance. *Evolution* **52**, 37–43.

Burke, J.M., Voss, T.J., and Arnold, M.L. (1998b) Genetic interactions and natural selection in Louisiana Iris hybrids. *Evolution* **52**, 1304–1310.

Burke, J.M., Bulger, M.R., Wesselingh, R.A., and Arnold, M.L. (2000a) Frequency and spatial patterning of clonal reproduction in Louisiana Iris hybrid populations. *Evolution* **54**, 137–144.

Burke, J.M., Wyatt, R., DePamphilis, C.W., and Arnold, M.L. (2000b) Nectar characteristics of interspecific hybrids and their parents in *Aesculus* and *Iris*. *Journal of the Torrey Botanical Society* **127**, 200–206.

Burke, J.M., Gardner, K.A., and Rieseberg, L.H. (2002) The potential for gene flow between cultivated and wild sunflower (*Helianthus annuus*) in the United States. *American Journal of Botany* **89**, 1550–1552.

Burke, J.M., Lai, Z., Salmaso, M., *et al.* (2004) Comparative mapping and rapid karyotypic evolution in the genus *Helianthus*. *Genetics* **167**, 449–457.

Bush, G.L. (1969) Sympatric host race formation and spe-ciation in frugivorous flies of the genus *Rhagoletis* (Diptera, Tephritidae). *Evolution* **23**, 237–251.

Butcher, P.A., Skinner, A.K., and Gardiner, C.A. (2005) Increased inbreeding and inter-species gene flow in remnant populations of the rare *Eucalyptus benthamii*. *Conservation Genetics* **6**, 213–226.

Bynum, E.L. (2002) Morphological variation within a macaque hybrid zone. *American Journal of Physical Anthropology* **118**, 45–49.

Bynum, E.L., Bynum, D.Z., and Supriatna, J. (1997) Confirmation and location of the hybrid zone between wild populations of *Macaca tonkeana* and *Macaca hecki* in central Sulawesi, Indonesia. *American Journal of Primatology* **43**, 181–209.

Byrne, M., MacDonald, B., and Coates, D. (2002) Phylogeographical patterns in chloroplast DNA varia-tion within the *Acacia acuminata* (Leguminosae: Mimosoideae) complex in Western Australia. *Journal of Evolutionary Biology* **15**, 576–587.

Cain, M.L., Andreasen, V., and Howard, D.J. (1999) Reinforcing selection is effective under a relatively broad set of conditions in a mosaic hybrid zone. *Evolution* **53**, 1343–1353.

Campbell, D. and Bernatchez, L. (2004) Generic scan using AFLP markers as a means to assess the role of directional selection in the divergence of sympatric whitefish ecotypes. *Molecular Biology and Evolution* **21**, 945–956.

Campbell, D.R. (1991) Comparing pollen dispersal and gene flow in a natural population. *Evolution* **45**, 1965–1968.

Campbell, D.R. (1996) Evolution of floral traits in a hermaphroditic plant: field measurements of heritabilities and genetic correlations. *Evolution* **50**, 1442–1453.

Campbell, D.R. (1997) Genetic and environmental variation in life-history traits of a monocarpic perennial: a decade-long field experiment. *Evolution* **51**, 373–382.

Campbell, D.R. (1998) Multiple paternity in fruits of *Ipomopsis aggregata* (Polemoniaceae). *American Journal of Botany* **85**, 1022–1027.

Campbell, D.R. (2003) Natural selection in *Ipomopsis* hybrid zones: implications for ecological speciation. *New Phytologist* **161**, 83–90.

Campbell, D.R. and Waser, N.M. (1989) Variation in pollen flow within and among populations of *Ipomopsis aggregata*. *Evolution* **43**, 1444–1455.

Campbell, D.R., and Dooley, J.L. (1992) The spatial scale of genetic differentiation in a hummingbird-pollinated plant: comparison with models of isolation by distance. *American Naturalist* **139**, 735–748.

Campbell, D.R. and Waser, N.M. (2001) Genotype-by-environment interaction and the fitness of plant hybrids in the wild. *Evolution* **55**, 669–676.

Campbell, D.R., Waser, N.M., Price, M.V., et al. (1991) Components of phenotypic selection: pollen export and flower corolla width in *Ipomopsis aggregata*. *Evolution* **45**, 1458–1467.

Campbell, D.R., Waser, N.M., and Price, M.V. (1994) Indirect selection of stigma position in *Ipomopsis aggregata* via a genetically correlated trait. *Evolution* **48**, 55–68.

Campbell, D.R., Waser, N.M., and Meléndez-Ackerman, E.J. (1997) Analyzing pollinator-mediated selection in a plant hybrid zone: hummingbird visitation patterns on three spatial scales. *American Naturalist* **149**, 295–315.

Campbell, D.R., Waser, N.M., and Wolf, P.G. (1998) Pollen transfer by natural hybrids and parental species in an *Ipomopsis* hybrid zone. *Evolution* **52**, 1602–1611.

Campbell, D.R., Crawford, M., Brody, A.K., and Forbis, T.A. (2002a) Resistance to pre-dispersal seed predators in a natural hybrid zone. *Oecologia* **131**, 436–443.

Campbell, D.R., Waser, N.M., and Pederson, G.T. (2002b) Predicting patterns of mating and potential hybridization from pollinator behavior. *American Naturalist* **159**, 438–450.

Campbell, D.R., Alarcón, R., and Wu, C.A. (2003) Reproductive isolation and hybrid pollen disadvantage in *Ipomopsis*. *Journal of Evolutionary Biology* **16**, 536–540.

Campbell, D.R., Galen, C., and Wu, C.A. (2005) Ecophysiology of first and second generation hybrids in a natural plant hybrid zone. *Oecologia* **144**, 214–225.

Caramelli, D., Lalueza-Fox, C., Vernesi, C., et al. (2003) Evidence for a genetic discontinuity between Neandertals and 24,000-year-old anatomically modern Europeans. *Proceedings of the National Academy of Sciences USA* **100**, 6593–6597.

Carleton, K.L., Parry, J.W.L., Bowmaker, J.K., et al. (2005) Colour vision and speciation in Lake Victoria cichlids of the genus *Pundamilia*. *Molecular Ecology* **14**, 4341–4353.

Carney, S.E. and Arnold, M.L. (1997) Differences in pollen-tube growth rate and reproductive isolation between Louisiana Irises. *Journal of Heredity* **88**, 545–549.

Carney, S.E., Cruzan, M.B., and Arnold, M.L. (1994) Reproductive interactions between hybridizing irises: analyses of pollen tube growth and fertilization success. *American Journal of Botany* **81**, 1169–1175.

Carney, S.E., Hodges, S.A., and Arnold, M.L. (1996) Effects of pollen-tube growth and ovule position on hybridization in the Louisiana Irises. *Evolution* **50**, 1871–1878.

Carney, S.E., Gardner, K.A., and Rieseberg, L.H. (2000) Evolutionary changes over the fifty-year history of a hybrid population of sunflowers (*Helianthus*). *Evolution* **54**, 462–474.

Carr, G.D. and Kyhos, D.W. (1981) Adaptive radiation in the Hawaiian Silversword alliance (Compositae-Madiinae) I. Cytogenetics of spontaneous hybrids. *Evolution* **35**, 543–556.

Carr, G.D., Robichaux, R.H., Witter, M.S., and Kyhos, D.W. (1989) Adaptive radiation of the Hawaiian Silversword alliance (Compositae-Madiinae): a comparison with Hawaiian Picture-Winged *Drosophila*. In L.V. Giddings, K.Y. Kaneshiro, and W.W. Anderson, eds. *Genetics, Speciation, and the Founder Principle*, pp. 79–97. Oxford University Press, Oxford.

Carr, S.M., Ballinger, S.W., Derr, J.N., et al. (1986) Mitochondrial DNA analysis of hybridization between sympatric white-tailed deer and mule deer in west Texas. *Proceedings of the National Academy of Sciences USA* **83**, 9576–9580.

Carr, S.M., Brothers, A.J., and Wilson, A.C. (1987) Evolutionary inferences from restriction maps of mitochondrial DNA from nine taxa of *Xenopus* frogs. *Evolution* **41**, 176–188.

Carreno, R.A., Martin, D.S., and Barta, J.R. (1999) *Cryptosporidium* is more closely related to the gregarines than to coccidian as shown by phylogenetic analysis of apicomplexan parasites inferred using small-subunit ribosomal RNA gene sequences. *Parasitology Research* **85**, 899–904.

Casjens, S. (2003) Prophages and bacterial genomics: what have we learned so far? *Molecular Microbiology* **49**, 277–300.

Castro, L.F.C. and Holland, P.W.H. (2003) Chromosomal mapping of ANTP class homeobox genes in amphioxus: piecing together ancestral genomes. *Evolution & Development* **5**, 459–465.

Castro, L.F.C., Furlong, R.F., and Holland, P.W.H. (2004) An antecedent of the MHC-linked genomic region in amphioxus. *Immunogenetics* **55**, 782–784.

Cathey, J.C., Bickham, J.W., and Patton, J.C. (1998) Introgressive hybridization and nonconcordant evolutionary history of maternal and paternal lineages in North American deer. *Evolution* **52**, 1224–1229.

Cavalier-Smith, T. (1991) Archamoebae: the ancestral eukaryotes? *BioSystems* **25**, 25–38.

Chain, P.S.G., Carniel, E., Larimer, F.W., *et al.* (2004) Insights into the evolution of *Yersinia pestis* through whole-genome comparison with *Yersinia pseudotuberculosis*. *Proceedings of the National Academy of Sciences USA* **101**, 13826–13831.

Chang, A.S. (2004) Conspecific sperm precedence in sister species of *Drosophila* with overlapping ranges. *Evolution* **58**, 781–789.

Chapman, M.A. and Burke, J.M. (2006) Letting the gene out of the bottle: the population genetics of GM crops. *New Phytologist*, **170**, 429–443.

Chapman, M.A., Forbes, D.G., and Abbott, R.J. (2005) Pollen competition among two species of *Senecio* (Asteraceae) that form a hybrid zone on Mt. Etna, Sicily. *American Journal of Botany* **92**, 730–735.

Chatti, N., Britton-Davidian, J., Catalan, J., *et al.* (2005) Reproductive trait divergence and hybrid fertility patterns between chromosomal races of the house mouse in Tunisia: analysis of wild and laboratory-bred males and females. *Biological Journal of the Linnean Society* **84**, 407–416.

Chen, F. and Lu, J. (2002) Genomic sequence and evolution of marine cyanophage P60: A new insight on lytic and lysogenic phages. *Applied and Environmental Microbiology* **68**, 2589–2594.

Chen, H., Deng, G., Li, Z., *et al.* (2004) The evolution of H5N1 influenza viruses in ducks in southern China. *Proceedings of the National Academy of Sciences USA* **101**, 10452–10457.

Chen, Z.J. and Pikaard, C.S. (1997) Transcriptional analysis of nucleolar dominance in polyploid plants: biased expression/silencing of progenitor rRNA genes is developmentally regulated in *Brassica*. *Proceedings of the National Academy of Sciences USA* **94**, 3442–3447.

Chiba, S. (2005) Appearance of morphological novelty in a hybrid zone between two species of land snail. *Evolution* **59**, 1712–1720.

Chien, M., Morozova, I., Shi, S., *et al.* (2004) The genomic sequence of the accidental pathogen *Legionella pneumophila*. *Science* **305**, 1966–1968.

Chin, M.P.S., Rhodes, T.D., Chen, J., *et al.* (2005) Identification of a major restriction in HIV-1 intersubtype recombination. *Proceedings of the National Academy of Sciences USA* **102**, 9002–9007.

Cho, Y., Qiu, Y.-L., Kuhlman, P., and Palmer, J.D. (1998) Explosive invasion of plant mitochondria by a group I intron. *Proceedings of the National Academy of Sciences USA* **95**, 14244–14249.

Christiansen, D.G., Fog, K., Pedersen, B.V., and Boomsma, J.J. (2005) Reproduction and hybrid load in all-hybrid populations of *Rana esculenta* water frogs in Denmark. *Evolution* **59**, 1348–1361.

Chu, J., Powers, E., and Howard, D.J. (1995) Gene exchange in a ground cricket hybrid zone. *Journal of Heredity* **86**, 17–21.

Clark, A.G., Begun, D.J., and Prout, T. (1999) Female x male interactions in *Drosophila* sperm competition. *Science* **283**, 217–220.

Clark, A.G., Dermitzakis, E.T., and Civetta, A. (2000) Nontransitivity of sperm precedence in *Drosophila*. *Evolution* **54**, 1030–1035.

Clark, C.G., and Roger, A.J. (1995) Direct evidence for secondary loss of mitochondria in *Entamoeba histolytica*. *Proceedings of the National Academy of Sciences USA* **92**, 6518–6521.

Clarke, K.E., Rinderer, T.E., Franck, P., *et al.* (2002) The africanization of honeybees (*Apis mellifera* L.) of the Yucatan: a study of a massive hybridization event across time. *Evolution* **56**, 1462–1474.

Cock, M.J.W. (1978) The assessment of preference. *Journal of Animal Ecology* **47**, 805–816.

Cohan, F.M. (2002) What are bacterial species? *Annual Review of Microbiology* **56**, 457–487.

Cohuet, A., Dia, I., Simard, F., *et al.* (2005) Gene flow between chromosomal forms of the malaria vector *Anopheles funestus* in Cameroon, Central Africa, and its relevance in malaria fighting. *Genetics* **169**, 301–311.

Colbourne, J.K., Crease, T.J., Weider, L.J., *et al.* (1998) Phylogenetics and evolution of a circumarctic species complex (Cladocera: *Daphnia pulex*). *Biological Journal of the Linnean Society* **65**, 347–365.

Cole, S.T., Brosch, R., Parkhill, J., *et al.* (1998) Deciphering the biology of *Mycobacterium tuberculosis* from the complete genome sequence. *Nature* **393**, 537–544.

Coluzzi, M., Sabatini, A., della Torre, A., *et al.* (2002) A polytene chromosome analysis of the *Anopheles gambiae* species complex. *Science* **298**, 1415–1418.

Comai, L. (2005) The advantages and disadvantages of being polyploid. *Nature Reviews Genetics* **6**, 836–846.

Comes, H.P. and Abbott, R.J. (1999) Population genetic structure and gene flow across arid versus mesic environments: a comparative study of two parapatric *Senecio* species from the near east. *Evolution* **53**, 36–54.

Comes, H.P. and Abbott, R.J. (2001) Molecular phylogeography, reticulation, and lineage sorting in Mediterranean *Senecio* sect *Senecio* (Asteraceae). *Evolution* **55**, 1943–1962.

Comstock, K.E., Georgiadis, N., Pecon-Slattery, J., *et al.* (2002) Patterns of molecular genetic variation among African elephant populations. *Molecular Ecology* **11**, 2489–2498.

Cornman, R.S., Burke, J.M., Wesselingh, R.A., and Arnold, M.L. (2004) Contrasting genetic structure of adults and progeny in a Louisiana Iris hybrid population. *Evolution* **58**, 2669–2681.

Coyne, J.A. and Orr, H.A. (2004) *Speciation*. Sinauer Associates, Sunderland, MA.

Cracraft, J. (1989) Speciation and its ontology: the empirical consequences of alternative species concepts for understanding patterns and processes of differentiation. In D. Otte and J.A. Endler, eds. *Speciation and its Consequences*, pp. 28–59. Sinauer Associates, Sunderland, MA.

Crease, T.J., Stanton, D.J., and Hebert, P.D.N. (1989) Polyphyletic origins of asexuality in *Daphnia pulex*. II. Mitochondrial-DNA variation. *Evolution* **43**, 1016–1026.

Cronquist, A. and Keck, D.D. (1957) A reconstitution of the genus *Machaeranthera*. *Brittonia* **9**, 231–239.

Crow, K.D., Stadler, P.F., Lynch, V.J., *et al.* (2006) The 'fish-specific' Hox cluster duplication is coincident with the origin of teleosts. *Molecular Biology and Evolution* **23**, 121–136.

Cruzan, M.B. and Arnold, M.L. (1993) Ecological and genetic associations in an *Iris* hybrid zone. *Evolution* **47**, 1432–1445.

Cruzan, M.B. and Arnold, M.L. (1994) Assortative mating and natural selection in an *Iris* hybrid zone. *Evolution* **48**, 1946–1958.

Cruzan, M.B., Arnold, M.L., Carney, S.E., and Wollenberg, K.R. (1993) cpDNA inheritance in interspecific crosses and evolutionary inference in Louisiana Irises. *American Journal of Botany* **80**, 344–350.

Cuellar, H.S. (1971) Levels of genetic compatibility of *Rana areolata* with southwestern members of the *Rana pipiens* complex (Anura: Ranidae). *Evolution* **25**, 399–409.

Culver, M., Johnson, W.E., Pecon-Slattery, J., and O'Brien, S.J. (2000) Genomic ancestry of the American Puma (*Puma concolor*). *Journal of Heredity* **91**, 186–197.

Cummings, C.L., Alexander, H.M., Snow, A.A., *et al.* (2002) Fecundity selection in a sunflower crop-wild study: can ecological data predict crop allele changes? *Ecological Applications* **12**, 1661–1671.

Czesak, M.E., Knee, M.J., Gale, R.G., *et al.* (2004) Genetic architecture of resistance to aphids and mites in a willow hybrid system. *Heredity* **93**, 619–626.

Daly, J.C., Wilkinson, P., and Shaw, D.D. (1981) Reproductive isolation in relation to allozymic and chromosomal differentiation in the grasshopper *Caledia captiva*. *Evolution* **35**, 1164–1179.

Dambroski, H.R., Linn, C., Jr, Berlocher, S.H., *et al.* (2005) The genetic basis for fruit odor discrimination in *Rhagoletis* flies and its significance for sympatric host shifts. *Evolution* **59**, 1953–1964.

Dance, D.A.B. (1991) Melioidosis: the tip of the iceberg? *Clinical Microbiology Reviews* **4**, 52–60.

D'Antonio, C.M., Jackson, N.E., Horvitz, C.C., and Hedberg, R. (2004) Invasive plants in wildland ecosystems: merging the study of invasion processes with management needs. *Frontiers in Ecology and the Environment* **2**, 513–521.

Darbyshire, S.J., Cayouette, J., and Warwick, S.I. (1992) The intergeneric hybrid origin of *Poa labradorica* (Poaceae). *Plant Systematics and Evolution* **181**, 57–76.

Darwin, C. (1845) *The Voyage of the Beagle*, 2nd edn. PF Collier & Son, New York.

Darwin, C. (1859) *On the Origin of Species by Means of Natural Selection or the Preservation of Favoured Races in the Struggle for Life*. John Murray, London.

Daubin, V. and Ochman, H. (2004) Bacterial genomes as new gene homes: the genealogy of ORFans in *E. coli*. *Genome Research* **14**, 1036–1042.

David, J., Lemeunier, F., Tsacas, L., and Bocquet, C. (1974) Hybridation d'une nouvelle espèce, *Drosophila mauritiana* avec *D. melanogaster* et *D. simulans*. *Annales de Genetique* **17**, 235–241.

Davis, C.C. and Wurdack, K.J. (2004) Host-to-parasite gene transfer in flowering plants: phylogenetic evidence from Malpighiales. *Science* **305**, 676–678.

Davis, C.C., Anderson, W.R., and Wurdack, K.J. (2005) Gene transfer from a parasitic flowering plant to a fern. *Proceedings of the Royal Society of London Series B* **272**, 2237–2242.

Davis, J.C. and Petrov, D.A. (2005) Do disparate mechanisms of duplication add similar genes to the genome? *Trends in Genetics* **21**, 548–551.

Davis, W.T. (1892) Interesting oaks recently discovered on Staten Island. *Bulletin of the Torrey Botanical Club* **19**, 301–303.

de Barros Lopes, M., Bellon, J.R., Shirley, N.J., and Ganter, P.F. (2002) Evidence for multiple interspecific hybridization in *Saccharomyces sensu stricto*. *FEMS Yeast Research* **1**, 323–331.

De Bodt, S., Maere, S., and Van de Peer, Y. (2005) Genome duplication and the origin of angiosperms. *Trends in Ecology and Evolution* **20**, 591–597.

Debruyne, R. (2005) A case study of apparent conflict between molecular phylogenies: The interrelationships of African elephants. *Cladistics* **21**, 31–50.

de Kort, S.R., den Hartog, P.M., and ten Cate, C. (2002a) Diverge or merge? The effect of sympatric occurrence on the territorial vocalizations of the vinaceous dove *Streptopelia vinacea* and the ring-necked dove *S. capicola*. *Journal of Avian Biology* **33**, 150–158.

de Kort, S.R., den Hartog, P.M., and ten Cate, C. (2002b) Vocal signals, isolation and hybridization in the vinaceous dove (*Streptopelia vinacea*) and the ring-necked dove (*S. capicola*). *Behavioral Ecology and Sociobiology* **51**, 378–385.

della Torre, A., Merzagora, L., Powell, J.R., and Coluzzi, M. (1997) Selective introgression of paracentric inversions between two sibling species of the *Anopheles gambiae* complex. *Genetics* **146**, 239–244.

Delnari, D., Colson, I., Grammenoudi, S., *et al.* (2003) Engineering evolution to study speciation in yeasts. *Nature* **422**, 68–72.

DeMarais, B.D., Dowling, T.E., Douglas, M.E., *et al.* (1992) Origin of *Gila seminuda* (Teleostei: Cyprinidae) through introgressive hybridization: implications for evolution and conservation. *Proceedings of the National Academy of Sciences USA* **89**, 2747–2751.

Dennis, E.S., Peacock, W.J., White, M.J.D., *et al.* (1981) Cytogenetics of the parthenogenetic grasshopper *Warramaba virgo* and its bisexual relatives. VII. Evidence from repeated DNA sequences for a dual origin of *W. virgo*. *Chromosoma* **82**, 453–469.

Densmore, L.D., III, Moritz, C.C., Wright, J.W., and Brown, W.M. (1989a) Mitochondrial-DNA analyses and the origin and relative age of parthenogenetic lizards (genus *Cnemidophorus*) IV. Nine *sexlineatus*-group unisexuals. *Evolution* **43**, 969–983.

Densmore, L.D., III, Wright, J.W., and Brown, W.M. (1989b) Mitochondrial-DNA analyses and the origin and relative age of parthenogenetic lizards (genus *Cnemidophorus*) II. *C. neomexicanus* and the *C. tesselatus* complex. *Evolution* **43**, 943–957.

De Sá, R.O. and Hillis, D.M. (1990) Phylogenetic relationships of the pipid frogs *Xenopus* and *Silurana*: an integration of ribosomal DNA and morphology. *Molecular Biology and Evolution* **7**, 365–376.

Dessauer, H.C. and Cole, C.J. (1989) Diversity between and within nominal forms of unisexual teiid lizards. In R.M. Dawley and J.P. Bogart, eds. *Evolution and Ecology of Unisexual Vertebrates*, pp. 49–71. New York State Museum, New York.

de Souza, F.S.J., Bumaschny, V.F., Low, M.J., and Rubinstein, M. (2005) Subfunctionalization of expression and peptide domains following the ancient duplication of the proopiomelanocortin gene in teleost fishes. *Molecular Biology and Evolution* **22**, 2417–2427.

Detwiler, K.M., Burrell, A.S., and Jolly, C.J. (2005) Conservation implications of hybridization in African cercopithecine monkeys. *International Journal of Primatology* **26**, 661–684.

Diaz, A. and Macnair, M.R. (1999) Pollen tube competition as a mechanism of prezygotic reproductive isolation between *Mimulus nasutus* and its presumed progenitor *M. guttatus*. *New Phytologist* **144**, 471–478.

Dirks, W., Reid, D.J., Jolly, C.J., *et al.* (2002) Out of the mouths of baboons: stress, life history, and dental development in the Awash National Park hybrid zone, Ethiopia. *American Journal of Physical Anthropology* **118**, 239–252.

Dobzhansky, T. (1935) A critique of the species concept in biology. *Philosophy of Science* **2**, 344–355.

Dobzhansky, T. (1937) *Genetics and the Origin of Species*. Columbia University Press, New York.

Dobzhansky, T. (1940) Speciation as a stage in evolutionary divergence. *American Naturalist* **74**, 312–321.

Dobzhansky, T. (1970) *Genetics of the Evolutionary Process*. Columbia University Press, New York.

Dobzhansky, T. (1973) Is there gene exchange between *Drosophila pseudoobscura* and *Drosophila persimilis* in their natural habitats? *American Naturalist* **107**, 312–314.

Dod, B., Jermiin, L.S., Boursot, P., *et al.* (1993) Counterselection on sex chromosomes in the *Mus musculus* European hybrid zone. *Journal of Evolutionary Biology* **6**, 529–546.

Doherty, J.A. and Howard, D.J. (1996) Lack of preference for conspecific calling songs in female crickets. *Animal Behaviour* **51**, 981–990.

Doiron, S., Bernatchez, L., and Blier, P.U. (2002) A comparative mitogenomic analysis of the potential adaptive value of arctic charr mtDNA introgression in brook charr populations (*Salvelinus fontinalis* Mitchill). *Molecular Biology and Evolution* **19**, 1902–1909.

Donnelly, M.J., Pinto, J., Girod, R., *et al.* (2004) Revisiting the role of introgression *vs* shared ancestral polymorphisms as key processes shaping genetic diversity in the recently separated sibling species of the *Anopheles gambiae* complex. *Heredity* **92**, 61–68.

Donoghue, P.C.J. and Purnell, M.A. (2005) Genome duplication, extinction and vertebrate evolution. *Trends in Ecology and Evolution* **20**, 312–319.

Doolittle, R.F., Feng D.-F., Tsang, S., *et al.* (1996) Determining divergence times of the major kingdoms of living organisms with a protein clock. *Science* **271**, 470–477.

Doolittle, W.F. (1999) Phylogenetic classification and the universal tree. *Science* **284**, 2124–2128.

Doolittle, W.F., Boucher, Y., Nesbø, C.L., *et al.* (2003) How big is the iceberg of which organellar genes in nuclear genomes are but the tip? *Philosophical Transactions of the Royal Society of London Series B* **358**, 39–58.

Dorado, O., Rieseberg, L.H., and Arias, D.M. (1992) Chloroplast DNA introgression in Southern California sunflowers. *Evolution* **46**, 566–572.

Dover, G. (1982) Molecular drive: a cohesive mode of species evolution. *Nature* **299**, 111–116.

Dover, G.A. and Tautz, D. (1986) Conservation and divergence in multigene families: alternatives to selection and drift. *Philosophical Transactions of the Royal Society of London Series B* **312**, 275–289.

Dowling, T.E. and DeMarais, B.D. (1993) Evolutionary significance of introgressive hybridization in cyprinid fishes. *Nature* **362**, 444–446.

Doyle, J.J. and Brown, A.H.D. (1989) 5S nuclear ribosomal gene variation in the *Glycine tomentella* polyploid complex (Leguminosae). *Systematic Botany* **14**, 398–407.

Doyle, J.J., Doyle, J.L., Grace, J.P., and Brown, A.H.D. (1990) Reproductively isolated polyploid races of *Glycine tabacina* (Leguminosae) had different chloroplast genome donors. *Systematic Botany* **15**, 173–181.

Doyle, J.J., Doyle, J.L., Brown, A.H.D., and Pfeil, B.E. (2000) Confirmation of shared and divergent genomes in the *Glycine tabacina* polyploid complex (Leguminosae) using histone H3-D sequences. *Systematic Botany* **25**, 437–448.

Doyle, J.J., Doyle, J.L., Brown, A.H.D., and Palmer, R.G. (2002) Genomes, multiple origins, and lineage recombination in the *Glycine tomentella* (Leguminosae) polyploid complex: Histone H3-D gene sequences. *Evolution* **56**, 1388–1402.

Doyle, J.J., Doyle, J.L., Rauscher, J.T., and Brown, A.H.D. (2003) Diploid and polyploid reticulate evolution throughout the history of the perennial soybeans (*Glycine* subgenus *Glycine*). *New Phytologist* **161**, 121–132.

Doyle, J.J., Doyle, J.L., Rauscher, J.T., and Brown, A.H.D. (2004) Evolution of the perennial soybean polyploid

complex (*Glycine* subgenus *Glycine*): a study of contrasts. *Biological Journal of the Linnean Society* **82**, 583–597.

Doyle, M.J. and Brown, A.H.D. (1985) Numerical analysis of isozyme variation in *Glycine tomentella*. *Biochemical Systematics and Ecology* **13**, 413–419.

Drummond, D.A., Bloom, J.D., Adami, C., *et al.* (2005) Why highly expressed proteins evolve slowly. *Proceedings of the National Academy of Sciences USA* **102**, 14338–14343.

Duarte, J.M., Cui, L., Wall, P.K., *et al.* (2006) Expression pattern shifts following duplication indicative of subfunctionalization and neofunctionalization in regulatory genes of *Arabidopsis*. *Molecular Biology and Evolution* **23**, 469–478.

Dufresne, A., Salanoubat, M., Partensky, F., *et al.* (2003) Genome sequence of the cyanobacterium *Prochlorococcus marinus* SS120, a nearly minimal oxyphototrophic genome. *Proceedings of the National Academy of Sciences USA* **100**, 10020–10025.

Dufresne, F. and Hebert, P.D.N. (1994) Hybridization and origins of polyploidy. *Proceedings of the Royal Society of London Series B* **258**, 141–146.

Dufresne, F. and Hebert, P.D.N. (1997) Pleistocene glaciations and polyphyletic origins of polyploidy in an arctic cladoceran. *Proceedings of the Royal Society of London Series B* **264**, 201–206.

Dumolin-Lapègue, S., Demesure, B., Fineschi, S., *et al.* (1997) Phylogeographic structure of white oaks throughout the European continent. *Genetics* **146**, 1475–1487.

Dumolin-Lapègue, S., Pemonge, M.-H., and Petit, R.J. (1998) Association between chloroplast and mitochondrial lineages in oaks. *Molecular Biology and Evolution* **15**, 1321–1331.

Dumolin-Lapègue, S., Kremer, A., and Petit, R.J. (1999) Are chloroplast and mitochondrial DNA variation species independent in oaks? *Evolution* **53**, 1406–1413.

Dungey, H.S., Potts, B.M., Whitham, T.G., and Li, H.-F. (2000) Plant genetics affects arthropod community richness and composition: evidence from a synthetic eucalypt hybrid population. *Evolution* **54**, 1938–1946.

Dvorak, J. and Akhunov, E.D. (2005) Tempos of gene locus deletions and duplications and their relationship to recombination rate during diploid and polyploid evolution in the *Aegilops-Triticum* alliance. *Genetics* **171**, 323–332.

Eckenwalder, J.E. (1984a) Natural intersectional hybridization between North American species of *Populus* (Salicaceae) in sections *Aigeiros* and *Tacamahaca*. I. Population studies of *P. x parryi*. *Canadian Journal of Botany* **62**, 317–324.

Eckenwalder, J.E. (1984b) Natural intersectional hybridization between North American species of

Populus (Salicaceae) in sections *Aigeiros* and *Tacamahaca*. II. Taxonomy. *Canadian Journal of Botany* **62**, 325–335.

Eckenwalder, J.E. (1984c) Natural intersectional hybridization between North American species of *Populus* (Salicaceae) in sections *Aigeiros* and *Tacamahaca*. III. Paleobotany and evolution. *Canadian Journal of Botany* **62**, 336–342.

Eggert, L.S., Rasner, C.A., and Woodruff, D.S. (2002) The evolution and phylogeography of the African elephant inferred from mitochondrial DNA sequence and nuclear microsatellite markers. *Proceedings of the Royal Society of London Series B* **269**, 1993–2006.

Ellstrand, N.C. (2003) *Dangerous Liaisons? When Cultivated Plants Mate With Their Wild Relatives*. Johns Hopkins University Press, Baltimore.

Ellstrand, N.C. and Hoffman, C.A. (1990) Hybridization as an avenue of escape for engineered genes—strategies for risk reduction. *BioScience* **40**, 438–442.

Ellstrand, N.C. and Schierenbeck, K.A. (2000) Hybridization as a stimulus for the evolution of invasiveness in plants? *Proceedings of the National Academy of Sciences USA* **97**, 7043–7050.

El-Sayed, N.M., Myler, P.J., Blandin, G., *et al.* (2005) Comparative genomics of trypanosomatid parasitic protozoa. *Science* **309**, 404–409.

Emms, S.K. and Arnold, M.L. (1997) The effect of habitat on parental and hybrid fitness: reciprocal transplant experiments with Louisiana Irises. *Evolution* **51**, 1112–1119.

Emms, S.K. and Arnold, M.L. (2000) Site-to-site differences in pollinator visitation patterns in a Louisiana Iris hybrid zone. *Oikos* **91**, 568–578.

Emms, S.K., Hodges, S.A., and Arnold, M.L. (1996) Pollen-tube competition, siring success, and consistent asymmetric hybridization in Louisiana Irises. *Evolution* **50**, 2201–2206.

Evans, B.J., Morales, J.C., Picker, M.D., *et al.* (1997) Comparative molecular phylogeography of two *Xenopus* species, *X. gilli* and *X. laevis*, in the southwestern Cape Province, South Africa. *Molecular Ecology* **6**, 333–343.

Evans, B.J., Morales, J.C., Supriatna, J., and Melnick, D.J. (1999) Origin of the Sulawesi macaques (Cercopithecidae: *Macaca*) as suggested by mitochondrial DNA phylogeny. *Biological Journal of the Linnean Society* **66**, 539–560.

Evans, B.J., Supriatna, J., and Melnick, D.J. (2001) Hybridization and population genetics of two macaque species in Sulawesi, Indonesia. *Evolution* **55**, 1686–1702.

Evans, B.J., Supriatna, J., Andayani, N., and Melnick, D.J. (2003) Diversification of Sulawesi macaque monkeys: decoupled evolution of mitochondrial and autosomal DNA. *Evolution* **57**, 1931–1946.

Evans, B.J., Kelley, D.B., Tinsley, R.C., *et al.* (2004) A mitochondrial DNA phylogeny of African clawed frogs: phylogeography and implications for polyploid evolution. *Molecular Phylogenetics and Evolution* **33**, 197–213.

Evans, B.J., Kelley, D.B., Melnick, D.J., and Cannatella, D.C. (2005) Evolution of RAG-1 in polyploid clawed frogs. *Molecular Biology and Evolution* **22**, 1193–1207.

Farrar, M.D., Ingham, E., and Holland, K.T. (2000) Heat shock proteins and inflammatory acne vulgaris: molecular cloning, overexpression and purification of a *Propionibacterium acnes* GroEL and DnaK homologue. *FEMS Microbiology Letters* **191**, 183–186.

Feder, J.L., Stolz, U., Lewis, K.M., *et al.* (1997) The effects of winter length on the genetics of apple and hawthorn races of *Rhagoletis pomonella* (Diptera: Tephritidae). *Evolution* **51**, 1862–1876.

Feder, J.L., Berlocher, S.H., Roethele, J.B., *et al.* (2003) Allopatric genetic origins for sympatric host-plant shifts and race formation in *Rhagoletis*. *Proceedings of the National Academy of Sciences USA* **100**, 10314–10319.

Feil, E.J., Cooper, J.E., Grundmann, H., *et al.* (2003) How clonal is *Staphylococcus aureus*? *Journal of Bacteriology* **185**, 3307–3316.

Feil, E.J., Enright, M.C., and Spratt, B.G. (2000) Estimating the relative contributions of mutation and recombination to clonal diversification: a comparison between *Neisseria meningitidis* and *Streptococcus pneumonidae*. *Research in Microbiology* **151**, 465–469.

Ferguson, D. and Sang, T. (2001) Speciation through homoploid hybridization between allotetraploids in peonies (*Paeonia*). *Proceedings of the National Academy of Sciences USA* **98**, 3915–3919.

Ferris, C., Oliver, R.P., Davy, A.J., and Hewitt, G.M. (1993) Native oak chloroplasts reveal an ancient divide across Europe. *Molecular Ecology* **2**, 337–344.

Ferris, C., King, R.A., and Gray, A.J. (1997) Molecular evidence for the maternal parentage in the hybrid origin of *Spartina anglica* CE Hubbard. *Molecular Ecology* **6**, 185–187.

Feschotte, C. and Wessler, S.R. (2002) *Mariner*-like transposases are widespread and diverse in flowering plants. *Proceedings of the National Academy of Sciences USA* **99**, 280–285.

Feschotte, C., Jiang, N., and Wessler, S.R. (2002) Plant transposable elements: where genetics meets genomics. *Nature Reviews Genetics* **3**, 329–341.

Fields, B.S., Benson, R.F., and Besser, R.E. (2002) *Legionella* and Legionnaires' disease: 25 years of investigation. *Clinical Microbiology Reviews* **15**, 506–526.

Figueroa-Bossi, N. and Bossi, L. (1999) Inducible prophages contribute to *Salmonella* virulence in mice. *Molecular Microbiology* **33**, 167–176.

Filchak, K.E., Feder, J.L., Roethele, J.B., and Stolz, U. (1999) A field test for host-plant dependent selection on larvae of the apple maggot fly, *Rhagoletis pomonella*. *Evolution* **53**, 187–200.

Filchak, K.E., Roethele, J.B., and Feder, J.L. (2000) Natural selection and sympatric divergence in the apple maggot *Rhagoletis pomonella*. *Nature* **407**, 739–742.

Filée, J., Forterre, P., and Laurent, J. (2003) The role played by viruses in the evolution of their hosts: A view based on informational protein phylogenies. *Research in Microbiology* **154**, 237–243.

Filée, J., Tétart, F., Suttle, C.A., and Krisch, H.M. (2005) Marine T4-type bacteriophages, a ubiquitous component of the dark matter of the biosphere. *Proceedings of the National Academy of Sciences USA* **102**, 12471–12476.

Finlayson, C. (2005) Biogeography and evolution of the genus *Homo*. *Trends in Ecology and Evolution* **20**, 457–463.

Fischer, G., James, S.A., Roberts, I.N., *et al.* (2000) Chromosomal evolution in *Saccharomyces*. *Nature* **405**, 451–454.

Fischer, W.J., Koch, W.A., and Elepfandt, A. (2000) Sympatry and hybridization between the clawed frogs *Xenopus laevis laevis* and *Xenopus muelleri* (Pipidae). *Journal of Zoology* **252**, 99–107.

Fitzpatrick, B.M. (2004) Rates of evolution of hybrid inviability in birds and mammals. *Evolution* **58**, 1865–1870.

Floate, K.D. (2004) Extent and patterns of hybridization among the three species of *Populus* that constitute the riparian forest of southern Alberta, Canada. *Canadian Journal of Botany* **82**, 253–264.

Floate, K.D. and Whitham, T.G. (1995) Insects as traits in plant systematics: their use in discriminating between hybrid cottonwoods. *Canadian Journal of Botany* **73**, 1–13.

Floate, K.D., Kearsley, M.J.C., and Whitham, T.G. (1993) Elevated herbivory in plant hybrid zones: *Chrysomela confluens*, *Populus* and phenological sinks. *Ecology* **74**, 2056–2065.

Fonseca, D.M., Keyghobadi, N., Malcolm, C.A., *et al.* (2004) Emerging vectors in the *Culex pipiens* complex. *Science* **303**, 1535–1538.

Force, A., Lynch, M., Pickett, F.B., *et al.* (1999) Preservation of duplicate genes by complementary, degenerative mutations. *Genetics* **151**, 1531–1545.

Forterre, P. (1999) Displacement of cellular proteins by functional analogues from plasmids or viruses could explain puzzling phylogenies of many DNA informational proteins. *Molecular Microbiology* **33**, 457–465.

Fosberg, F.R. (1960) Introgression in *Artocarpus* (Moraceae) in Micronesia. *Brittonia* **12**, 101–113.

Foster, R.C. (1937) A cyto-taxonomic survey of the North American species of *Iris*. *Contributions from the Gray Herbarium*, no. CXIX, pp. 3–80.

Franck, P., Garnery, L., Solignac, M., and Cornuet, J.-M. (1998) The origin of West European subspecies of honeybees *Apis mellifera*: new insights from microsatellite and mitochondrial data. *Evolution* **52**, 1119–1134.

Franck, P., Garnery, L., Celebrano, G., *et al.* (2000) Hybrid origins of honeybees from Italy (*Apis mellifera ligustica*) and Sicily (*A. m. sicula*). *Molecular Ecology* **9**, 907–921.

Fraser, D.W., Tsai, T.R., Orenstein, W., *et al.* (1977) Legionnaires' disease: description of an epidemic of pneumonia. *New England Journal of Medicine* **297**, 1189–1197.

Freeland, J.R. and Boag, P.T. (1999) The mitochondrial and nuclear genetic homogeneity of the phenotypically diverse Darwin's ground finches. *Evolution* **53**, 1553–1563.

Freeman, D.C., Turner, W.A., McArthur, E.D., and Graham, J.H. (1991) Characterization of a narrow hybrid zone between two subspecies of Big Sagebrush (*Artemisia tridentata*: Asteraceae). *American Journal of Botany* **78**, 805–815.

Freeman, D.C., Graham, J.H., Byrd, D.W., *et al.* (1995) Narrow hybrid zone between two subspecies of Big Sagebrush, *Artemisia tridentata* (Asteraceae). *American Journal of Botany* **82**, 1144–1152.

Freeman, D.C., Wang, H., Sanderson, S., and McArthur, E.D. (1999) Characterization of a narrow hybrid zone between two subspecies of Big Sagebrush (*Artemisia tridentata*, Asteraceae): VII. Community and demographic analyses. *Evolutionary Ecology Research* **1**, 487–502.

Freeman, J.S., Jackson, H.D., Steane, D.A., *et al.* (2001) Chloroplast DNA phylogeography of *Eucalyptus globulus*. *Australian Journal of Botany* **49**, 585–596.

Frieman, M., Chen, Z.J., Saez-Vasquez, J., *et al.* (1999) RNA polymerase I transcription in a *Brassica* interspecific hybrid and its progenitors: tests of transcription factor involvement in nucleolar dominance. *Genetics* **152**, 451–460.

Fritz, R.S. (1999) Resistance of hybrid plants to herbivores: genes, environment, or both? *Ecology* **80**, 382–391.

Fritz, R.S., Nichols-Orians, C.M., and Brunsfeld, S.J. (1994) Interspecific hybridization of plants and resistance to herbivores: hypotheses, genetics, and variable responses in a diverse herbivore community. *Oecologia* **97**, 97–106.

Fritz, R.S., Moulia, C., and Newcombe, G. (1999) Resistance of hybrid plants and animals to herbivores, pathogens, and parasites. *Annual Review of Ecology and Systematics* **30**, 565–591.

Fritz, R.S., Hochwender, C.G., Brunsfeld, S.J., and Roche, B.M. (2003) Genetic architecture of susceptibility to herbivores in hybrid willows. *Journal of Evolutionary Biology* **16**, 1115–1126.

Frost, J.S. and Bagnara, J.T. (1976) A new species of leopard frog (*Rana pipiens* complex) from northwestern Mexico. *Copeia* **1976**, 332–338.

Frost, J.S. and Platz, J.E. (1983) Comparative assessment of modes of reproductive isolation among four species of leopard frogs (*Rana pipiens* complex). *Evolution* **37**, 66–78.

Frost, L.S., Ippen-Ihler, K., and Skurray, R.A. (1994) Analysis of the sequence and gene products of the transfer region of the F sex factor. *Microbiological Reviews* **58**, 162–210.

Frost, L.S., Leplae, R., Summers, A.O., and Toussaint, A. (2005) Mobile genetic elements: the agents of open source evolution. *Nature Reviews Microbiology* **3**, 722–732.

Fulnecek, J., Lim, K.Y., Leitch, A.R., *et al.* (2002) Evolution and structure of 5S rDNA loci in allotetraploid *Nicotiana tabacum* and its putative parental species. *Heredity* **88**, 19–25.

Furlong, R.F. and Holland, P.W.H. (2002) Were vertebrates octoploid? *Philosophical Transactions of the Royal Society of London* **357**, 531–544.

Furlong, R.F. and Holland, P.W.H. (2004) Polyploidy in vertebrate ancestry: Ohno and beyond. *Biological Journal of the Linnean Society* **82**, 425–430.

Futuyma, D.J. (1998) *Evolutionary Biology*, 3rd edn. Sinauer Associates, Sunderland, MA.

Gallagher, K.G., Schierenbeck, K.A., and D'Antonio, C.M. (1997) Hybridization and introgression in *Carpobrotus* spp. (Aizoaceae) in California II. Allozyme evidence. *American Journal of Botany* **84**, 905–911.

Gallardo, M.H. and Kirsch, J.A.W. (2001) Molecular relationships among Octodontidae (Mammalia: Rodentia: Caviomorpha). *Journal of Mammalian Evolution* **8**, 73–89.

Gallardo, M.H., Bickham, J.W., Honeycutt, R.L., *et al.* (1999) Discovery of tetraploidy in a mammal. *Nature* **401**, 341.

Gallardo, M.H., Bickham, J.W., Kausel, G., *et al.* (2003) Gradual and quantum genome size shifts in the hystricognath rodents. *Journal of Evolutionary Biology* **16**, 163–169.

Gallardo, M.H., Kausel, G., Jiménez, A., *et al.* (2004) Whole-genome duplications in South American desert rodents (Octodontidae). *Biological Journal of the Linnean Society* **82**, 443–451.

Gamieldien, J., Ptitsyn, A., and Hide, W. (2002) Eukaryotic genes in *Mycobacterium tuberculosis* could have a role in pathogenesis and immunomodulation. *Trends in Genetics* **18**, 5–8.

Gao, F., Bailes, E., Robertson, D.L., *et al.* (1999) Origin of HIV-1 in the chimpanzee *Pan troglodytes troglodytes*. *Nature* **397**, 436–441.

Garrigan, D., Mobasher, Z., Severson, T., *et al.* (2005) Evidence for archaic Asian ancestry on the human X chromosome. *Molecular Biology and Evolution* **22**, 189–192.

Gaunt, M.W., Yeo, M., Frame, I.A., *et al.* (2003) Mechanism of genetic exchange in American trypanosomes. *Nature* **421**, 936–939.

Ge, F., Wang, L.-S., and Kim, J. (2005) The cobweb of life revealed by genome-scale estimates of horizontal gene transfer. *PLoS Biology* **3**, e316.

Gentile, G., della Torre, A., Maegga, B., *et al.* (2002) Genetic differentiation in the African malaria vector, *Anopheles gambiae* s.s., and the problem of taxonomic status. *Genetics* **161**, 1561–1578.

Gerstel, D.U. and Sisson, V.A. (1995) Tobacco—*Nicotiana tabacum* (Solanaceae). In J. Smartt and N.W. Simmonds, eds. *Evolution of Crop Plants*, 2nd edn, pp. 458–463. Longman Scientific & Technical, Harlow.

Gevers, D., Cohan, F.M., Lawrence, J.G., *et al.* (2005) Re-evaluating prokaryotic species. *Nature Reviews Microbiology* **3**, 733–739.

Geyer, L.B. and Palumbi, S.R. (2005) Conspecific sperm precedence in two species of tropical sea urchins. *Evolution* **59**, 97–105.

Ghedin, E., Sengamalay, N.A., Shumway, M., *et al.* (2005) Large-scale sequencing of human influenza reveals the dynamic nature of viral genome evolution. *Nature* **437**, 1162–1166.

Gibbs, M.J., Armstrong, J.S., and Gibbs, A.J. (2001) Recombination in the hemagglutinin gene of the 1918 'Spanish Flu'. *Science* **293**, 1842–1845.

Gibbs, M.J., Armstrong, J.S., and Gibbs, A.J. (2002) Response to Worobey, *et al. Science* **296**, 211a.

Gillett, G.W. (1966) Hybridization and its taxonomic implications in the *Scaevola gaudichaudiana* complex of the Hawaiian Islands. *Evolution* **20**, 506–516.

Glémet, H., Blier, P., and Bernatchez, L. (1998) Geographical extent of arctic char (*Salvelinus alpinus*) mtDNA introgression in brook char populations (*S. fontinalis*) from eastern Québec, Canada. *Molecular Ecology* **7**, 1655–1662.

Gogarten, J.P., and Townsend, J.P. (2005) Horizontal gene transfer, genome innovation, and evolution. *Nature Reviews Microbiology* **3**, 679–687.

Gogarten, J.P., Doolittle, W.F., and Lawrence, J.G. (2002) Prokaryotic evolution in light of gene transfer. *Molecular Biology and Evolution* **19**, 2226–2238.

Gollmann, G., Roth, P., and Hodl, W. (1988) Hybridization between the fire-bellied toads *Bombina bombina* and *Bombina variegata* in the Karst regions of Slovakia and Hungary: morphological and allozyme evidence. *Journal of Evolutionary Biology* **1**, 3–14.

Gómez, A., Carvalho, G.R., and Lunt, D.H. (2000) Phylogeography and regional endemism of a passively dispersing zooplankter: Mitochondrial DNA variation in rotifer resting egg banks. *Proceedings of the Royal Society of London Series B* **267**, 2189–2197.

Goodfriend, G.A. and Gould, S.J. (1996) Paleontology and chronology of two evolutionary transitions by hybridization in the Bahamian land snail *Cerion*. *Science* **274**, 1894–1897.

Goremykin, V.V., Hirsch-Ernst, K.I., Wölfl, S., and Hellwig, F.H. (2003) Analysis of the *Amborella trichopoda* chloroplast genome sequence suggests that *Amborella* is not a basal angiosperm. *Molecular Biology and Evolution* **20**, 1499–1505.

Götherström, A., Anderung, C., Hellborg, L., *et al.* (2005) Cattle domestication in the Near East was followed by hybridization with aurochs bulls in Europe. *Proceedings of the Royal Society of London Series B* **272**, 2345–2350.

Graham, J.H., Freeman, D.C., and McArthur, E.D. (1995) Narrow hybrid zone between two subspecies of Big Sagebrush (*Artemisia tridentata*: Asteraceae) II. Selection gradients and hybrid fitness. *American Journal of Botany* **82**, 709–716.

Grant, B.R. and Grant, P.R. (1993) Evolution of Darwin's finches caused by a rare climatic event. *Proceedings of the Royal Society of London Series B* **251**, 111–117.

Grant, B.R. and Grant, P.R. (1996) High survival of Darwin's Finch hybrids: effects of beak morphology and diets. *Ecology* **77**, 500–509.

Grant, P.R. (1993) Hybridization of Darwin's finches on Isla Daphne Major, Galápagos. *Philosophical Transactions of the Royal Society of London Series B.* **340**, 127–139.

Grant, P.R. and Grant, B.R. (1992) Hybridization of bird species. *Science* **256**, 193–197.

Grant, P.R. and Grant, B.R. (2002) Unpredictable evolution in a 30-year study of Darwin's finches. *Science* **296**, 707–711.

Grant, P.R., Grant, B.R., Keller, L.F., *et al.* (2003) Inbreeding and interbreeding in Darwin's finches. *Evolution* **57**, 2911–2916.

Grant, P.R., Grant, B.R., Markert, J.A., *et al.* (2004) Convergent evolution of Darwin's finches caused by introgressive hybridization and selection. *Evolution* **58**, 1588–1599.

Grant, P.R., Grant, B.R., and Petren, K. (2005) Hybridization in the recent past. *American Naturalist* **166**, 56–67.

Grant, V. (1981) *Plant Speciation*. Columbia University Press, New York.

Grant, V. and Grant, K.A. (1971) Dynamics of clonal microspecies in cholla cactus. *Evolution* **25**, 144–155.

Gregory, P.G. and Howard, D.J. (1994) A postinsemination barrier to fertilization isolates two closely related ground crickets. *Evolution* **48**, 705–710.

Greig, D., Borts, R.H., Louis, E.J., and Travisano, M. (2001) Epistasis and hybrid sterility in *Saccharomyces*. *Proceedings of the Royal Society of London Series B* **269**, 1167–1171.

Greig, D., Louis, E.J., Borts, R.H., and Travisano, M. (2002) Hybrid speciation in experimental populations of yeast. *Science* **298**, 1773–1775.

Griffin, A.R., Burgess, I.P., and Wolf, L. (1988) Patterns of natural and manipulated hybridisation in the genus *Eucalyptus* L'Hérit—A review. *Australian Journal of Botany* **36**, 41–66.

Griffiths, H.I. and Butlin, R.K. (1995) A timescale for sex versus parthenogenesis: evidence from subfossil ostracods. *Proceedings of the Royal Society of London Series B* **260**, 65–71.

Groeters, F.R. and Shaw, D.D. (1992) Association between latitudinal variation for embryonic development time and chromosome structure in the grasshopper *Caledia captiva* (Orthoptera: Acrididae). *Evolution* **46**, 245–257.

Gross, B.L., Kane, N.C., Lexer, C., *et al.* (2004) Reconstructing the origin of *Helianthus deserticola*: survival and selection on the desert floor. *American Naturalist* **164**, 145–156.

Groth, C., Hansen, J., and Piskur, J. (1999) A natural chimeric yeast containing genetic material from three species. *International Journal of Systematic Bacteriology* **49**, 1933–1938.

Gu, X., Wang, Y., and Gu, J. (2002) Age distribution of human gene families shows significant roles of both large- and small-scale duplications in vertebrate evolution. *Nature Genetics* **31**, 205–209.

Gu, X., Zhang, Z., and Huang, W. (2005) Rapid evolution of expression and regulatory divergences after yeast gene duplication. *Proceedings of the National Academy of Sciences USA* **102**, 707–712.

Guan, Y., Shortridge, K.F., Krauss, S., and Webster, R.G. (1999) Molecular characterization of H9N2 influenza viruses: were they the donors of the 'internal' genes of H5N1 viruses in Hong Kong? *Proceedings of the National Academy of Sciences USA* **96**, 9363–9367.

Guan, Y., Peiris, J.S.M., Lipatov, A.S., *et al.* (2002) Emergence of multiple genotypes of H5N1 avian influenza viruses in Hong Kong SAR. *Proceedings of the National Academy of Sciences USA* **99**, 8950–8955.

Haavie, J., Borge, T., Bures, S., *et al.* (2004) Flycatcher song in allopatry and sympatry—convergence, divergence and reinforcement. *Journal of Evolutionary Biology* **17**, 227–237.

Haesler, M.P. and Seehausen, O. (2005) Inheritance of female mating preference in a sympatric sibling species pair of

Lake Victoria cichlids: implications for speciation. *Proceedings of the Royal Society Series B* **272**, 237–245.

Haig, S.M., Mullins, T.D., and Forsman, E.D. (2004a) Subspecific relationships and genetic structure in the spotted owl. *Conservation Genetics* **5**, 683–705.

Haig, S.M., Mullins, T.D., Forsman, E.D., *et al.* (2004b) Genetic identification of spotted owls, barred owls, and their hybrids: legal implications of hybrid identity. *Conservation Biology* **18**, 1347–1357.

Hails, R.S. and Morley, K. (2005) Genes invading new populations: a risk assessment perspective. *Trends in Ecology and Evolution* **20**, 245–252.

Haldane, J.B.S. (1922) Sex ratio and unisexual sterility in hybrid animals. *Journal of Genetics* **12**, 101–109.

Haldane, J.B.S. (1990) *The Causes of Evolution*. Princeton University Press, Princeton, NJ.

Hall, S.E., Luo, S., Hall, A.E., and Preuss, D. (2005) Differential rates of local and global homogenization in centromere satellites from *Arabidopsis* relatives. *Genetics* **170**, 1913–1927.

Hamzeh, M. and Dayanandan, S. (2004) Phylogeny of *Populus* (Salicaceae) based on nucleotide sequences of chloroplast *trnT-trnF* region and nuclear rDNA. *American Journal of Botany* **91**, 1398–1408.

Han, F., Fedak, G., Guo, W., and Liu, B. (2005) Rapid and repeatable elimination of a parental genome-specific DNA repeat (pGc1R-1a) in newly synthesized wheat allopolyploids. *Genetics* **170**, 1239–1245.

Hansen, J. and Kielland-Brandt, M.C. (1994) *Saccharomyces carlsbergensis* contains two functional *MET2* alleles similar to homologues from *S. cerevisiae* and *S. monacensis*. *Gene* **140**, 33–40.

Hapke, A., Zinner, D., and Zischler, H. (2001) Mitochondrial DNA variation in Eritrean hamadryas baboons (*Papio hamadryas hamadryas*): life history influences population genetic structure. *Behavioral Ecology and Sociobiology* **50**, 483–492.

Hardig, T.M., Brunsfeld, S.J., Fritz, R.S., *et al.* (2000) Morphological and molecular evidence for hybridization and introgression in a willow (*Salix*) hybrid zone. *Molecular Ecology* **9**, 9–24.

Harris, E.E. and Disotell, T.R. (1998) Nuclear gene trees and the phylogenetic relationships of the mangabeys (Primates: Papionini). *Molecular Biology and Evolution* **15**, 892–900.

Harrison, R.G. (1986) Pattern and process in a narrow hybrid zone. *Heredity* **56**, 337–349.

Harrison, R.G. (1990) Hybrid zones: windows on evolutionary process. *Oxford Surveys in Evolutionary Biology* **7**, 69–128.

Hartwell, L.H., Hood, L., Goldberg, M.L., *et al.* (2004) *Genetics: From Genes to Genomes*, 2nd edn. McGraw-Hill, Boston.

Haufler, C.H., Soltis, D.E., and Soltis, P.S. (1995a) Phylogeny of the *Polypodium vulgare* complex: insights from chloroplast DNA restriction site data. *Systematic Botany* **20**, 110–119.

Haufler, C.H., Windham, M.D., and Rabe, E.W. (1995b) Reticulate evolution in the *Polypodium vulgare* complex. *Systematic Botany* **20**, 89–109.

Hawks, J.D. and Wolpoff, M.H. (2001) The accretion model of Neandertal evolution. *Evolution* **55**, 1474–1485.

Hayasaka, K., Fujii, K., and Horai, S. (1996) Molecular phylogeny of macaques: implications of nucleotide sequences from an 896-base pair region of mitochondrial DNA. *Molecular Biology and Evolution* **13**, 1044–1053.

Hayes, C.G. (2001) West Nile virus: Uganda, 1937, to New York City, 1999. *Annals of the New York Academy of Sciences* **951**, 25–37.

Hebert, P.D.N., and Finston, T.L. (2001) Macrogeographic patterns of breeding system diversity in the *Daphnia pulex* group from the United States and Mexico. *Heredity* **87**, 153–161.

Hedrén, M. (2003) Plastid DNA variation in the *Dactylorhiza incarnata/maculata* polyploid complex and the origin of allotetraploid *D. sphagnicola* (Orchidaceae). *Molecular Ecology* **12**, 2669–2680.

Hegde, S.G., and Waines, J.G. (2004) Hybridization and introgression between bread wheat and wild and weedy relatives in North America. *Crop Science* **44**, 1145–1155.

Heiser, C.B., Jr (1947) Hybridization between the sunflower species *Helianthus annuus* and *H. petiolaris*. *Evolution* **1**, 249–262.

Heiser, C.B., Jr (1951a) Hybridization in the annual sunflowers: *Helianthus annuus* X *H. argophyllus*. *American Naturalist* **85**, 65–72.

Heiser, C.B., Jr (1951b) Hybridization in the annual sunflowers: *Helianthus annuus* X *H. debilis* var *cucumerifolius*. *Evolution* **5**, 42–51.

Heiser, C.B., Jr (1954) Variation and subspeciation in the common sunflower, *Helianthus annuus*. *American Midland Naturalist* **51**, 287–305.

Heiser, C.B., Jr (1958) Three new annual sunflowers (*Helianthus*) from the southwestern United States. *Rhodora* **60**, 272–283.

Heiser, C.B., Jr (1965) Species crosses in *Helianthus*: delimitation of 'Sections'. *Annals of the Missouri Botanical Garden* **52**, 364–370.

Heiser, C.B., Jr (1979) Origins of some cultivated New World plants. *Annual Review of Ecology and Systematics* **10**, 309–326.

Heiser, C.B., Jr, Smith, D.M., Clevenger, S.B., and Martin, W.C., Jr (1969) The North American sunflowers (*Helianthus*). *Memoirs of the Torrey Botanical Club* **22**, 1–213.

Hellriegel, B., and Reyer H.-U. (2000) Factors influencing the composition of mixed populations of a hemiclonal hybrid and its sexual host. *Journal of Evolutionary Biology* **13**, 906–918.

Hendrix, R.W., Lawrence, J.G., Hatfull, G.F., and Casjens, S. (2000) The origins and ongoing evolution of viruses. *Trends in Microbiology* **8**, 504–508.

Hennig, W. (1966) *Phylogenetic Systematics*. University of Illinois Press, Urbana, IL.

Hepp, G.R., Novak, J.M., Scribner, K.T., and Stangel, P.W. (1988) Genetic distance and hybridization of black ducks and mallards: a morph of a different color? *Auk* **105**, 804–807.

Hess, W.R., Rocap, G., Ting, C.S., *et al.* (2001) The photosynthesis apparatus of *Prochlorococcus*: insights through comparative genomics. *Photosynthesis Research* **70**, 53–71.

Hewitt, G.M. (1988) Hybrid zones-natural laboratories for evolutionary studies. *Trends in Ecology and Evolution* **3**, 158–167.

Hey, J. and Nielsen, R. (2004) Multilocus methods for estimating population sizes, migration rates and divergence time, with applications to the divergence of *Drosophila pseudoobscura* and *D. persimilis*. *Genetics* **167**, 747–760.

Hey, J., Waples, R.S., Arnold, M.L., *et al.* (2003) Understanding and confronting species uncertainty in biology and conservation. *Trends in Ecology and Evolution* **18**, 597–603.

Hey, J., Won, Y.-J., Sivasundar, A.S., *et al.* (2004) Using nuclear haplotypes with microsatellites to study gene flow between recently separated cichlid species. *Molecular Ecology* **13**, 909–919.

Hibben, F.C. (1992) *Indian Hunts and Indian Hunters of the Old West*. Safari Press, Long Beach, CA.

Hillis, D.M., Moritz, C., Porter, C.A., and Baker, R.J. (1991) Evidence for biased gene conversion in concerted evolution of ribosomal DNA. *Science* **251**, 308–310.

Hinchliffe, S.J., Isherwood, K.E., Stabler, R.A., *et al.* (2003) Application of DNA microarrays to study the evolutionary genomics of *Yersinia pestis* and *Yersinia pseudotuberculosis*. *Genome Research* **13**, 2018–2029.

Hinnebusch, B.J., Rudolph, A.E., Cherepanov, P., *et al.* (2002) Role of Yersinia Murine Toxin in survival of *Yersinia pestis* in the midgut of the flea vector. *Science* **296**, 733–735.

Hodges, S.A., Burke, J.M., and Arnold, M.L. (1996) Natural formation of *Iris* hybrids: experimental evidence on the establishment of hybrid zones. *Evolution* **50**, 2504–2509.

Holden, M.T.G., Feil, E.J., Lindsay, J.A., *et al.* (2004a) Complete genomes of two clinical *Staphylococcus aureus* strains: evidence for the rapid evolution of virulence and drug resistance. *Proceedings of the National Academy of Sciences USA* **101**, 9786–9791.

Holden, M.T.G., Titball, R.W., Peacock, S.J., *et al.* (2004b) Genomic plasticity of the causative agent of Melioidosis, *Burkholderia pseudomallei*. *Proceedings of the National Academy of Sciences USA* **101**, 14240–14245.

Holliday, T.W. (2003) Species concepts, reticulation, and human evolution. *Current Anthropology* **44**, 653–660.

Holmes, E.C. (2004) The phylogeography of human viruses. *Molecular Ecology* **13**, 745–756.

Honeycutt, R.L. and Wilkinson, P. (1989) Electrophoretic variation in the parthenogenetic grasshopper *Warramaba virgo* and its sexual relatives. *Evolution* **43**, 1027–1044.

Honeycutt, R.L., Rowe, D.L., and Gallardo, M.H. (2003) Molecular systematics of the South American caviomorph rodents: relationships among species and genera in the family Octodontidae. *Molecular Phylogenetics and Evolution* **26**, 476–489.

Hong, D.-Y., Pan, K.-Y., and Rao, G.-Y. (2001) Cytogeography and taxonomy of the *Paeonia obovata* polyploid complex (Paeoniaceae). *Plant Systematics and Evolution* **227**, 123–136.

Hoot, S.B., Napier, N.S., and Taylor, W.C. (2004) Revealing unknown or extinct lineages within *Isoëtes* (Isoëtaceae) using DNA sequences from hybrids. *American Journal of Botany* **91**, 899–904.

Hopkins, C.D. and Bass, A.H. (1981) Temporal coding of species recognition signals in an electric fish. *Science* **212**, 85–87.

Hoskin, C.J., Higgie, M., McDonald, K.R., and Moritz, C. (2005) Reinforcement drives rapid allopatric speciation. *Nature* **437**, 1353–1356.

Hotz, H., Beerli, P., and Spolsky, C. (1992) Mitochondrial DNA reveals formation of nonhybrid frogs by natural matings between hemiclonal hybrids. *Molecular Biology and Evolution* **9**, 610–620.

Houck, M.A., Clark, J.B., Peterson, K.R., and Kidwell, M.G. (1991) Possible horizontal transfer of *Drosophila* genes by the mite *Proctolaelaps regalis*. *Science* **253**, 1125–1129.

Howard, D.J. (1982) *Speciation and Coexistence in a Group of Closely Related Ground Crickets*. PhD Dissertation, Yale University, New Haven, CT.

Howard, D.J. (1983) Electrophoretic survey of eastern North American *Allonemobius* (Orthoptera: Gryllidae): evolutionary relationships and the discovery of three new species. *Annals of the Entomological Society of America* **76**, 1014–1021.

Howard, D.J. (1986) A zone of overlap and hybridization between two ground cricket species. *Evolution* **40**, 34–43.

Howard, D.J. (1993) Reinforcement: origin, dynamics, and fate of an evolutionary hypothesis. In R.G. Harrison, ed.

Hybrid Zones and the Evolutionary Process, pp. 46–69. Oxford University Press, Oxford.

Howard, D.J. and Furth, D.G. (1986) Review of the *Allonemobius fasciatus* (Orthoptera: Gryllidae) complex with the description of two new species separated by electrophoresis, songs, and morphometrics. *Annals of the Entomological Society of America* **79**, 472–481.

Howard, D.J. and Gregory, P.G. (1993) Post-insemination signaling systems and reinforcement. *Philosophical Transactions of the Royal Society of London Series B* **340**, 231–236.

Howard, D.J. and Harrison, R.G. (1984a) Habitat segregation in ground crickets: experimental studies of adult survival, reproductive success, and oviposition preference. *Ecology* **65**, 61–68.

Howard, D.J. and Harrison, R.G. (1984b) Habitat segregation in ground crickets: the role of interspecific competition and habitat selection. *Ecology* **65**, 69–76.

Howard, D.J. and Waring, G.L. (1991) Topographic diversity, zone width, and the strength of reproductive isolation in a zone of overlap and hybridization. *Evolution* **45**, 1120–1135.

Howard, D.J., Waring, G.L., Tibbets, C.A., and Gregory, P.G. (1993) Survival of hybrids in a mosaic hybrid zone. *Evolution* **47**, 789–800.

Howard, D.J., Preszler, R.W., Williams, J., *et al.* (1997) How discrete are oak species? Insights from a hybrid zone between *Quercus grisea* and *Quercus gambelii*. *Evolution* **51**, 747–755.

Howard, D.J., Gregory, P.G., Chu, J., and Cain, M.L. (1998) Conspecific sperm precedence is an effective barrier to hybridization between closely related species. *Evolution* **52**, 511–516.

Howard, D.J., Marshall, J.L., Hampton, D.D., *et al.* (2002) The genetics of reproductive isolation: a retrospective and prospective look with comments on ground crickets. *American Naturalist* **159**, S8–S21.

Howarth, D.G. and Baum, D.A. (2002) Phylogenetic utility of a nuclear intron from nitrate reductase for the study of closely related plant species. *Molecular Phylogenetics and Evolution* **23**, 525–528.

Howarth, D.G. and Baum, D.A. (2005) Genealogical evidence of homoploid hybrid speciation in an adaptive radiation of *Scaevola* (Goodeniaceae) in the Hawaiian Islands. *Evolution* **59**, 948–961.

Howarth, D.G., Gustafsson, M.H.G., Baum, D.A., and Motley, T.J. (2003) Phylogenetics of the genus *Scaevola* (Goodeniaceae): implication for dispersal patterns across the Pacific Basin and colonization of the Hawaiian Islands. *American Journal of Botany* **90**, 915–923.

Hu, W.-S., and Temin, H.M. (1990) Genetic consequences of packaging two RNA genomes in one retroviral particle: pseudodiploidy and high rate of genetic recombination. *Proceedings of the National Academy of Sciences USA* **87**, 1556–1560.

Huang, J., Mullapudi, N., Lancto, C.A., *et al.* (2004) Phylogenomic evidence supports past endosymbiosis, intracellular and horizontal gene transfer in *Cryptosporidium parvum*. *Genome Biology* **5**, R88.

Hubbs, C.L. and Hubbs, L.C. (1932) Apparent parthenogenesis in nature, in a form of fish of hybrid origin. *Science* **76**, 628–630.

Hughes, A.L. and Friedman, R. (2004) Patterns of sequence divergence in 5′ intergenic spacers and linked coding regions in 10 species of pathogenic bacteria reveal distinct recombinational histories. *Genetics* **168**, 1795–1803.

Hughes, A.L., Friedman, R., and Murray, M. (2002) Genomewide pattern of synonymous nucleotide substitution in two complete genomes of *Mycobacterium tuberculosis*. *Emerging Infectious Diseases* **8**, 1342–1346.

Hulse-Post, D.J., Sturm-Ramirez, K.M., Humberd, J., *et al.* (2005) Role of domestic ducks in the propagation and biological evolution of highly pathogenic H5N1 influenza viruses in Asia. *Proceedings of the National Academy of Sciences USA* **102**, 10682–10687.

Hunter, N., Chambers, S.R., Louis, E.J., and Borts, R.H. (1996) The mismatch repair system contributes to meiotic sterility in an interspecific yeast hybrid. *EMBO Journal* **15**, 1726–1733.

Huson, D.H. and Bryant, D. (2006) Application of phylogenetic networks in evolutionary studies. *Molecular Biology and Evolution* **23**, 254–267.

Huston, C.D. (2004) Parasite and host contributions to the pathogenesis of amebic colitis. *Trends in Parasitology* **20**, 23–26.

Hutchison, D.W. and Templeton, A.R. (1999) Correlation of pairwise genetic and geographic distance measures: inferring the relative influences of gene flow and drift on the distribution of genetic variability. *Evolution* **53**, 1898–1914.

Hutchinson, J.F. (2001) The biology and evolution of HIV. *Annual Review of Anthropology* **30**, 85–108.

Intrieri, M.C. and Buiatti, M. (2001) The horizontal transfer of *Agrobacterium rhizogenes* genes and the evolution of the genus *Nicotiana*. *Molecular Phylogenetics and Evolution* **20**, 100–110.

Irwin, D.E. (2002) Phylogeographic breaks without geographic barriers to gene flow. *Evolution* **56**, 2383–2394.

Jackson, H.D., Steane, D.A., Potts, B.M., and Vaillancourt, R.E. (1999) Chloroplast DNA evidence for reticulate evolution in *Eucalyptus* (Myrtaceae). *Molecular Ecology* **8**, 739–751.

James, J.K. and Abbott, R.J. (2005) Recent, allopatric, homoploid hybrid speciation: the origin of *Senecio squalidus* (Asteraceae) in the British Isles from a hybrid zone on Mt. Etna, Sicily. *Evolution* **59**, 2533–2547.

Janko, K., Kotlík, P., and Ráb, P. (2003) Evolutionary history of asexual hybrid loaches (*Cobitis*: Teleostei) inferred from phylogenetic analysis of mitochondrial DNA variation. *Journal of Evolutionary Biology* **16**, 1280–1287.

Janko, K., Culling, M.A., Ráb, P., and Kotlík, P. (2005) Ice age cloning—comparison of the quaternary evolutionary histories of sexual and clonal forms of spiny loaches (*Cobitis*; Teleostei) using the analysis of mitochondrial DNA variation. *Molecular Ecology* **14**, 2991–3004.

Jiang, C.-X., Wright, R.J., El-Zik, K., and Paterson, A.H. (1998) Polyploid formation created unique avenues for response to selection in *Gossypium* (cotton). *Proceedings of the National Academy of Sciences USA* **95**, 4419–4424.

Jiménez, P., López de Heredia, U., Collada, C., *et al.* (2004) High variability of chloroplast DNA in three Mediterranean evergreen oaks indicates complex evolutionary history. *Heredity* **93**, 510–515.

Johannesson, H., Townsend, J.P., Hung, C.-Y., *et al.* (2005) Concerted evolution in the repeats of an immunomodulating cell surface protein, *SOWgp*, of the human pathogenic fungi *Coccidioides immitis* and *C. posadasii*. *Genetics* **171**, 109–117.

Johnsgard, P.A. (1967) Sympatry changes and hybridization incidence in mallards and black ducks. *American Midland Naturalist* **77**, 51–63.

Johnson, A.P., Aucken, H.M., Cavendish, S., *et al.* (2001) Dominance of EMRSA-15 and -16 among MRSA causing nosocomial bacteraemia in the UK: analysis of isolates from the European Antimicrobial Resistance Surveillance System (EARSS). *Journal of Antimicrobial Chemotherapy* **48**, 141–156.

Johnson, S.G. (2005) Mode of origin differentially influences the fitness of parthenogenetic freshwater snails. *Proceedings of the Royal Society of London Series B* **272**, 2149–2153.

Johnson, T.C., Scholz, C.A., Talbot, M.R., *et al.* (1996) Late Pleistocene desiccation of Lake Victoria and rapid evolution of cichlid fishes. *Science* **273**, 1091–1093.

Johnson, W.E. and O'Brien, S.J. (1997) Phylogenetic reconstruction of the Felidae using 16S rRNA and NADH-5 mitochondrial genes. *Journal of Molecular Evolution* **44** (suppl. 1), S98–S116.

Johnston, J.A., Wesselingh, R.A., Bouck, A.C., *et al.* (2001) Intimately linked or hardly speaking? The relationship between genotype and environmental gradients in a Louisiana Iris hybrid population. *Molecular Ecology* **10**, 673–681.

Johnston, J.A., Arnold, M.L., and Donovan, L.A. (2003) High hybrid fitness at seed and seedling life history stages in Louisiana Irises. *Journal of Ecology* **91**, 438–446.

Johnston, J.A., Donovan, L.A., and Arnold, M.L. (2004) Novel phenotypes among early generation hybrids of two Louisiana Iris species: flooding experiments. *Journal of Ecology* **92**, 967–976.

Johnston, J.R., Baccari, C., and Mortimer, R.K. (2000) Genotypic characterization of strains of commercial wine yeasts by tetrad analysis. *Research in Microbiology* **151**, 583–590.

Jolly, C.J. (2001) A proper study for mankind: analogies from the papionin monkeys and their implications for human evolution. *Yearbook of Physical Anthropology* **44**, 177–204.

Jolly, C.J., Wooley-Barker, T., Beyene, S., Disotell, T.R., and Phillips-Conroy, J.E. (1997) Intergeneric hybrid baboons. *International Journal of Primatology* **18**, 597–627.

Joyce, D.A., Lunt, D.H., Bills, R., *et al.* (2005) An extant cichlid fish radiation emerged in an extinct Pleistocene lake. *Nature* **435**, 90–95.

Juhala, R.J., Ford, M.E., Duda, R.L., *et al.* (2000) Genomic sequences of bacteriophages HK97 and HK022: pervasive genetic mosaicism in the lambdoid bacteriophages. *Journal of Molecular Biology* **299**, 27–51.

Kashkush, K., Feldman, M., and Levy, A.A. (2002) Gene loss, silencing and activation in a newly synthesized wheat allotetraploid. *Genetics* **160**, 1651–1659.

Kashkush, K., Feldman, M., and Levy, A.A. (2003) Transcriptional activation of retrotransposons alters the expression of adjacent genes in wheat. *Nature Genetics* **33**, 102–106.

Katz, L.A. (2002) Lateral gene transfers and the evolution of eukaryotes: theories and data. *International Journal of Systematic and Evolutionary Microbiology* **52**, 1893–1900.

Kearney, M.R. (2003) Why is sex so unpopular in the Australian desert? *Trends in Ecology and Evolution* **18**, 605–607.

Kearney, M. (2005) Hybridization, glaciation and geographical parthenogenesis. *Trends in Ecology and Evolution* **20**, 495–502.

Kearney, M. and Shine, R. (2004a) Developmental success, stability, and plasticity in closely related parthenogenetic and sexual lizards (*Heteronotia*, Gekkonidae). *Evolution* **58**, 1560–1572.

Kearney, M. and Shine, R. (2004b) Morphological and physiological correlates of hybrid parthenogenesis. *American Naturalist* **164**, 803–813.

Kearney, M., Wahl, R., and Autumn, K. (2005) Increased capacity for sustained locomotion at low temperature in parthenogenetic geckos of hybrid origin. *Physiological and Biochemical Zoology* **78**, 316–324.

Keim, P., Paige, K.N., Whitham, T.G., and Lark, K.G. (1989) Genetic analysis of an interspecific hybrid swarm of *Populus*: occurrence of unidirectional introgression. *Genetics* **123**, 557–565.

Kellis, M., Birren, B.W., and Lander, E.S. (2004) Proof and evolutionary analysis of ancient genome duplication in the yeast *Saccharomyces cerevisiae*. *Nature* **428**, 617–624.

Kellogg, E.A., Appels, R., and Mason-Gamer, R.J. (1996) When genes tell different stories: the diploid genera of Triticeae (Gramineae). *Systematic Botany* **21**, 321–347.

Kelly, J.K. and Noor, M.A.F. (1996) Speciation by reinforcement: a model derived from studies of *Drosophila*. *Genetics* **143**, 1485–1497.

Kennedy, A.J. (1995) Cacao—*Theobroma cacao* (Sterculiaceae). In J. Smartt and N.W. Simmonds, eds. *Evolution of Crop Plants*, 2nd edn, pp. 472–475. Longman Scientific & Technical, Harlow.

Kentner, E.K. and Mesler, M.R. (2000) Evidence for natural selection in a fern hybrid zone. *American Journal of Botany* **87**, 1168–1174.

Key, K.H.L. (1968) The concept of stasipatric speciation. *Systematic Zoology* **17**, 14–22.

Kiang, Y.T. and Hamrick, J.L. (1978) Reproductive isolation in the *Mimulus guttatus—M. nasutus* complex. *American Midland Naturalist* **100**, 269–276.

Kidwell, M.G. (1993) Lateral transfer in natural populations of eukaryotes. *Annual Review of Genetics* **27**, 235–256.

Kidwell, M.G. and Lisch, D.R. (2000) Transposable elements and host evolution. *Trends in Ecology and Evolution* **15**, 95–99.

Kim, S.-C. and Rieseberg, L.H. (1999) Genetic architecture of species differences in annual sunflowers: implications for adaptive trait introgression. *Genetics* **153**, 965–977.

Kim, S.-C. and Rieseberg, L.H. (2001) The contribution of epistasis to species differences in annual sunflowers. *Molecular Ecology* **10**, 683–690.

King, L.M. (1993) Origins of genotypic variation in North American Dandelions inferred from ribosomal DNA and chloroplast DNA restriction enzyme analysis. *Evolution* **47**, 136–151.

Kinsella, R.J., Fitzpatrick, D.A., Creevey, C.J., and McInerney, J.O. (2003) Fatty acid biosynthesis in *Mycobacterium tuberculosis*: lateral gene transfer, adaptive evolution, and gene duplication. *Proceedings of the National Academy of Sciences USA* **100**, 10320–10325.

Kirkpatrick, J.B. and Fowler, M. (1998) Locating likely glacial forest refugia in Tasmania using palynological and ecological information to test alternative climactic models. *Biological Conservation* **85**, 171–182.

Kitamura, S., Inoue, M., Shikazono, N., and Tanaka, A. (2001) Relationships among *Nicotiana* species revealed by the 5S rDNA spacer sequence and fluorescence

in situ hybridization. *Theoretical and Applied Genetics* **103**, 678–686.

Knight, M.E., Turner, G.F., Rico, C., *et al.* (1998) Microsatellite paternity analysis on captive Lake Malawi cichlids supports reproductive isolation by direct mate choice. *Molecular Ecology* **7**, 1605–1610.

Knowles, L.L. and Maddison, W.P. (2002) Statistical phylogeography. *Molecular Ecology* **11**, 2623–2635.

Kobasa, D., Takada, A., Shinya, K., *et al.* (2004) Enhanced virulence of influenza A viruses with the haemagglutinin of the 1918 virus. *Nature* **431**, 703–707.

Kobel, H.R. and Du Pasquier, L. (1986) Genetics of polyploid *Xenopus*. *Trends in Genetics* **2**, 310–315.

Kocher, T.D. and Sage, R.D. (1986) Further genetic analyses of a hybrid zone between leopard frogs (*Rana pipiens* complex) in central Texas. *Evolution* **40**, 21–33.

Kohlmann, B., Nix, H., and Shaw, D.D. (1988) Environmental predictions and distributional limits of chromosomal taxa in the Australian grasshopper *Caledia captiva* (F.). *Oecologia* **75**, 483–493.

Konopka, R.J. and Benzer, S. (1971) Clock mutants of *Drosophila melanogaster*. *Proceedings of the National Academy of Sciences USA* **68**, 2112–2116.

Konstantinidis, K. and Tiedje, J.M. (2005) Genomic insights that advance the species definition for prokaryotes. *Proceedings of the National Academy of Sciences USA* **102**, 2567–2572.

Kornfield, I. and Smith, P.F. (2000) African cichlid fishes: model systems for evolutionary biology. *Annual Review of Ecology and Systematics* **31**, 163–196.

Kornkven, A.B., Watson, L.E., and Estes, J.R. (1999) Molecular phylogeny of *Artemisia* section *Tridentatae* (Asteraceae) based on chloroplast DNA restriction site variation. *Systematic Botany* **24**, 69–84.

Kovarík, A., Matyasek, R., Lim, K.Y., *et al.* (2004) Concerted evolution of 18–5.8–26S rDNA repeats in *Nicotiana* allotetraploids. *Biological Journal of the Linnean Society* **82**, 615–625.

Kovarík, A., Pires, J.C., Leitch, A.R., *et al.* (2005) Rapid concerted evolution of nuclear ribosomal DNA in two *Tragopogon* allopolyploids of recent and recurrent origin. *Genetics* **169**, 931–944.

Krings, M., Stone, A., Schmitz, R.W., *et al.* (1997) Neandertal DNA sequences and the origin of modern humans. *Cell* **90**, 19–30.

Kroll, J.S., Wilks, K.E., Farrant, J.L., and Langford, P.R. (1998) Natural genetic exchange between *Haemophilus* and *Neisseria*: intergeneric transfer of chromosomal genes between major human pathogens. *Proceedings of the National Academy of Sciences USA* **95**, 12381–12385.

Kruse, K.C. and Dunlap, D.G. (1976) Serum albumins and hybridization in two species of the *Rana pipiens* complex in the north central United States. *Copeia* **1976**, 394–396.

Kruuk, L.E.B., Gilchrist, J.S., and Barton, N.H. (1999) Hybrid dysfunction in fire-bellied toads (*Bombina*). *Evolution* **53**, 1611–1616.

Kuiken, T., Rimmelzwaan, G., van Riel, D., *et al.* (2004) Avian H5N1 influenza in cats. *Science* **306**, 241.

Kurland, C.G. (2005) What tangled web: barriers to rampant horizontal gene transfer. *BioEssays* **27**, 741–747.

Kwan, T., Liu, J., DuBow, M., Gros, P., and Pelletier, J. (2005) The complete genomes and proteomes of 27 *Staphylococcus aureus* bacteriophages. *Proceedings of the National Academy of Sciences USA* **102**, 5174–5179.

Kyriacou, C.P. and Hall, J.C. (1980) Circadian rhythm mutations in *Drosophila melanogaster* affect short-term fluctuations in the male's courtship song. *Proceedings of the National Academy of Sciences USA* **77**, 6729–6733.

Lack, D. (1947) *Darwin's finches*. Cambridge University Press, Cambridge.

Lampert, K.P., Lamatsch, D.K., Epplen, J.T., and Schartl, M. (2005) Evidence for a monophyletic origin of triploid clones of the Amazon molly, *Poecilia formosa*. *Evolution* **59**, 881–889.

Lanzaro, G.C., Touré, Y.T., Carnahan, J., *et al.* (1998) Complexities in the genetic structure of *Anopheles gambiae* populations in west Africa as revealed by microsatellite DNA analysis. *Proceedings of the National Academy of Sciences USA* **95**, 14260–14265.

Lashermes, P., Combes, M.-C., Robert, J., *et al.* (1999) Molecular characterisation and origin of the *Coffea arabica* L. genome. *Molecular and General Genetics* **261**, 259–266.

Laver, W.G., Bischofberger, N., and Webster, R.G. (2000) The origin and control of pandemic influenza. *Perspectives in Biology and Medicine* **43**, 173–192.

Lavoué, S., Bigorne, R., Lecointre, G., and Agnèse, J.-F. (2000) Phylogenetic relationships of mormyrid electric fishes (Mormyridae; Teleostei) inferred from cytochrome *b* sequences. *Molecular Phylogenetics and Evolution* **14**, 1–10.

Lavoué, S., Sullivan, J.P., and Hopkins, C.D. (2003) Phylogenetic utility of the first two introns of the S7 ribosomal protein gene in African electric fishes (Mormyroidea: Teleostei) and congruence with other molecular markers. *Biological Journal of the Linnean Society* **78**, 273–292.

Lawrence, J.G. and Ochman, H. (1998) Molecular archaeology of the *Escherichia coli* genome. *Proceedings of the National Academy of Sciences USA* **95**, 9413–9417.

Lawrence, J.G. and Hendrickson, H. (2003) Lateral gene transfer: when will adolescence end? *Molecular Microbiology* **50**, 739–749.

Lawrence, R., Potts, B.M., and Whitham, T.G. (2003) Relative importance of plant ontogeny, host genetic variation, and leaf age for a common herbivore. *Ecology* **84**, 1171–1178.

Lawton-Rauh, A., Robichaux, R.H., and Purugganan, M.D. (2003) Patterns of nucleotide variation in homoeologous regulatory genes in the allotetraploid Hawaiian Silversword alliance (Asteraceae). *Molecular Ecology* **12**, 1301–1313.

Leander, B.S., Clopton, R.E., and Keeling, P.J. (2003) Phylogeny of gregarines (Apicomplexa) as inferred from small-subunit rDNA and β-tubulin. *International Journal of Systematic and Evolutionary Microbiology* **53**, 345–354.

Le Comber, S.C. and Smith, C. (2004) Polyploidy in fishes: patterns and processes. *Biological Journal of the Linnean Society* **82**, 431–442.

Lee, J.-Y., Mummenhoff, K., and Bowman, J.L. (2002) Allopolyploidization and evolution of species with reduced floral structures in *Lepidium* L. (Brassicaceae). *Proceedings of the National Academy of Sciences USA* **99**, 16835–16840.

Lehman, N., Eisenhawer, A., Hansen, K., *et al.* (1991) Introgression of coyote mitochondrial DNA into sympatric North American Gray Wolf populations. *Evolution* **45**, 104–119.

Leitch, I.J. and Bennett, M.D. (2004) Genome downsizing in polyploid plants. *Biological Journal of the Linnean Society* **82**, 651–663.

Lemeunier, F. and Ashburner, M. (1984) Relationships within the *melanogaster* species subgroup of the genus *Drosophila* (*Sophophora*) IV. The chromosomes of two new species. *Chromosoma* **89**, 343–351.

Lenz, L.W. (1959) Hybridization and speciation in the Pacific Coast irises. *Aliso* **4**, 237–309.

Leonard, J.A., Wayne, R.K., Wheeler, J., *et al.* (2002) Ancient DNA evidence for Old World origin of New World dogs. *Science* **298**, 1613–1616.

León-Avila, G. and Tovar, J. (2004) Mitosomes of *Entamoeba histolytica* are abundant mitochondrion-related remnant organelles that lack a detectable organellar genome. *Microbiology* **150**, 1245–1250.

Levin, D.A. (2000) *The Origin, Expansion, and Demise of Plant Species*. Oxford University Press, Oxford.

Levitan, D.R. (2002) The relationship between conspecific fertilization success and reproductive isolation among three congeneric sea urchins. *Evolution* **56**, 1599–1609.

Levy, A.A. and Feldman, M. (2004) Genetic and epigenetic reprogramming of the wheat genome upon allopolyploidization. *Biological Journal of the Linnean Society* **82**, 607–613.

Lewontin, R.C. and Birch, L.C. (1966) Hybridization as a source of variation for adaptation to new environments. *Evolution* **20**, 315–336.

Lexer, C., Welch, M.E., Durphy, J.L., and Rieseberg, L.H. (2003a) Natural selection for salt tolerance quantitative trait loci (QTLs) in wild sunflower hybrids: implications for the origin of *Helianthus paradoxus*, a diploid hybrid species. *Molecular Ecology* **12**, 1225–1235.

Lexer, C., Welch, M.E., Raymond, O., and Rieseberg, L.H. (2003b) The origin of ecological divergence in *Helianthus paradoxus* (Asteraceae): selection on transgressive characters in a novel hybrid habitat. *Evolution* **57**, 1989–2000.

Lexer, C., Fay, M.F., Joseph, J.A., *et al.* (2005) Barriers to gene flow between two ecologically divergent *Populus* species, *P. alba* (white poplar) and *P. tremula* (European aspen): the role of ecology and life history in gene introgression. *Molecular Ecology* **14**, 1045–1057.

Li, K.S., Guan, Y., Wang, J., *et al.* (2004) Genesis of a highly pathogenic and potentially pandemic H5N1 influenza virus in eastern Asia. *Nature* **430**, 209–213.

Li, M.-S., Farrant, J.L., Langford, P.R., and Kroll, J.S. (2003) Identification and characterization of genomic loci unique to the Brazilian purpuric fever clonal group of *H. influenzae* biogroup aegyptius: functionality explored using meningococcal homology. *Molecular Microbiology* **47**, 1101–1111.

Li, W.-H., Yang, J., and Gu, X. (2005) Expression divergence between duplicate genes. *Trends in Genetics* **21**, 602–607.

Lim, K.Y., Matyásek, R., Lichtenstein, C.P., and Leitch, A.R. (2000) Molecular cytogenetic analyses and phylogenetic studies in the *Nicotiana* section Tomentosae. *Chromosoma* **109**, 245–258.

Lim, K.Y., Matyásek, R., Kovarík, A., and Leitch, A.R. (2004) Genome evolution in allotetraploid *Nicotiana*. *Biological Journal of the Linnean Society* **82**, 599–606.

Lindell, D., Sullivan, M.B., Johnson, Z.I., *et al.* (2004) Transfer of photosynthesis genes to and from *Prochlorococcus* viruses. *Proceedings of the National Academy of Sciences USA* **101**, 11013–11018.

Linn, C.E., Jr, Feder, J.L., Nojima, S., *et al.* (2003) Fruit odor discrimination and sympatric host race formation in *Rhagoletis*. *Proceedings of the National Academy of Sciences USA* **100**, 11490–11493.

Linn, C.E., Jr, Dambroski, H.R., Feder, J.L., *et al.* (2004) Postzygotic isolating factor in sympatric speciation in *Rhagoletis* flies: reduced response of hybrids to parental host-fruit odors. *Proceedings of the National Academy of Sciences USA* **101**, 17753–17758.

Linnaeus, C. (1760) *Disquisitio de Sexu Plantarum*. St. Petersburg.

Littlejohn, M.J. and Oldham, R.S. (1968) *Rana pipiens* complex: mating call structure and taxonomy. *Science* **162**, 1003–1005.

Liu, B., Brubaker, C.L., Mergeai, G., *et al.* (2001) Polyploid formation in cotton is not accompanied by rapid genomic changes. *Genome* **44**, 321–330.

Liu, H., Nolla, H.A., and Campbell, L. (1997) *Prochlorococcus* growth rate and contribution to primary production in the equatorial and subtropical North Pacific Ocean. *Aquatic Microbial Ecology* **12**, 39–47.

Liu, J., Xiao, H., Lei, F., *et al.* (2005) Highly pathogenic H5N1 influenza virus infection in migratory birds. *Science* **309**, 1206.

Liu, S.-L. and Sanderson, K.E. (1998) Homologous recombination between *rrn* operons rearranges the chromosome in host-specialized species of *Salmonella*. *FEMS Microbiology Letters* **164**, 275–281.

Liu, Z.-L., Zhang, D., Hong, D.-Y., and Wang, X.-R. (2003a) Chromosomal localization of 5S and 18S-5.8S-25S ribosomal DNA sites in five Asian pines using fluorescence in situ hybridization. *Theoretical and Applied Genetics* **106**, 198–204.

Liu, Z.-L., Zhang, D., Wang, X.-Q., *et al.* (2003b) Intragenomic and interspecific 5S rDNA sequence variation in five Asian pines. *American Journal of Botany* **90**, 17–24.

Llopart, A., Lachaise, D., and Coyne, J.A. (2005) Multilocus analysis of introgression between two sympatric sister species of *Drosophila*: *Drosophila yakuba* and *D. santomea*. *Genetics* **171**, 197–210.

Lochouarn, L., Dia, I., Boccolini, D., *et al.* (1998) Bionomical and cytogenetic heterogeneities of *Anopheles funestus* in Senegal. *Transactions of the Royal Society of Tropical Medicine and Hygiene* **92**, 607–612.

Loftus, B., Anderson, I., Davies, R., *et al.* (2005) The genome of the protist parasite *Entamoeba histolytica*. *Nature* **433**, 865–868.

Lotsy, J.P. (1916) *Evolution by Means of Hybridization*. Martinus Nijhoff, The Hague.

Lotsy, J.P. (1931) On the species of the taxonomist in its relation to evolution. *Genetica* **13**, 1–16.

Lowe, A.J. and Abbott, R.J. (2000) Routes of origin of two recently evolved hybrid taxa: *Senecio vulgaris* var *hibernicus* and York radiate groundsel (Asteraceae). *American Journal of Botany* **87**, 1159–1167.

Lowe, A.J. and Abbott, R.J. (2004) Reproductive isolation of a new hybrid species, *Senecio eboracensis* Abbott & Lowe (Asteraceae). *Heredity* **92**, 386–395.

Lowe, P.R. (1936) The finches of the Galápagos in relation to Darwin's conception of species. *Ibis* **6**, 310–321.

Lu, G., Basley, D.J., and Bernatchez, L. (2001) Contrasting patterns of mitochondrial DNA and microsatellite introgressive hybridization between lineages of lake whitefish (*Coregonus clupeaformis*); relevance for speciation. *Molecular Ecology* **10**, 965–985.

Ludwig, F., Rosenthal, D.M., Johnston, J.A., *et al.* (2004) Selection on leaf ecophysiological traits in a desert hybrid *Helianthus* species and early-generation hybrids. *Evolution* **58**, 2682–2692.

Lukhtanov, V.A., Kandul, N.P., Plotkin, J.B., *et al.* (2005) Reinforcement of pre-zygotic isolation and karyotype evolution in *Agrodiaetus* butterflies. *Nature* **436**, 385–389.

Lynch, M. (1984) The genetic structure of a cyclical parthenogen. *Evolution* **38**, 186–203.

MacCallum, C.J., Nürnberger, B., Barton, N.H., and Szymura, J.M. (1998) Habitat preference in the *Bombina* hybrid zone in Croatia. *Evolution* **52**, 227–239.

Machado, C.A. and Ayala, F.J. (2001) Nucleotide sequences provide evidence of genetic exchange among distantly related lineages of *Trypanosoma cruzi*. *Proceedings of the National Academy of Sciences USA* **98**, 7396–7401.

Machado, C.A. and Hey, J. (2003) The causes of phylogenetic conflict in a classic *Drosophila* species group. *Proceedings of the Royal Society of London Series B* **270**, 1193–1202.

Machado, C.A., Kliman, R.M., Markert, J.A., and Hey, J. (2002) Inferring the history of speciation from multilocus DNA sequence data: the case of *Drosophila pseudoobscura* and close relatives. *Molecular Biology and Evolution* **19**, 472–488.

Maddison, W.P. (1997) Gene trees in species trees. *Systematic Biology* **46**, 523–536.

Madern, D., Cai, X., Abrahamsen, M.S., and Zhu, G. (2004) Evolution of *Cryptosporidium parvum* lactate dehydrogenase from malate dehydrogenase by a very recent event of gene duplication. *Molecular Biology and Evolution* **21**, 489–497.

Madlung, A., Tyagi, A.P., Watson, B., *et al.* (2005) Genomic changes in synthetic *Arabidopsis* polyploids. *The Plant Journal* **41**, 221–230.

Maere, S., De Bodt, S., Raes, J., *et al.* (2005) Modeling gene and genomic duplications in eukaryotes. *Proceedings of the National Academy of Sciences USA* **102**, 5454–5459.

Mai, Z., Ghosh, S., Frisardi, M., *et al.* (1999) Hsp60 is targeted to a cryptic mitochondrion-derived organelle ('Crypton') in the microaerophilic protozoan parasite *Entamoeba histolytica*. *Molecular and Cellular Biology* **19**, 2198–2205.

Mallet, J. (2005) Hybridization as an invasion of the genome. *Trends in Ecology and Evolution* **20**, 229–237.

Mank, J.E., Carlson, J.E., and Brittingham, M.C. (2004) A century of hybridization: decreasing genetic distance between American black ducks and mallards. *Conservation Genetics* **5**, 395–403.

Mann, N.H., Cook, A., Millard, A., *et al.* (2003) Bacterial photosynthesis genes in a virus. *Nature* **424**, 741.

Marchant, A.D. and Shaw, D.D. (1993) Contrasting patterns of geographic variation shown by mtDNA and karyotype organization in two subspecies of *Caledia captiva* (Orthoptera). *Molecular Biology and Evolution* **10**, 855–872.

Marchant, A.D., Arnold, M.L., and Wilkinson, P. (1988) Gene flow across a chromosomal tension zone I. Relics of ancient hybridization. *Heredity* **61**, 321–328.

Marcus, S.L., Brumell, J.H., Pfeifer, C.G., and Finlay, B.B. (2000) *Salmonella* pathogenicity islands: big virulence in small packages. *Microbes and Infection* **2**, 145–156.

Marinoni, G., Manuel, M., Petersen, R.F., *et al.* (1999) Horizontal transfer of genetic material among *Saccharomyces* yeasts. *Journal of Bacteriology* **181**, 6488–6496.

Marshall, J.L. (2004) The *Allonemobius-Wolbachia* host-endosymbiont system: evidence for rapid speciation and against reproductive isolation driven by cytoplasmic incompatibility. *Evolution* **58**, 2409–2425.

Marshall, J.L., Arnold, M.L., and Howard, D.J. (2002) Reinforcement: the road not taken. *Trends in Ecology and Evolution* **17**, 558–563.

Martin, N.H., Bouck, A.C., and Arnold, M.L. (2005) Loci affecting long-term hybrid survivability in Louisiana Irises: implications for reproductive isolation and introgression. *Evolution* **59**, 2116–2124.

Martin, N.H., Bouck, A.C., and Arnold, M.L. (2006) Detecting adaptive trait introgression in Louisiana Irises. *Genetics* **172**, 2481–2489.

Martin, W. (2005) Lateral gene transfer and other possibilities. *Heredity* **94**, 565–566.

Martinsen, G.D., Whitham, T.G., Turek, R.J., and Keim, P. (2001) Hybrid populations selectively filter gene introgression between species. *Evolution* **55**, 1325–1335.

Masneuf, I., Hansen, J., Groth, C., *et al.* (1998) New hybrids between *Saccharomyces* sensu stricto yeast species found among wine and cider production strains. *Applied and Environmental Microbiology* **64**, 3887–3892.

Mason-Gamer, R.J. (2005) The β-amylase genes of grasses and a phylogenetic analysis of the Triticeae (Poaceae). *American Journal of Botany* **92**, 1045–1058.

Masta, S.E., Laurent, M., and Routman, E.J. (2003) Population genetic structure of the toad *Bufo woodhousii*: an empirical assessment of the effects of haplotype extinction on nested cladistic analysis. *Molecular Ecology* **12**, 1541–1554.

Masterson, J. (1994) Stomatal size in fossil plants: evidence for polyploidy in majority of angiosperms. *Science* **264**, 421–424.

Mayer, W.E., Tichy, H., and Klein, J. (1998) Phylogeny of African cichlid fishes as revealed by molecular markers. *Heredity* **80**, 702–714.

Maynard Smith, J. (1992) Age and the unisexual lineage. *Nature* **356**, 661–662.

Mayr, E. (1942) *Systematics and the Origin of Species.* Columbia University Press, New York.

Mayr, E. (1946) Experiments on sexual isolation in *Drosophila* VI. Isolation between *Drosophila pseudoobscura* and *Drosophila persimilis* and their hybrids. *Proceedings of the National Academy of Sciences USA* **32**, 57–59.

Mayr, E. (1963) *Animal Species and Evolution.* Belknap Press, Cambridge, Massachusetts.

McArthur, E.D. and Sanderson, S.C. (1999) Cytogeography and chromosome evolution of subgenus *Tridentatae* of *Artemisia* (Asteraceae). *American Journal of Botany* **86**, 1754–1775.

McArthur, E.D., Welch, B.L., and Sanderson, S.C. (1988) Natural and artificial hybridization between Big Sagebrush (*Artemisia tridentata*) subspecies. *Journal of Heredity* **79**, 268–276.

McArthur, E.D., Mudge, J., Van Buren, R., *et al.* (1998) Randomly amplified polymorphic DNA analysis (RAPD) of *Artemisia* subgenus *Tridentatae* species and hybrids. *Great Basin Naturalist* **58**, 12–27.

McClelland, M., Sanderson, K.E., Spieth, J., *et al.* (2001) Complete genome sequence of *Salmonella enterica* serovar Typhimurium LT2. *Nature* **413**, 852–856.

McClintock, B. (1984) The significance of responses of the genome to challenge. *Science* **226**, 792–801.

McCracken, K.G., Johnson, W.P., and Sheldon, F.H. (2001) Molecular population genetics, phylogeography, and conservation biology of the mottled duck (*Anas fulvigula*). *Conservation Genetics* **2**, 87–102.

McDade, L.A. (1992) Hybrids and phylogenetic systematics II. The impact of hybrids on cladistic analysis. *Evolution* **46**, 1329–1346.

McDonald, D.B., Clay, R.P., Brumfield, R.T., and Braun, M.J. (2001) Sexual selection on plumage and behavior in an avian hybrid zone: experimental tests of male-male interactions. *Evolution* **55**, 1443–1451.

McElroy, D.M., Shoemaker, J.A., and Douglas, M.E. (1997) Discriminating *Gila robusta* and *Gila cypha*: risk assessment and the Endangered Species Act. *Ecological Applications* **7**, 958–967.

McGraw, E.A., Li, J., Selander, R.K., and Whittam, T.S. (1999) Molecular evolution and mosaic structure of α, β, and γ intimins of pathogenic *Escherichia coli. Molecular Biology and Evolution* **16**, 12–22.

McKinnon, G.E., Vaillancourt, R.E., Jackson, H.D., and Potts, B.M. (2001) Chloroplast sharing in the Tasmanian eucalypts. *Evolution* **55**, 703–711.

McKinnon, G.E., Vaillancourt, R.E., Steane, D.A., and Potts, B.M. (2004) The rare silver gum, *Eucalyptus cordata*, is leaving its trace in the organellar gene pool of *Eucalyptus globulus. Molecular Ecology* **13**, 3751–3762.

Melayah, D., Lim, K.Y., Bonnivard, E., *et al.* (2004) Distribution of the Tnt1 retrotransposon family in the amphidiploid tobacco (*Nicotiana tabacum*) and its wild *Nicotiana* relatives. *Biological Journal of the Linnean Society* **82**, 639–649.

Meléndez-Ackerman, E. (1997) Patterns of color and nectar variation across an *Ipomopsis* (Polemoniaceae) hybrid zone. *American Journal of Botany* **84**, 41–47.

Meléndez-Ackerman, E. and Campbell, D.R. (1998) Adaptive significance of flower color and inter-trait correlations in an *Ipomopsis* hybrid zone. *Evolution* **52**, 1293–1303.

Meléndez-Ackerman, E., Campbell, D.R., and Waser, N.M. (1997) Hummingbird behavior and mechanisms of selection on flower color in *Ipomopsis. Ecology* **78**, 2532–2541.

Melo-Ferreira, J., Boursot, P., Suchentrunk, F., *et al.* (2005) Invasion from the cold past: extensive introgression of mountain hare (*Lepus timidus*) mitochondrial DNA into three other hare species in northern Iberia. *Molecular Ecology* **14**, 2459–2464.

Messina, F.J., Richards, J.H., and McArthur, E.D. (1996) Variable responses of insects to hybrid versus parental sagebrush in common gardens. *Oecologia* **107**, 513–521.

Meyer, A., Kocher, T.D., Basasibwaki, P., and Wilson, A.C. (1990) Monophyletic origin of Lake Victoria cichlid fishes suggested by mitochondrial DNA sequences. *Nature* **347**, 550–553.

Michel, A.P., Ingrasci, M.J., Schemerhorn, B.J., *et al.* (2005) Rangewide population genetic structure of the African malaria vector *Anopheles funestus. Molecular Ecology* **14**, 4235–4248.

Miglia, K.J., McArthur, E.D., Moore, W.S., *et al.* (2005) Nine-year reciprocal transplant experiment in the gardens of the basin and mountain big sagebrush (*Artemisia tridentata*: Asteraceae) hybrid zone of Salt Creek Canyon: the importance of multiple-year tracking of fitness. *Biological Journal of the Linnean Society* **86**, 213–225.

Millard, A., Clokie, M.R.J., Shub, D.A., and Mann, N.H. (2004) Genetic organization of the *psbAD* region in phages infecting marine *Synechococcus* strains. *Proceedings of the National Academy of Sciences USA* **101**, 11007–11012.

Miller, E.S., Heidelberg, J.F., Eisen, J.A., *et al.* (2003a) Complete genome sequence of the broad-host-range vibriophage KVP40: comparative genomics of a T4-related bacteriophage. *Journal of Bacteriology* **185**, 5220–5233.

Miller, E.S., Kutter, E., Mosig, G., *et al.* (2003b) Bacteriophage T4 genome. *Microbiology and Molecular Biology Reviews* **67**, 86–156.

Mira, A., Ochman, H., and Moran, N.A. (2001) Deletional bias and the evolution of bacterial genomes. *Trends in Genetics* **17**, 589–596.

Miskin, J.E., Farrell, A.M., Cunliffe, W.J., and Holland, K.T. (1997) *Propionibacterium acnes*, a resident of lipid-rich human skin, produces a 33 kDa extracellular lipase encoded by *gehA*. *Microbiology* **143**, 1745–1755.

Moncada, P. and McCouch, S. (2004) Simple sequence repeat diversity in diploid and tetraploid *Coffea* species. *Genome* **47**, 501–509.

Moore, W.S. (1971) An evaluation of narrow hybrid zones in vertebrates. *Quarterly Review of Biology* **52**, 263–277.

Moore, W.S. and Buchanan, D.B. (1985) Stability of the Northern Flicker hybrid zone in historical times: implications for adaptive speciation theory. *Evolution* **39**, 135–151.

Morales, J.C. and Melnick, D.J. (1998) Phylogenetic relationships of the macaques (Cercopithecidae: *Macaca*), as revealed by high resolution restriction site mapping of mitochondrial ribosomal genes. *Journal of Human Evolution* **34**, 1–23.

Moran, C. (1979) The structure of the hybrid zone in *Caledia captiva*. *Heredity* **42**, 13–32.

Moran, C. and Shaw, D.D. (1977) Population cytogenetics of the genus *Caledia* (Orthoptera: Acridinae). *Chromosoma* **63**, 181–204.

Moran, C., Wilkinson, P., and Shaw, D.D. (1980) Allozyme variation across a narrow hybrid zone in the grasshopper, *Caledia captiva*. *Heredity* **44**, 69–81.

Moreira, D. (2000) Multiple independent horizontal transfers of informational genes from bacteria to plasmids and phages: implications for the origin of bacterial replication machinery. *Molecular Microbiology* **35**, 1–5.

Morgan, D.R. (1993) A molecular systematic study and taxonomic revision of *Psilactis* (Asteraceae: Astereae). *Systematic Botany* **18**, 290–308.

Morgan, D.R. (1997) Reticulate evolution in *Machaeranthera* (Asteraceae). *Systematic Botany* **22**, 599–615.

Morgan, D.R. (2003) nrDNA external transcribed spacer (ETS) sequence data, reticulate evolution, and the systematics of *Machaeranthera* (Asteraceae). *Systematic Botany* **28**, 179–190.

Morgan, D.R. and Simpson, B.B. (1992) A systematic study of *Machaeranthera* (Asteraceae) and related groups using restriction site analysis of chloroplast DNA. *Systematic Botany* **17**, 511–531.

Morgan-Richards, M. and Trewick, S.A. (2005) Hybrid origin of a parthenogenetic genus? *Molecular Ecology* **14**, 2133–2142.

Moritz, C. (1983) Parthenogenesis in the endemic Australian lizard *Heteronotia binoei* (Gekkonidae). *Science* **220**, 735–737.

Moritz, C. (1991) The origin and evolution of parthenogenesis in *Heteronotia binoei* (Gekkonidae): evidence for recent and localized origins of widespread clones. *Genetics* **129**, 211–219.

Moritz, C., Donnellan, S., Adams, M., and Baverstock, P.R. (1989a) The origin and evolution of parthenogenesis in *Heteronotia binoei* (Gekkonidae): extensive genotypic diversity among parthenogens. *Evolution* **43**, 994–1003.

Moritz, C., Wright, J.W., and Brown, W.M. (1989a) Mitochondrial-DNA analyses and the origin and relative age of parthenogenetic lizards (genu *Cnemidophorus*) III. *C. velox* and *C. exsangus*. *Evolution* **43**, 958–968.

Moritz, C., Wright, J.W., and Brown, W.M. (1992) Mitochondrial-DNA analyses and the origin and relative age of parthenogenetic *Cnemidophorus*: phylogenetic constraints on hybrid origins. *Evolution* **46**, 184–192.

Motamayor, J.C., Risterucci, A.M., Lopez, C.F., Ortiz, C.F., Moreno, A., and Lanaud, C. (2002) Cacao domestication I: the origin of the cacao cultivated by the Mayas. *Heredity* **89**, 380–386.

Motamayor, J.C., Risterucci, A.M., Heath, M., and Lanaud, C. (2003) Cacao domestication II: progenitor germplasm of the Trinitario cacao cultivar. *Heredity* **91**, 322–330.

Mousseau, T.A. and Howard, D.J. (1998) Genetic variation in cricket calling song across a hybrid zone between two sibling species. *Evolution* **52**, 1104–1110.

Mower, J.P., Stefanovic, S., Young, G.J., and Palmer, J.D. (2004) Gene transfer from parasitic to host plants. *Nature* **432**, 165–166.

Moya, A., Holmes, E.C., and González-Candelas, F. (2004) The population genetics and evolutionary epidemiology of RNA viruses. *Nature Reviews Microbiology* **2**, 1–10.

Muir, G. and Schlötterer, C. (2005) Evidence for shared ancestral polymorphism rather than recurrent gene flow at microsatellite loci differentiating two hybridizing oaks (*Quercus* spp). *Molecular Ecology* **14**, 549–561.

Muir, G., Fleming, C.C., and Schlötterer, C. (2000) Species status of hybridizing oaks. *Nature* **405**, 1016.

Muir, G., Fleming, C.C., and Schlötterer, C. (2001) Three divergent rDNA clusters predate the species divergence in *Quercus petraea* (Matt) Liebl and *Quercus robur* L. *Molecular Biology and Evolution* **18**, 112–119.

Mulley, J. and Holland, P. (2004) Small genome, big insights. *Nature* **431**, 916–917.

Murdoch, D.A., Banatvala, N.A., Bone, A., *et al.* (1998) Epidemic ciprofloxacin-resistant *Salmonella typhi* in Tajikistan. *Lancet* **351**, 339.

Nagai, H., and Roy, C.R. (2003) Show me the substrates: modulation of host cell function by type IV secretion systems. *Cellular Microbiology* **5**, 373–383.

Naisbit, R.E., Jiggins, C.D., Linares, M., *et al.* (2002) Hybrid sterility, Haldane's Rule and speciation in *Heliconius cydno* and *H. melpomene. Genetics* **161**, 1517–1526.

Nason, J.D., Ellstrand, N.C., and Arnold, M.L. (1992) Patterns of hybridization and introgression in populations of oaks, manzanitas and irises. *American Journal of Botany* **79**, 101–111.

Natcheva, R. and Cronberg, N. (2004) What do we know about hybridization among bryophytes in nature? *Canadian Journal of Botany* **82**, 1687–1704.

Navarro, A. and Barton, N.H. (2003) Chromosomal speciation and molecular divergence—accelerated evolution in rearranged chromosomes. *Science* **300**, 321–324.

Neale, D.B. and Sederoff, R.R. (1989) Paternal inheritance of chloroplast DNA and maternal inheritance of mitochondrial DNA in loblolly pine. *Theoretical and Applied Genetics* **77**, 212–216.

Nei, M. (1978) Estimation of average heterozygosity and genetic distance from a small number of individuals. *Genetics* **89**, 583–590.

Newman, T.K., Jolly, C.J., and Rogers, J. (2004) Mitochondrial phylogeny and systematics of baboons (*Papio*). *American Journal of Physical Anthropology* **124**, 17–27.

Nielson, M., Lohman, K., and Sullivan, J. (2001) Phylogeography of the tailed frog (*Ascaphus truei*): implications for the biogeography of the Pacific Northwest. *Evolution* **55**, 147–160.

Nielsen, R. and Wakeley, J. (2001) Distinguishing migration from isolation: a Markov Chain Monte Carlo approach. *Genetics* **158**, 885–896.

Nierman, W.C., DeShazer, D., Kim, H.S., *et al.* (2004) Structural flexibility in the *Burkholderia mallei* genome. *Proceedings of the National Academy of Sciences USA* **101**, 14246–14251.

Nixon, K.C. and Wheeler, Q.D. (1990) An amplification of the phylogenetic species concept. *Cladistics* **6**, 211–223.

Noor, M.A.F. (1995) Speciation driven by natural selection in *Drosophila. Nature* **375**, 674–675.

Noor, M.A.F. (1999) Reinforcement and other consequences of sympatry. *Heredity* **83**, 503–508.

Noor, M.A.F., Johnson, N.A., and Hey, J. (2000) Gene flow between *Drosophila pseudoobscura* and *D. persimilis. Evolution* **54**, 2174–2175.

Normile, D. and Enserink, M. (2004) Avian influenza makes a comeback, reviving pandemic worries. *Science* **305**, 321.

NSW National Parks and Wildlife Service (2000) Threatened species information—*Eucalyptus benthamii*. National Parks and Wildlife Service, Sydney.

Nürnberger, B., Barton, N., MacCallum, C., *et al.* (1995) Natural selection on quantitative traits in the *Bombina* hybrid zone. *Evolution* **49**, 1224–1238.

Nürnberger, B., Barton, N.H., Kruuk, L.E.B., and Vines, T.H. (2005) Mating patterns in a hybrid zone of fire-bellied toads (*Bombina*): inferences from adult and full-sib genotypes. *Heredity* **94**, 247–257.

Nyakaana, S. and Arctander, P. (1999) Population genetic structure of the African elephant in Uganda based on variation at mitochondrial and nuclear loci: evidence for male-biased gene flow. *Molecular Ecology* **8**, 1105–1115.

Nyakaana, S., Abe, E.L., Arctander, P., and Siegismund, H.R. (2001) DNA evidence for elephant social behaviour breakdown in Queen Elizabeth National Park, Uganda. *Animal Conservation* **4**, 231–237.

Nyakaana, S., Arctander, P., and Siegismund, H.R. (2002) Population structure of the African savannah elephant inferred from mitochondrial control region sequences and nuclear microsatellite loci. *Heredity* **89**, 90–98.

Nystrom, P., Phillips-Conroy, J.E., and Jolly, C.J. (2004) Dental microwear in anubis and hybrid baboons (*Papio hamadryas*, sensu lato) living in Awash National Park, Ethiopia. *American Journal of Physical Anthropology* **125**, 279–291.

Oborník, M. and Green, B.R. (2005) Mosaic origin of the heme biosynthesis pathway in photosynthetic eukaryotes. *Molecular Biology and Evolution* **22**, 2343–2353.

O'Brien, S.J. and Mayr, E. (1991) Bureaucratic mischief: recognizing endangered species and subspecies. *Science* **251**, 1187–1188.

O'Brien, S.J., Roelke, M.E., Yuhki, N., *et al.* (1990) Genetic introgression within the Florida Panther *Felis concolor coryi. National Geographic Research* **6**, 485–494.

Ochman, H., Lawrence, J.G., and Groisman, E.A. (2000) Lateral gene transfer and the nature of bacterial innovation. *Nature* **405**, 299–304.

Ochman, H., Lerat, E., and Daubin, V. (2005) Examining bacterial species under the specter of gene transfer and exchange. *Proceedings of the National Academy of Sciences USA* **102**, 6595–6599.

O'hUigin, C., Satta, Y., Takahata, N., and Klein, J. (2002) Contribution of homoplasy and of ancestral polymorphism to the evolution of genes in anthropoid primates. *Molecular Biology and Evolution* **19**, 1501–1513.

Olendzenski, L., Zhaxybayeva, O., and Gogarten, J.P. (2004) A brief history of views of prokaryotic evolution and taxonomy. In C.M. Fraser, T. Read, and K.E. Nelson, eds. *Microbial Genomes*, pp. 143–154. Humana Press, Totowa, NJ.

Olmstead, R.G. (1995) Species concepts and plesiomorphic species. *Systematic Botany* **20**, 623–630.

Onyabe, D.Y. and Conn, J.E. (2001) Genetic differentiation of the malaria vector *Anopheles gambiae* across Nigeria suggests that selection limits gene flow. *Heredity* **87**, 647–658.

Orians, C.M., Bolnick, D.I., Roche, B.M., *et al.* (1999) Water availability alters the relative performance of *Salix sericea*, *Salix eriocephala*, and their F_1 hybrids. *Canadian Journal of Botany* **77**, 514–522.

Orr, H.A. (1990) 'Why polyploidy is rarer in animals than in plants' revisited. *American Naturalist* **136**, 759–770.

Ortiz-Barrientos, D. and Noor, M.A.F. (2005) Evidence for a one-allele assortative mating locus. *Science* **310**, 1467.

Osada, N. and Wu C-I (2005) Inferring the mode of speciation from genomic data: a study of the great apes. *Genetics* **169**, 259–264.

Ovchinnikov, I.V., Götherström, A., Romanova, G.P., *et al.* (2000) Molecular analysis of Neanderthal DNA from the northern Caucasus. *Nature* **404**, 490–493.

Ownbey, M. (1950) Natural hybridization and amphidiploidy in the genus *Tragopogon*. *American Journal of Botany* **37**, 487–499.

Pääbo, S. (2003) The mosaic that is our genome. *Nature* **421**, 409–412.

Palenik, B., Brahamsha, B., Larimer, F.W., *et al.* (2003) The genome of a motile marine *Synechococcus*. *Nature* **424**, 1037–1042.

Palmer, E.J. (1948) Hybrid oaks of North America. *Journal of the Arnold Arboretum* **29**, 1–48.

Palopoli, M.F. and Wu C-I (1994) Genetics of hybrid male sterility between *Drosophila* sibling species: a complex web of epistasis is revealed in interspecific studies. *Genetics* **138**, 329–341.

Panopoulou, G. and Poustka, A.J. (2005) Timing and mechanism of ancient vertebrate genome duplications—the adventure of a hypothesis. *Trends in Genetics* **21**, 559–567.

Paoletti, M., Buck, K.W., and Brasier, C.M. (2006) Selective acquisition of novel mating type and vegetative incompatibility genes via interspecies gene transfer in the globally invading eukaryote *Ophiostoma novo-ulmi*. *Molecular Ecology* **15**, 249–262.

Paraskevis, D., Lemey, P., Salemi, M., *et al.* (2003) Analysis of the evolutionary relationships of HIV-1 and SIVcpz sequences using Bayesian inference: implications for the origin of HIV-1. *Molecular Biology and Evolution* **20**, 1986–1996.

Parker, E.D., Jr and Selander, R.K. (1976) The organization of genetic diversity in the parthenogenetic lizard *Cnemidophorus tesselatus*. *Genetics* **84**, 791–805.

Parkhill, J., Dougan, G., James, K.D., *et al.* (2001) Complete genome sequence of a multiple drug resistant *Salmonella enterica* serovar Typhi CT18. *Nature* **413**, 848–852.

Parkhill, J., Wren, B.W., Thomson, N.R., *et al.* (2001) Genome sequence of *Yersinia pestis*, the causative agent of plague. *Nature* **413**, 523–527.

Parris, M.J. (1999) Hybridization in leopard frogs (*Rana pipiens* complex): larval fitness components in single-genotype populations and mixtures. *Evolution* **53**, 1872–1883.

Parris, M.J. (2000) Experimental analysis of hybridization in leopard frogs (Anurae: Ranidae): larval performance in desiccating environments. *Copeia* **2000**, 11–19.

Parris, M.J. (2001a) High larval performance of leopard frog hybrids: effects of environment-dependent selection. *Ecology* **82**, 3001–3009.

Parris, M.J. (2001b) Hybridization in leopard frogs (*Rana pipiens* complex): variation in interspecific hybrid larval fitness components along a natural contact zone. *Evolutionary Ecology Research* **3**, 91–105.

Parris, M.J., Semlitsch, R.D., and Sage, R.D. (1999) Experimental analysis of the evolutionary potential of hybridization in leopard frogs (Anura: Ranidae). *Journal of Evolutionary Biology* **12**, 662–671.

Parry, C., Wain, J., Chinh, N.T., *et al.* (1998) Quinolone-resistant *Salmonella typhi*. *Lancet* **351**, 1289.

Parsons, T.J., Olson, S.L., and Braun, M.J. (1993) Unidirectional spread of secondary sexual plumage traits across an avian hybrid zone. *Science* **260**, 1643–1646.

Partensky, F., Hess, W.R., and Vaulot, D. (1999) *Prochlorococcus*, a marine photosynthetic prokaryote of global significance. *Microbiology and Molecular Biology Reviews* **63**, 106–127.

Paterson, A.H., Freeling, M., and Sasaki, T. (2005) Grains of knowledge: genomics of model cereals. *Genome Research* **15**, 1643–1650.

Paterson, H.E.H. (1985) The recognition concept of species. In E.S. Vrba, ed. *Species and Speciation*, pp. 21–29. Transvaal Museum Monograph no. 4. Transvaal Museum, Pretoria.

Paulo, O.S., Jordan, W.C., Bruford, M.W., and Nichols, R.A. (2002) Using nested clade analysis to assess the history of colonization and the persistence of populations of an Iberian lizard. *Molecular Ecology* **11**, 809–819.

Payseur, B.A. and Nachman, M.W. (2005) The genomics of speciation: Investigating the molecular correlates of X chromosome introgression across the hybrid zone between *Mus domesticus* and *Mus musculus*. *Biological Journal of the Linnean Society* **84**, 523–534.

Payseur, B.A., Krenz, J.G., and Nachman, M.W. (2004) Differential patterns of introgression across the X chromosome in a hybrid zone between two species of house mice. *Evolution* **58**, 2064–2078.

Pedulla, M.L., Ford, M.E., Houtz, J.M., *et al.* (2003) Origins of highly mosaic mycobacteriophage genomes. *Cell* **113**, 171–182.

Pereyra, R., Taylor, M.I., Turner, G.F., and Rico, C. (2004) Variation in habitat preference and population structure among three species of the Lake Malawi cichlid genus *Protomelas*. *Molecular Ecology* **13**, 2691–2697.

Pérombelon, M.C.M. (2002) Potato diseases caused by soft rot erwinias: an overview of pathogenesis. *Plant Pathology* **51**, 1–12.

Perry, R.D. and Fetherston, J.D. (1997) *Yersinia pestis*—etiologic agent of plague. *Clinical Microbiology Reviews* **10**, 35–66.

Petit, R.J., Pineau, E., Demesure, B., *et al.* (1997) Chloroplast DNA footprints of postglacial recolonization by oaks. *Proceedings of the National Academy of Sciences USA* **94**, 9996–10001.

Petit, R.J., Bodénès, C., Ducousso, A., *et al.* (2003) Hybridization as a mechanism of invasion in oaks. *New Phytologist* **161**, 151–164.

Petren, K., Grant, P.R., Grant, B.R., and Keller, L.F. (2005) Comparative landscape genetics and the adaptive radiation of Darwin's finches: the role of peripheral isolation. *Molecular Ecology* **14**, 2943–2957.

Pfenninger, M. and Posada, D. (2002) Phylogeographic history of the land snail *Candidula unifasciata* (Helicellinae, Stylommatophora): fragmentation, corridor migration, and secondary contact. *Evolution* **56**, 1776–1788.

Pigeon, D., Chouinard, A., and Bernatchez, L. (1997) Multiple modes of speciation involved in the parallel evolution of sympatric morphotypes of lake whitefish (*Coregonus clupeaformis*, Salmonidae). *Evolution* **51**, 196–205.

Pigliucci, M. (2003) Species as family resemblance concepts: the (dis-)solution of the species problem? *BioEssays* **25**, 596–602.

Pinceel, J., Jordaens, K., and Backeljau, T. (2005) Extreme mtDNA divergences in a terrestrial slug (Gastropoda, Pulmonata, Arionidae): accelerated evolution, allopatric divergence and secondary contact. *Journal of Evolutionary Biology* **18**, 1264–1280.

Pinto, M.A., Rubink, W.L., Coulson, R.N., *et al.* (2004) Temporal pattern of africanization in a feral honeybee population from Texas inferred from mitochondrial DNA. *Evolution* **58**, 1047–1055.

Pinto, M.A., Rubink, W.L., Patton, J.C., *et al.* (2005) Africanization in the United States: replacement of feral European honeybees (*Apis mellifera* L.) by an African hybrid swarm. *Genetics* **170**, 1653–1665.

Pires, J.C., Zhao, J., Schranz, M.E., *et al.* (2004) Flowering time divergence and genomic rearrangements in resynthesized *Brassica* polyploids (Brassicaceae). *Biological Journal of the Linnean Society* **82**, 675–688.

Plénet, S., Pagano, A., Joly, P., and Fouillet, P. (2000) Variation of plastic responses to oxygen availability within the hybridogenetic *Rana esculenta* complex. *Journal of Evolutionary Biology* **13**, 20–28.

Plénet, S., Joly, P., Hervant, F., *et al.* (2005) Are hybridogenetic complexes structured by habitat in water frogs? *Journal of Evolutionary Biology* **18**, 1575–1586.

Pongratz, N., Storhas, M., Carranza, S., and Michiels, N.K. (2003) Phylogeography of competing sexual and parthenogenetic forms of a freshwater flatworm: patterns and explanations. *BMC Evolutionary Biology* **3**, 23.

Pontes, O., Neves, N., Silva, M., *et al.* (2004) Chromosomal locus rearrangements are a rapid response to formation of the allotetraploid *Arabidopsis suecica* genome. *Proceedings of the National Academy of Sciences USA* **101**, 18240–18245.

Porwollik, S., Wong, R.M.-Y., and McClelland, M. (2002) Evolutionary genomics of *Salmonella*: gene acquisitions revealed by microarray analysis. *Proceedings of the National Academy of Sciences USA* **99**, 8956–8961.

Powell, J.R. (1983) Interspecific cytoplasmic gene flow in the absence of nuclear gene flow: evidence from *Drosophila*. *Proceedings of the National Academy of Sciences USA* **80**, 492–495.

Powell, J.R. (1991) Monophyly/paraphyly/polyphyly and gene/species trees: an example from *Drosophila*. *Molecular Biology and Evolution* **8**, 892–896.

Prager, E.M. and Wilson, A.C. (1975) Slow evolutionary loss of the potential for interspecific hybridization in birds: a manifestation of slow regulatory evolution. *Proceedings of the National Academy of Sciences USA* **72**, 200–204.

Price, C.S.C., Kim, C.H., Posluszny, J., and Coyne, J.A. (2000) Mechanisms of conspecific sperm precedence in *Drosophila*. *Evolution* **54**, 2028–2037.

Prince, V.E. and Pickett, F.B. (2002) Splitting pairs: the diverging fates of duplicated genes. *Nature Reviews Genetics* **3**, 827–837.

Promislow, D.E.L., Jung, C.F., and Arnold, M.L. (2001) Age-specific fitness components in hybrids of *Drosophila pseudoobscura* and *D. persimilis*. *Journal of Heredity* **92**, 30–37.

Pröschold, T., Harris, E.H., and Coleman, A.W. (2005) Portrait of a species: *Chlamydomonas reinhardtii*. *Genetics* **170**, 1601–1610.

Pyne, S., Skiena, S., and Futcher, B. (2005) Copy correction and concerted evolution in the conservation of yeast genes. *Genetics* **170**, 1501–1513.

Rabe, E.W. and Haufler, C.H. (1992) Incipient polyploid speciation in the maidenhair fern (*Adiantum pedatum*; Adiantaceae)? *American Journal of Botany* **79**, 701–707.

Raina, S.N., Mukai, Y., and Yamamoto, M. (1998) In situ hybridization identifies the diploid progenitor species of *Coffea arabica* (Rubiaceae). *Theoretical and Applied Genetics* **97**, 1204–1209.

Ramsey, J., Bradshaw, H.D., Jr, and Schemske, D.W. (2003) Components of reproductive isolation between the monkeyflowers *Mimulus lewisii* and *M. cardinalis* (Phrymaceae). *Evolution* **57**, 1520–1534.

Randolph, L.F. (1934) Chromosome numbers in native American and introduced species and cultivated varieties of *Iris. Bulletin of the American Iris Society* **52**, 61–66.

Randolph, L.F. (1966) *Iris nelsonii*, a new species of Louisiana iris of hybrid origin. *Baileya* **14**, 143–169.

Randolph, L.F., Mitra, J., and Nelson, I.S. (1961) Cytotaxonomic studies of Louisiana irises. *Botanical Gazette* **123**, 125–133.

Randolph, L.F., Nelson, I.S., and Plaisted, R.L. (1967) Negative evidence of introgression affecting the stability of Louisiana *Iris* species. *Cornell University Agricultural Experiment Station Memoir* **398**, 1–56.

Raufaste, N., Orth, A., Belkhir, K., *et al.* (2005) Inferences of selection and migration in the Danish house mouse hybrid zone. *Biological Journal of the Linnean Society* **84**, 593–616.

Rauscher, J.T., Doyle, J.J., and Brown, A.H.D. (2004) Multiple origins and nrDNA Internal Transcribed Spacer homeologue evolution in the *Glycine tomentella* (Leguminosae) allopolyploid complex. *Genetics* **166**, 987–998.

Ravid, K., Lu, J., Zimmet, J.M., and Jones, M.R. (2002) Roads to polyploidy: the megakaryocyte example. *Journal of Cellular Physiology* **190**, 7–20.

Ren, N. and Timko, M.P. (2001) AFLP analysis of genetic polymorphism and evolutionary relationships among cultivated and wild *Nicotiana* species. *Genome* **44**, 559–571.

Reyna-Lopez, G.E., Simpson, J., and Ruiz-Herrera, J. (1997) Differences in DNA methylation patterns are detectable during the dimorphic transition of fungi by amplification of restriction polymorphisms. *Molecular and General Genetics* **253**, 703–710.

Rhode, J.M. and Cruzan, M.B. (2005) Contributions of heterosis and epistasis to hybrid fitness. *American Naturalist* **166**, E124–E139.

Rhymer, J.M. and Simberloff, D. (1996) Extinction by hybridization and introgression. *Annual Review of Ecology and Systematics* **27**, 83–109.

Richard, F., Lombard, M., and Dutrillaux, B. (2000) Phylogenetic origin of human chromosomes 7, 16, and 19 and their homologs in placental mammals. *Genome Research* **10**, 644–651.

Richards, A.J. (1970) Hybridization in *Taraxacum. New Phytologist* **69**, 1103–1121.

Richt, J.A., Lager, K.M., Janke, B.H., *et al.* (2003) Pathogenic and antigenic properties of phylogenetically distinct reassortant H3N2 swine influenza viruses cocirculating in the United States. *Journal of Clinical Microbiology* **41**, 3198–3205.

Riddle, N.C. and Birchler, J.A. (2003) Effects of reunited diverged regulatory hierarchies in allopolyploids and species hybrids. *Trends in Genetics* **19**, 597–600.

Rieseberg, L.H. (1991) Homoploid reticulate evolution in *Helianthus* (Asteraceae): evidence from ribosomal genes. *American Journal of Botany* **78**, 1218–1237.

Rieseberg, L.H. (1997) Hybrid origins of plant species. *Annual Review of Ecology and Systematics* **28**, 359–389.

Rieseberg, L.H. and Soltis, D.E. (1991) Phylogenetic consequences of cytoplasmic gene flow in plants. *Evolutionary Trends in Plants* **5**, 65–84.

Rieseberg, L.H. and Livingstone, K. (2003) Chromosomal speciation in primates. *Science* **300**, 267–268.

Rieseberg, L.H., Soltis, D.E., and Palmer, J.D. (1988) A molecular reexamination of introgression between *Helianthus annuus* and *H. bolanderi* (Compositae). *Evolution* **42**, 227–238.

Rieseberg, L.H., Beckstorm-Sternberg, S., and Doan, K. (1990a) *Helianthus annuus* ssp *texanus* has chloroplast DNA and nuclear ribosomal RNA genes of *Helianthus debilis* ssp *cucumerifolius. Proceedings of the National Academy of Sciences USA* **87**, 593–597.

Rieseberg, L.H., Carter, R., and Zona, S. (1990b) Molecular tests of the hypothesized hybrid origin of two diploid *Helianthus* species (Asteraceae). *Evolution* **44**, 1498–1511.

Rieseberg, L.H., Chan, H.C.R., and Spore, C. (1993) Genomic map of a diploid hybrid species. *Heredity* **70**, 285–293.

Rieseberg, L.H., Van Fossen, C., and Desrochers, A.M. (1994) Hybrid speciation accompanied by genomic reorganization in wild sunflowers. *Nature* **375**, 313–316.

Rieseberg, L.H., Desrochers, A.M., and Youn, S.J. (1995) Interspecific pollen competition as a reproductive barrier between sympatric species of *Helianthus* (Asteraceae). *American Journal of Botany* **82**, 515–519.

Rieseberg, L.H., Sinervo, B., Linder, C.R., *et al.* (1996) Role of gene interactions in hybrid speciation: evidence from ancient and experimental hybrids. *Science* **272**, 741–745.

Rieseberg, L.H., Whitton, J., and Gardner, K. (1999) Hybrid zones and the genetic architecture of a barrier to gene flow between two sunflower species. *Genetics* **152**, 713–727.

Rieseberg, L.H., Raymond, O., Rosenthal, D.M., *et al.* (2003) Major ecological transitions in wild sunflowers facilitated by hybridization. *Science* **301**, 1211–1216.

Riley, H.P. (1938) A character analysis of colonies of *Iris fulva*, *Iris hexagona* var. *giganticaerulea* and natural hybrids. *American Journal of Botany* **25**, 727–738.

Riley, H.P. (1939) Pollen fertility in *Iris* and its bearing on the hybrid origin of some of Small's 'species'. *Journal of Heredity* **30**, 481–483.

Riley, H.P. (1942) Development of the embryo sac of *Iris fulva* and *I. hexagona* var *giganticaerulea*. *Transactions of the American Microscopical Society* **61**, 328–335.

Riley, H.P. (1943a) Cell size in developing ovaries of *Iris hexagona* var *giganticaerulea*. *American Journal of Botany* **30**, 356–361.

Riley, H.P. (1943b) Development and relative growth in ovaries of *Iris fulva* and *I. hexagona* var *giganticaerulea*. *American Journal of Botany* **29**, 323–331.

Riordan, C.E., Ault, J.G., Langreth, S.G., and Keithly, J.S. (2003) *Cryptosporidium parvum* Cpn60 targets a relict organelle. *Current Genetics* **44**, 138–147.

Robertson, H.M. (1983) Mating behavior and the evolution of *Drosophila mauritiana*. *Evolution* **37**, 1283–1293.

Robertson, W.R.B. (1916) Chromosome studies. I. Taxonomic relationships shown in the chromosomes of Tettigidae and Acrididae. V-shaped chromosomes and their significance in Acrididae, Locustidae and Gryllidae: chromosomes and variation. *Journal of Morphology* **27**, 179–331.

Robichaux, R.H. (1984) Variation in the tissue water relations of two sympatric Hawaiian *Dubautia* species and their natural hybrid. *Oecologia* **65**, 75–81.

Robichaux, R.H., Carr, G.D., Liebman, M., and Pearcy, R.W. (1990) Adaptive radiation of the Hawaiian Silversword alliance (Compositae-Madiinae): ecological, morphological, and physiological diversity. *Annals of the Missouri Botanical Garden* **77**, 64–72.

Robinson, T., Johnson, N.A., and Wade, M.J. (1994) Postcopulatory, prezygotic isolation: intraspecific and interspecific sperm precedence in *Tribolium* spp., flour beetles. *Heredity* **73**, 155–159.

Roca, A.L., Georgiadis, N., Pecon-Slattery, J., and O'Brien, S.J. (2001) Genetic evidence for two species of elephant in Africa. *Science* **293**, 1473–1477.

Roca, A.L., Georgiadis, N., and O'Brien, S.J. (2005) Cytonuclear genomic dissociation in African elephant species. *Nature Genetics* **37**, 96–100.

Rocap, G., Larimer, F.W., Lamerdin, J., *et al.* (2003) Genome divergence in two *Prochlorococcus* ecotypes reflects oceanic niche differentiation. *Nature* **424**, 1042–1047.

Roff, D.A., Mousseau, T.A., and Howard, D.J. (1999) Variation in genetic architecture of calling song among populations of *Allonemobius socius*, *A. fasciatus* and a hybrid population: drift or selection? *Evolution* **53**, 216–224.

Rogers, S.M. and Bernatchez, L. (2005) Integrating QTL mapping and genome scans towards the characterization of candidate loci under parallel selection in the lake whitefish (*Coregonus clupeaformis*). *Molecular Ecology* **14**, 351–361.

Rohwer, F., Segall, A., Steward, G., *et al.* (2000) The complete genomic sequence of the marine phage Roseophage SIO1 shares homology with nonmarine phages. *Limnology and Oceanography* **45**, 408–418.

Rong, J., Bowers, J.E., Schulze, S.R., *et al.* (2005) Comparative genomics of *Gossypium* and *Arabidopsis*: unraveling the consequences of both ancient and recent polyploidy. *Genome Research* **15**, 1198–1210.

Roosevelt, T. (1996) *Hunting Trips of a Ranchman*. Random House, New York.

Roques, S., Sévigny, J.-M., and Bernatchez, L. (2001) Evidence for broadscale introgressive hybridization between two redfish (genus *Sebastes*) in the North-west Atlantic: a rare marine example. *Molecular Ecology* **10**, 149–165.

Rosen, D.E. (1978) Vicariant patterns and historical explanation in biogeography. *Systematic Zoology* **27**, 159–188.

Rosenthal, D.M., Rieseberg, L.H., and Donovan, L.A. (2005) Re-creating ancient hybrid species' complex phenotypes from early-generation synthetic hybrids: three examples using wild sunflowers. *American Naturalist* **166**, 26–41.

Rosselló-Mora, R. and Amann, R. (2001) The species concept for prokaryotes. *FEMS Microbiology Reviews* **25**, 39–67.

Ruas, P.M., Ruas, C.F., Rampim, L., *et al.* (2003) Genetic relationships in *Coffea* species and parentage determination of interspecific hybrids using ISSR (Inter- Simple Sequence Repeat) markers. *Genetics and Molecular Biology* **26**, 319–327.

Rüber, L., Verheyen, E., and Meyer, A. (1999) Replicated evolution of trophic specializations in an endemic cichlid fish lineage from Lake Tanganyika. *Proceedings of the National Academy of Sciences USA* **96**, 10230–10235.

Rüber, L., Meyer, A., Sturmbauer, C., and Verheyen, E. (2001) Population structure in two sympatric species of the Lake Tanganyika cichlid tribe Eretmodini: evidence for introgression. *Molecular Ecology* **10**, 1207–1225.

Ruvolo, M., Disotell, T.R., Allard, M.W., *et al.* (1991) Resolution of the African hominoid trichotomy by use of a mitochondrial gene sequence. *Proceedings of the National Academy of Sciences USA* **88**, 1570–1574.

Sætre, G.-P., Moum, T., Bures, S., *et al.* (1997) A sexually selected character displacement in flycatchers reinforces premating isolation. *Nature* **387**, 589–592.

Sætre, G.-P., Borge, T., Lindell, J., *et al.* (2001) Speciation, introgressive hybridization and nonlinear rate of molecular evolution in flycatchers. *Molecular Ecology* **10**, 737–749.

Sætre, G.-P., Borge, T., Lindroos, K., *et al.* (2003) Sex chromosome evolution and speciation in *Ficedula* flycatchers. *Proceedings of the Royal Society of London Series B* **270**, 53–59.

Sage, R.D. and Selander, R.K. (1979) Hybridization between species of the *Rana pipiens* complex in central Texas. *Evolution* **33**, 1069–1088.

Sage, R.D., Atchley, W.R., and Capanna, E. (1993) House mice as models in systematic biology. *Systematic Biology* **42**, 523–561.

Salem, A.-H., Ray, D.A., Xing, J., *et al.* (2003) Alu elements and hominid phylogenetics. *Proceedings of the National Academy of Sciences USA* **100**, 12787–12791.

Salik, J. and Pfeffer, E. (1999) The interplay of hybridization and clonal reproduction in the evolution of willows. *Plant Ecology* **141**, 163–178.

Salmon, A., Ainoche, M.L., and Wendel, J.F. (2005) Genetic and epigenetic consequences of recent hybridization and polyploidy in *Spartina* (Poaceae). *Molecular Ecology* **14**, 1163–1175.

Salzburger, W., Baric, S., and Sturmbauer, C. (2002) Speciation via introgressive hybridization in East African cichlids? *Molecular Ecology* **11**, 619–625.

Sanderson, N., Szymura, J.M., and Barton, N.H. (1992) Variation in mating call across the hybrid zone between the fire-bellied toads *Bombina bombina* and *B. variegata*. *Evolution* **46**, 595–607.

Sang, T., Crawford, D.J., and Stuessy, T.F. (1995) Documentation of reticulate evolution in peonies (*Paeonia*) using internal transcribed spacer sequences of nuclear ribosomal DNA: implications for biogeography and concerted evolution. *Proceedings of the National Academy of Sciences USA* **92**, 6813–6817.

Sang, T., Crawford, D.J., and Stuessy, T.F. (1997a) Chloroplast DNA phylogeny, reticulate evolution, and biogeography of *Paeonia* (Paeoniaceae). *American Journal of Botany* **84**, 1120–1136.

Sang, T., Donoghue, M.J., and Zhang, D. (1997b) Evolution of alcohol dehydrogenase genes in peonies (*Paeonia*): phylogenetic relationships of putative nonhybrid species. *Molecular Biology and Evolution* **14**, 994–1007.

Sang, T., Pan, J., Zhang, D., *et al.* (2004) Origins of polyploids: an example from peonies (*Paeonia*) and a model for angiosperms. *Biological Journal of the Linnean Society* **82**, 561–571.

Sang, T. and Zhang, D. (1999) Reconstructing hybrid speciation using sequences of low copy nuclear genes: hybrid origins of five *Paeonia* species based on *Adh* gene phylogenies. *Systematic Botany* **24**, 148–163.

Santucci, F., Nascetti, G., and Bullini, L. (1996) Hybrid zones between two genetically differentiated forms of the pond frog *Rana lessonae* in southern Italy. *Journal of Evolutionary Biology* **9**, 429–450.

Sarich, V.M. and Wilson, A.C. (1967) Immunological time scale for hominid evolution. *Science* **158**, 1200–1203.

Såstad, S.M., Stenøien, H.K., Flatberg, K.I., and Bakken, S. (2001) The narrow endemic *Sphagnum troendelagicum* is an allopolyploid derivative of the widespread *S. balticum* and *S. tenellum*. *Systematic Botany* **26**, 66–74.

Savolainen, P., Zhang, Y.-p., Luo, J., *et al.* (2002) Genetic evidence for an East Asian origin of domestic dogs. *Science* **298**, 1610–1613.

Scali, V., Passamonti, M., Marescalchi, O., and Mantovani, B. (2003) Linkage between sexual and asexual lineages: genome evolution in *Bacillus* stick insects. *Biological Journal of the Linnean Society* **79**, 137–150.

Schaeffer, S.W. and Miller, E.L. (1993) Estimates of linkage disequilibrium and the recombination parameter determined from segregating nucleotide sites in the alcohol dehydrogenase region of *Drosophila pseudoobscura*. *Genetics* **135**, 541–552.

Scheen, A.-C., Elven, R., and Brochmann, C. (2002) A molecular-morphological approach solves taxonomic controversy in arctic *Draba* (Brassicaceae). *Canadian Journal of Botany* **80**, 59–71.

Schemske, D.W. (2000) Understanding the origin of species. *Evolution* **54**, 1069–1073.

Schilling, E.E. and Heiser, C.B., Jr (1981) Infrageneric classification of *Helianthus* (Compositae). *Taxon* **80**, 393–403.

Schlupp, I. (2005) The evolutionary ecology of gynogenesis. *Annual Review of Ecology, Evolution and Systematics* **36**, 399–417.

Schmeller, D.S., Seitz, A., Crivelli, A., and Veith, M. (2005) Crossing species' range borders: interspecies gene exchange mediated by hybridogenesis. *Proceedings of the Royal Society of London Series B* **272**, 1625–1631.

Schranz, M.E., Dobes, C., Koch, M.A., and Mitchell-Olds, T. (2005) Sexual reproduction, hybridization, apomixis, and polyploidization in the genus *Boechera* (Brassicaceae). *American Journal of Botany* **92**, 1797–1810.

Schwartz, M.K., Mills, L.S., McKelvey, K.S., *et al.* (2002) DNA reveals high dispersal synchronizing the population dynamics of Canada lynx. *Nature* **415**, 520–522.

Schwartz, M.K., Pilgrim, K.L., McKelvey, K.S., *et al.* (2004) Hybridization between Canada lynx and bobcats: genetic results and management implications. *Conservation Genetics* **5**, 349–355.

Schwarz, D., Matta, B.M., Shakir-Botteri, N.L., and McPheron, B.A. (2005) Host shift to an invasive plant triggers rapid animal hybrid speciation. *Nature* **436**, 546–549.

Schweitzer, J.A., Martinsen, G.D., and Whitham, T.G. (2002) Cottonwood hybrids gain fitness traits of both parents: a mechanism for their long-term persistence? *American Journal of Botany* **89**, 981–990.

Sebastiani, F., Barberio, C., Casalone, E., *et al.* (2002) Crosses between *Saccharomyces cerevisiae* and *Saccharomyces bayanus* generate fertile hybrids. *Research in Microbiology* **153**, 53–58.

Seehausen, O. (2004) Hybridization and adaptive radiation. *Trends in Ecology and Evolution* **19**, 198–107.

Seehausen, O. and van Alphen, J.J.M. (1998) The effect of male coloration on female mate choice in closely related Lake Victoria cichlids (*Haplochromis nyererei* complex). *Behavioral Ecology and Sociobiology* **42**, 1–8.

Seehausen, O., van Alphen, J.J.M., and Witte, F. (1997) Cichlid fish diversity threatened by eutrophication that curbs sexual selection. *Science* **277**, 1808–1811.

Seehausen, O., Koetsier, E., Schneider, M.V., *et al.* (2002) Nuclear markers reveal unexpected genetic variation and a Congolese-Nilotic origin of the Lake Victoria cichlid species flock. *Proceedings of the Royal Society of London Series B* **270**, 129–137.

Segal, G., Purcell, M., and Shuman, H.A. (1998) Host cell killing and bacterial conjugation require overlapping sets of genes within a 22-kb region of the *Legionella pneumophila* genome. *Proceedings of the National Academy of Sciences USA* **95**, 1669–1674.

Segal, G., Russo, J.J., and Shuman, H.A. (1999) Relationships between a new type IV secretion system and the *icm/dot* virulence system of *Legionella pneumophila*. *Molecular Microbiology* **34**, 799–809.

Semlitsch, R.D. (1993a) Adaptive genetic variation in growth and development of tadpoles of the hybridogenetic *Rana esculenta* complex. *Evolution* **47**, 1805–1818.

Semlitsch, R.D. (1993b) Asymmetric competition in mixed populations of tadpoles of the hybridogenetic *Rana esculenta* complex. *Evolution* **47**, 510–519.

Semlitsch, R.D. and Reyer, H.-U. (1992) Performance of tadpoles from the hybridogenetic *Rana esculenta* complex: interactions with pond drying and interspecific competition. *Evolution* **46**, 665–676.

Semlitsch, R.D., Schmiedehausen, S., Hotz, H., and Beerli, P. (1996) Genetic compatibility between sexual and clonal genomes in local populations of the hybridogenetic *Rana esculenta*. *Evolutionary Ecology* **10**, 531–543.

Semlitsch, R.D., Hotz, H., and Guex, G.-D. (1997) Competition among tadpoles of coexisting hemiclones of hybridogenetic *Rana esculenta*: support for the frozen niche variation model. *Evolution* **51**, 1249–1261.

Servedio, M.R. and Noor, M.A.F. (2003) The role of reinforcement in speciation: theory and data. *Annual Review of Ecology and Systematics* **34**, 339–364.

Shaked, H., Kashkush, K., Ozkan, H., *et al.* (2001) Sequence elimination and cytosine methylation are rapid and reproducible responses of the genome to wide hybridization and allopolyploidy in wheat. *The Plant Cell* **13**, 1749–1759.

Sharakhov, I.V., Sharakhova, M.V., Mbogo, C.M., *et al.* (2001) Linear and spatial organization of polytene chromosomes of the African malaria mosquito *Anopheles funestus*. *Genetics* **159**, 211–218.

Shaw, A.J. and Goffinet, B. (2000) Molecular evidence of reticulate evolution in the peatmosses (*Sphagnum*), including *S. ehyalinum sp. nov. The Bryologist* **103**, 357–374.

Shaw, A.J., Cox, C.J., and Boles, S.B. (2005) Phylogeny, species delimitation, and recombination in *Sphagnum* section *Acutifolia*. *Systematic Botany* **30**, 16–33.

Shaw, D.D. (1976) Population cytogenetics of the genus *Caledia* (Orthoptera: Acridinae) I. Inter- and intraspecific karyotype diversity. *Chromosoma* **54**, 221–243.

Shaw, D.D. and Wilkinson, P. (1978) 'Homologies' between non-homologous chromosomes in the grasshopper *Caledia captiva*. *Chromosoma* **68**, 241–259.

Shaw, D.D. and Coates, D.J. (1983) Chromosomal variation and the concept of the coadapted genome—a direct cytological assessment. In P.E. Brandham and M.D. Bennett, eds. *Kew Chromosome Conference, II.*, pp. 207–216. George Allen and Unwin, London.

Shaw, D.D., Webb, G.C., and Wilkinson, P. (1976) Population cytogenetics of the genus *Caledia* (Orthoptera: Acridinae) II. Variation in the pattern of C-banding. *Chromosoma* **56**, 169–190.

Shaw, D.D., Wilkinson, P., and Moran, C. (1979) A comparison of chromosomal and allozymal variation across a narrow hybrid zone in the grasshopper *Caledia captiva*. *Chromosoma* **75**, 333–351.

Shaw, D.D., Moran, C., and Wilkinson, P. (1980) Chromosomal reorganization, geographic differentiation and the mechanism of speciation in the genus *Caledia*. In R.L. Blackman, G.M. Hewitt and M. Ashburner, eds. *Insect Cytogenetics*, pp. 171–194. Blackwell Scientific Publications, Oxford.

Shaw, D.D., Wilkinson, P., and Coates, D.J. (1982) The chromosomal component of reproductive isolation in the grasshopper *Caledia captiva* II. The relative viabilities of recombinant and non-recombinant chromosomes during embryogenesis. *Chromosoma* **86**, 533–549.

Shaw, D.D., Wilkinson, P., and Coates, D.J. (1983) Increased chromosomal mutation rate after hybridization between two subspecies of grasshoppers. *Science* **220**, 1165–1167.

Shaw, D.D., Coates, D.J., Arnold, M.L., and Wilkinson, P. (1985) Temporal variation in the chromosomal structure of a hybrid zone and its relationship to karyotypic repatterning. *Heredity* **55**, 293–306.

Shaw, D.D., Coates, D.J., and Arnold, M.L. (1988) Complex patterns of chromosomal variation along a latitudinal cline in the grasshopper *Caledia captiva*. *Genome* **30**, 108–117.

Shaw, D.D., Marchant, A.D., Arnold, M.L., *et al.* (1990) The control of gene flow across a narrow hybrid zone: a selective role for chromosomal rearrangement? *Canadian Journal of Zoology* **68**, 1761–1769.

Shaw, D.D., Marchant, A.D., Contreras, N., *et al.* (1993) Genomic and environmental determinants of a narrow hybrid zone: cause or coincidence? In R.G. Harrison, ed. *Hybrid Zones and the Evolutionary Process*, pp. 165–195. Oxford University Press, Oxford.

Sherburne, C.K., Lawley, T.D., Gilmour, M.W., *et al.* (2000) The complete DNA sequence and analysis of R27, a large IncHI plasmid from *Salmonella typhi* that is temperature sensitive for transfer. *Nucleic Acids Research* **28**, 2177–2186.

Shoemaker, D.D., Ross, K.G., and Arnold, M.L. (1996) Genetic structure and evolution of a fire ant hybrid zone. *Evolution* **50**, 1958–1976.

Shotake, T., Nozawa, K., and Tanabe, Y. (1977) Blood protein variations in baboons I. Gene exchange and genetic distance between *Papio anubis*, *Papio hamadryas* and their hybrid. *Japanese Journal of Genetics* **52**, 223–237.

Shriner, D., Rodrigo, A.G., Nickle, D.C., and Mullins, J.I. (2004) Pervasive genomic recombination of HIV-1 *in vivo*. *Genetics* **167**, 1573–1583.

Shutt, T.E. and Gray, M.W. (2006) Bacteriophage origins of mitochondrial replication and transcription proteins. *Trends in Genetics* **22**, 90–95.

Silander, O.K., Weinreich, D.M., Wright, K.M., *et al.* (2005). Widespread genetic exchange among terrestrial bacteriophages. *Proceedings of the National Academy of Sciences USA* **102**, 19009–19014.

Simonson, A.B., Servin, J.A., Skophammer, R.G., *et al.* (2005) Decoding the genomic tree of life. *Proceedings of the National Academy of Sciences USA* **102**, 6608–6613.

Sites, J.W., Jr and Marshall, J.C. (2004) Operational criteria for delimiting species. *Annual Review of Ecology, Evolution and Systematics* **35**, 199–227.

Skalická, K., Lim, K.Y., Matyásek, R., *et al.* (2003) Rapid evolution of parental rDNA in a synthetic tobacco allotetraploid line. *American Journal of Botany* **90**, 988–996.

Skalická, K., Lim, K.Y., Matyásek, R., *et al.* (2005) Preferential elimination of repeated DNA sequences from the paternal, *Nicotiana tomentosiformis* genome donor of a synthetic, allotetraploid tobacco. *New Phytologist* **166**, 291–303.

Slightom, J.L., Blechl, A.E., and Smithies, O. (1980) Human fetal $^{G}\gamma$- and $^{A}\gamma$-globin genes: complete nucleotide sequences suggest that DNA can be exchanged between these duplicated genes. *Cell* **21**, 627–638.

Slotman, M., della Torre, A., and Powell, J.R. (2004) The genetics of inviability and male sterility in hybrids between *Anopheles gambiae* and *An. arabiensis*. *Genetics* **167**, 275–287.

Slotman, M., della Torre, A., and Powell, J.R. (2005) Female sterility in hybrids between *Anopheles gambiae* and *A. arabiensis*, and the causes of Haldane's rule. *Evolution* **59**, 1016–1026.

Small, J.K. and Alexander, E.J. (1931) Botanical interpretation of the Iridaceous plants of the Gulf States. *New York Botanical Garden Contribution* **327**, 325–357.

Smith, D.J., Lapedes, A.S., deJong, J.C., *et al.* (2004) Mapping antigenic and genetic evolution of influenza virus. *Science* **305**, 371–376.

Smith, E.B. (1968) Pollen competition and relatedness in *Haplopappus* section *Isopappus*. *Botanical Gazette* **129**, 371–373.

Smith, E.B. (1970) Pollen competition and relatedness in *Haplopappus* section *Isopappus* (Compositae) II. *American Journal of Botany* **57**, 874–880.

Smith, G.P. (1976) Evolution of repeated DNA sequences by unequal crossover. *Science* **191**, 528–535.

Smith, J.L. and Fonseca, D.M. (2004) Rapid assays for identification of members of the *Culex (Culex) pipiens* complex, their hybrids, and other sibling species (Diptera: Culicidae). *American Journal of Tropical Medicine and Hygiene* **70**, 339–345.

Smith, P.F. and Kornfield, I. (2002) Phylogeography of Lake Malawi cichlids of the genus *Pseudotropheus*: significance of allopatric colour variation. *Proceedings of the Royal Society of London Series B* **269**, 2495–2502.

Smith, P.F., Konings, A.D., and Kornfield, I. (2003) Hybrid origin of a cichlid population in Lake Malawi: implications for genetic variation and species diversity. *Molecular Ecology* **12**, 2497–2504.

Smith, R.L. and Sytsma, K.J. (1990) Evolution of *Populus nigra* (sect. *Aigeiros*): introgressive hybridization and the chloroplast contribution of *Populus alba* (sect. *Populus*). *American Journal of Botany* **77**, 1176–1187.

Smoot, L.M., Franke, D.D., McGillivary, G., and Actis, L.A. (2002) Genomic analysis of the F3031 Brazilian purpuric fever clone of *Haemophilus influenzae* biogroup aegyptius by PCR-based subtractive hybridization. *Infection and Immunity* **70**, 2694–2699.

Snow, A.A., Pilson, D., Rieseberg, L.H., *et al.* (2003) A Bt transgene reduces herbivory and enhances fecundity in wild sunflowers. *Ecological Applications* **13**, 279–286.

Solignac, M. and Monnerot, M. (1986) Race formation, speciation, and introgression within *Drosophila simulans*, *D. mauritiana*, and *D. sechellia* inferred from mitochondrial DNA analysis. *Evolution* **40**, 531–539.

Soltis, D.E. and Soltis, P.S. (1993) Molecular data and the dynamic nature of polyploidy. *Critical Reviews in Plant Sciences* **12**, 243–273.

Soltis, D.E. and Soltis, P.S. (1995) The dynamic nature of polyploid genomes. *Proceedings of the National Academy of Sciences USA* **92**, 8089–8091.

Soltis, D.E., Soltis, P.S., and Tate, J.A. (2003) Advances in the study of polyploidy since *Plant Speciation*. *New Phytologist* **161**, 173–191.

Soltis, D.E., Soltis, P.S., Pires, J.C., *et al.* (2004) Recent and recurrent polyploidy in *Tragopogon* (Asteraceae). Cytogenetic, genomic and genetic comparisons. *Biological Journal of the Linnean Society* **82**, 485–501.

Song, B.-H., Wang, X.-Q., Wang, X.-R., *et al.* (2002) Maternal lineages of *Pinus densata*, a diploid hybrid. *Molecular Ecology* **11**, 1057–1063.

Song, B.-H., Wang, X.-Q., Wang, X.-R., *et al.* (2003) Cytoplasmic composition in *Pinus densata* and population establishment of the diploid hybrid pine. *Molecular Ecology* **12**, 2995–3001.

Song, K., Lu, P., Tang, K., and Osborn, T.C. (1995) Rapid genome change in synthetic polyploids of *Brassica* and its implications for polyploid evolution. *Proceedings of the National Academy of Sciences USA* **92**, 7719–7723.

Sørensen, S.J., Bailey, M., Hansen, L.H., *et al.* (2005) Studying plasmid horizontal transfer *in situ*: a critical review. *Nature Reviews Microbiology* **3**, 700–710.

Souto, R.P. and Zingales, B. (1993) Sensitive detection and strain classification of *Trypanosoma cruzi* by amplification of a ribosomal RNA sequence. *Molecular and Biochemical Parasitology* **62**, 45–52.

Spielman, A. (2001) Structure and seasonality of Nearctic *Culex pipiens* populations. *Annals of the New York Academy of Sciences* **951**, 220–234.

Spies, T.A. and Barnes, B.V. (1981) A morphological analysis of *Populus alba*, *P. grandidentata* and their natural hybrids in southeastern Michigan. *Silvae Genetica* **30**, 102–106.

Spring, J. (1997) Vertebrate evolution by interspecific hybridisation—are we polyploid? *FEBS Letters* **400**, 2–8.

Sreevatsan, S., Pan, X., Stockbauer, K.E., *et al.* (1997) Restricted structural gene polymorphism in the *Mycobacterium tuberculosis* complex indicates evolutionarily recent global dissemination. *Proceedings of the National Academy of Sciences USA* **94**, 9869–9874.

Stackebrandt, E., Frederiksen, W., Garrity, G.M., *et al.* (2002) Report of the ad hoc committee for the re-evaluation of the species definition in bacteriology. *International Journal of Systematic and Evolutionary Microbiology* **52**, 1043–1047.

Stager, J.C., Day, J.J., and Santini, S. (2004) Comment on 'Origin of the superflock of cichlid fishes from Lake Victoria, East Africa'. *Science* **304**, 963b.

Stanley, S.L., Jr (2003) Amoebiasis. *Lancet* **361**, 1025–1034.

Stebbins, G.L., Jr (1938) Cytogenetic studies in *Paeonia* II. The cytology of the diploid species and hybrids. *Genetics* **23**, 83–110.

Stebbins, G.L., Jr (1942) The genetic approach to problems of rare and endemic species. *Madroño* **6**, 241–272.

Stebbins, G.L., Jr (1947) Types of polyploids: their classification and significance. *Advances in Genetics* **1**, 403–429.

Stebbins, G.L., Jr (1950) *Variation and Evolution in Plants*. Columbia University Press, New York.

Stebbins, G.L., Jr (1959) The role of hybridization in evolution. *Proceedings of the American Philosophical Society* **103**, 231–251.

Stebbins, G.L., Jr and Daly, K. (1961) Changes in the variation pattern of a hybrid population of *Helianthus* over an eight-year period. *Evolution* **15**, 60–71.

Stebbins, G.L., Jr, Matzke, E.B., and Epling, C. (1947) Hybridization in a population of *Quercus marilandica* and *Quercus ilicifolia*. *Evolution* **1**, 79–88.

Steiger, D.L., Nagai, C., Moore, P.H., *et al.* (2002) AFLP analysis of genetic diversity within and among *Coffea arabica* cultivars. *Theoretical and Applied Genetics* **105**, 209–215.

Steinhauer, D.A. and Skehel, J.J. (2002) Genetics of influenza viruses. *Annual Review of Genetics* **36**, 305–332.

Stöhr, K. (2002) Influenza—WHO cares. *Lancet Infectious Diseases* **2**, 517.

Storchová, R., Gregorová, S., Buckiová, D., *et al.* (2004) Genetic analysis of X-linked hybrid sterility in the house mouse. *Mammalian Genome* **15**, 515–524.

Strasburg, J.L. and Kearney, M. (2005) Phylogeography of sexual *Heteronotia binoei* (Gekkonidae) in the Australian arid zone: climatic cycling and repetitive hybridization. *Molecular Ecology* **14**, 2755–2772.

Streelman, J.T., Gmyrek, S.L., Kidd, M.R., *et al.* (2004) Hybridization and contemporary evolution in an introduced cichlid fish from Lake Malawi National Park. *Molecular Ecology* **13**, 2471–2479.

Streiff, R., Ducousso, A., Lexer, C., *et al.* (1999) Pollen dispersal inferred from paternity analysis in a mixed oak stand of *Quercus robur* L. and *Q. petraea* (Matt) Liebl. *Molecular Ecology* **8**, 831–841.

Streiff, R., Veyrier, R., Audiot, P., *et al.* (2005) Introgression in natural populations of bioindicators: a case study of

Carabus splendens and *Carabus punctatoauratus. Molecular Ecology* **14**, 3775–3786.

Striepen, B., White, M.W., Li, C., *et al.* (2002) Genetic complementation in apicomplexan parasites. *Proceedings of the National Academy of Sciences USA* **99**, 6304–6309.

Striepen, B., Pruijssers, A.J.P., Huang, J., *et al.* (2004) Gene transfer in the evolution of parasite nucleotide biosynthesis. *Proceedings of the National Academy of Sciences USA* **101**, 3154–3159.

Stump, A.D., Shoener, J.A., Costantini, C., *et al.* (2005) Sex-linked differentiation between incipient species of *Anopheles gambiae. Genetics* **169**, 1509–1519.

Sturgill-Koszycki, S. and Swanson, M.S. (2000) *Legionella pneumophila* replication vacuoles mature into acidic, endocytic organelles. *Journal of Experimental Medicine* **192**, 1261–1272.

Sturmbauer, C., Baric, S., Salzburger, W., *et al.* (2001) Lake level fluctuations synchronize genetic divergences of cichlid fishes in African lakes. *Molecular Biology and Evolution* **18**, 144–154.

Subbarao, K., Klimov, A., Katz, J., *et al.* (1998) Characterization of an avian influenza A (H5N1) virus isolated from a child with a fatal respiratory illness. *Science* **279**, 393–396.

Suehs, C.M., Affre, L., and Médail, F. (2004a) Invasion dynamics of two alien *Carpobrotus* (Aizoaceae) taxa on a Mediterranean island: I. Genetic diversity and introgression. *Heredity* **92**, 31–40.

Suehs, C.M., Affre, L., and Médail, F. (2004b) Invasion dynamics of two alien *Carpobrotus* (Aizoaceae) taxa on a Mediterranean island: II. Reproductive strategies. *Heredity* **92**, 550–556.

Suehs, C.M., Affre, L., and Médail, F. (2005) Unexpected insularity effects in invasive plant mating systems: the case of *Carpobrotus* (Aizoaceae) taxa in the Mediterranean Basin. *Biological Journal of the Linnean Society* **85**, 65–79.

Sugawara, T., Terai, Y., Imai, H., *et al.* (2005) Parallelism of amino acid changes at the RH1 affecting spectral sensitivity among deep-water cichlids from Lake Tanganyika and Malawi. *Proceedings of the National Academy of Sciences USA* **102**, 5448–5453.

Sugino, R.P. and Innan, H. (2005) Estimating the time to the whole-genome duplication and the duration of concerted evolution via gene conversion in yeast. *Genetics* **171**, 63–69.

Sullivan, J.P., Lavoué, S., and Hopkins, C.D. (2000) Molecular systematics of the African electric fishes (Mormyroidea: Teleostei) and a model for the evolution of their electric organs. *Journal of Experimental Biology* **203**, 665–683.

Sullivan, J.P., Lavoué, S., and Hopkins, C.D. (2002) Discovery and phylogenetic analysis of a riverine species flock of African electric fishes (Mormyridae: Teleostei). *Evolution* **56**, 597–616.

Sullivan, J.P., Lavoué, S., Arnegard, M.E., and Hopkins, C.D. (2004) AFLPs resolve phylogeny and reveal mitochondrial introgression within a species flock of African electric fish (Mormyroidea: Teleostei) *Evolution* **58**, 825–841.

Sullivan, M.B., Waterbury, J.B., and Chisholm, S.W. (2003) Cyanophages infecting the oceanic cyanobacterium *Prochlorococcus. Nature* **424**, 1047–1051.

Sweigart, A.L. and Willis, J.H. (2003) Patterns of nucleotide diversity in two species of *Mimulus* are affected by mating system and asymmetric introgression. *Evolution* **57**, 2490–2506.

Szmulewicz, M.N., Andino, L.M., Reategui, E.P., *et al.* (1999) An *Alu* insertion polymorphism in a baboon hybrid zone. *American Journal of Physical Anthropology* **109**, 1–8.

Szymura, J.M. (1976) New data on the hybrid zone between *Bombina bombina* and *Bombina variegata* (Anura, Discoglossidae). *Bulletin Academia Polonica Scientatis Class II* **24**, 355–363.

Szymura, J.M. (1993) Analysis of hybrid zones with *Bombina*. In R.G. Harrison, ed. *Hybrid Zones and the Evolutionary Process*, pp. 261–289. Oxford University Press, Oxford.

Szymura, J.M. and Barton, N.H. (1986) Genetic analysis of a hybrid zone between the fire-bellied toads, *Bombina bombina* and *B. variegata*, near Cracow in southern Poland. *Evolution* **40**, 1141–1159.

Szymura, J.M. and Barton, N.H. (1991) The genetic structure of the hybrid zone between the fire-bellied toads *Bombina bombina* and *B. variegata*: comparisons between transects and between loci. *Evolution* **45**, 237–261.

Szymura, J.M., Spolsky, C., and Uzzell, T. (1985) Concordant change in mitochondrial and nuclear genes in a hybrid zone between two frog species (genus *Bombina*). *Experientia* **41**, 1469–1470.

Szymura, J.M., Uzzell, T., and Spolsky, C. (2000) Mitochondrial DNA variation in the hybridizing fire-bellied toads, *Bombina bombina* and *B. variegata. Molecular Ecology* **9**, 891–899.

Takahashi, K., Terai, Y., Nishida, M., and Okada, N. (1998) A novel family of short interspersed repetitive elements (SINEs) from cichlids: the patterns of insertion of SINEs at orthologous loci support the proposed monophyly of four major groups of cichlid fishes in Lake Tanganyika. *Molecular Biology and Evolution* **15**, 391–407.

Takahashi, K., Nishida, M., Yuma, M., and Okada, N. (2001) Retroposition of the AFC family of SINEs (Short Interspersed Repetitive Elements) before and during the adaptive radiation of cichlid fishes in Lake Malawi and related inferences about phylogeny. *Journal of Molecular Evolution* **53**, 496–507.

Takahata, N., Lee, S.-H., and Satta, Y. (2001) Testing multiregionality of modern human origins. *Molecular Biology and Evolution* **18**, 172–183.

Tao, Y. and Hartl, D.L. (2003) Genetic dissection of hybrid incompatibilities between *Drosophila simulans* and *D. mauritiana*. III. Heterogeneous accumulation of hybrid incompatibilities, degree of dominance, and implications for Haldane's Rule. *Evolution* **57**, 2580–2598.

Taubenberger, J.K., Reid, A.H., Lourens, R.M., *et al.* (2005) Characterization of the 1918 influenza virus polymerase genes. *Nature* **437**, 889–893.

Taverne, L. (1972) Ostéologie des genres *Mormyrus* Linné, *Mormyrops* Müller, *Hyperopisus* Gill, *Myomyrus* Boulenger, *Stomatorhinus* Boulenger et *Gymnarchus* Cuvier. Considérations générales sur la systématique des Poissons de l'ordre des Mormyriformes. *Annales du Musée Royal de l'Afrique Centrale, Sciences Zoologiques* **200**, 1–194.

Taylor, D.J., Finston, T.L., and Hebert, P.D.N. (1998) Biogeography of a widespread freshwater crustacean: pseudocongruence and cryptic endemism in the North American *Daphnia laevis* complex. *Evolution* **52**, 1648–1670.

Taylor, D.J., Sprenger, H.L., and Ishida, S. (2005) Geographic and phylogenetic evidence for dispersed nuclear introgression in a daphniid with sexual propagules. *Molecular Ecology* **14**, 525–537.

Tegelström, H. and Gelter, H.P. (1990) Haldane's rule and sex biased gene flow between two hybridizing flycatcher species (*Ficedula albicollis* and *F. hypoleuca*, Aves: Muscicapidae). *Evolution* **44**, 2012–2021.

Templeton, A.R. (1981) Mechanisms of speciation—a population genetic approach. *Annual Review of Ecology and Systematics* **12**, 23–48.

Templeton, A.R. (1983) Phylogenetic inference from restriction endonuclease cleavage site maps with particular reference to the evolution of humans and apes. *Evolution* **37**, 221–244.

Templeton, A.R. (1989) The meaning of species and speciation: a genetic perspective. In D. Otte and J.A. Endler, eds. *Speciation and its Consequences*, pp. 3–27. Sinauer Associates, Sunderland, MA.

Templeton, A.R. (1993) The 'Eve' hypotheses: a genetic critique and reanalysis. *American Anthropologist* **95**, 51–72.

Templeton, A.R. (1998) Nested clade analyses of phylogeographic data: testing hypotheses about gene flow and population history. *Molecular Ecology* **7**, 381–397.

Templeton, A.R. (2001) Using phylogeographic analyses of gene trees to test species status and processes. *Molecular Ecology* **10**, 779–791.

Templeton, A.R. (2002) Out of Africa again and again. *Nature* **416**, 45–51.

Templeton, A.R. (2004a) Statistical phylogeography: methods of evaluating and minimizing inference errors. *Molecular Ecology* **13**, 789–809.

Templeton, A.R. (2004b) Using haplotype trees for phylogeographic and species inferences in fish populations. *Environmental Biology of Fishes* **69**, 7–20.

Templeton, A.R. (2005) Haplotype trees and modern human origins. *Yearbook of Physical Anthropology* **48**, 33–59.

Templeton, A.R., Boerwinkle, E., and Sing, C.F. (1987) A cladistic analysis of phenotypic associations with haplotypes inferred from restriction endonuclease mapping. I. Basic theory and an analysis of alcohol dehydrogenase activity in *Drosophila*. *Genetics* **117**, 343–351.

Templeton, A.R., Crandall, K.A., and Sing, C.F. (1992) A cladistic analysis of phenotypic associations with haplotypes inferred from restriction endonuclease mapping and DNA sequence data. III. Cladogram estimation. *Genetics* **132**, 619–633.

Templeton, A.R., Routman, E., and Phillips, C.A. (1995) Separating population structure from population history: a cladistic analysis of the geographical distribution of mitochondrial DNA haplotypes in the tiger salamander, *Ambystoma tigrinum*. *Genetics* **140**, 767–782.

Terai, Y., Mayer, W.E., Klein, J., *et al.* (2002) The effect of selection on a long wavelength-sensitive (*LWS*) opsin gene of Lake Victoria cichlid fishes. *Proceedings of the National Academy of Sciences USA* **99**, 15501–15506.

Terai, Y., Morikawa, N., Kawakami, K., and Okada, N. (2003) The complexity of alternative splicing of *hagoromo* mRNAs is increased in an explosively speciated lineage in East African cichlids. *Proceedings of the National Academy of Sciences USA* **100**, 12798–12803.

Theisen, B.F., Christensen, B., and Arctander, P. (1995) Origin of clonal diversity in triploid parthenogenetic *Trichoniscus pusillus pusillus* (Isopoda, Crustacea) based upon allozyme and nucleotide sequence data. *Journal of Evolutionary Biology* **8**, 71–80.

Thomas, C.M. and Nielsen, K.M. (2005) Mechanisms of, and barriers to, horizontal gene transfer between bacteria. *Nature Reviews Microbiology* **3**, 711–721.

Thompson, J.D. (1991) The biology of an invasive plant: what makes *Spartina anglica* so successful? *BioScience* **41**, 393–401.

Thompson, J.D., McNeilly, T., and Gray, A.J. (1991a) Population variation in *Spartina anglica* CE Hubbard I. Evidence from a common garden experiment. *New Phytologist* **117**, 115–128.

Thompson, J.D., McNeilly, T., and Gray, A.J. (1991b) Population variation in *Spartina anglica* CE Hubbard II.

Reciprocal transplants among three successional populations. *New Phytologist* **117**, 129–139.

Thompson, J.D., McNeilly, T., and Gray, A.J. (1991c) Population variation in *Spartina anglica* CE Hubbard III. Response to substrate variation in a glasshouse experiment. *New Phytologist* **117**, 141–152.

Thulin, C.-G. and Tegelström, H. (2002) Biased geographical distribution of mitochondrial DNA that passed the species barrier from mountain hares to brown hares (genus *Lepus*): an effect of genetic incompatibility and mating behaviour? *Journal of the Zoological Society of London* **258**, 299–306.

Thulin, C.-G., Jaarola, M., and Tegelström, H. (1997) The occurrence of mountain hare mitochondrial DNA in wild brown hares. *Molecular Ecology* **6**, 463–467.

Tibayrenc, M. and Ayala, F.J. (1988) Isozyme variability in *Trypanosoma cruzi*, the agent of Chagas' disease: genetical, taxonomical, and epidemiological significance. *Evolution* **42**, 277–292.

Tibayrenc, M., Ward, P., Moya, A., and Ayala, F.J. (1986) Natural populations of *Trypanosoma cruzi*, the agent of Chagas disease, have a complex multiclonal structure. *Proceedings of the National Academy of Sciences USA* **83**, 115–119.

Tibayrenc, M., Neubauer, K., Barnabé, C., et al. (1993) Genetic characterization of six parasitic protozoa: parity between random-primer DNA typing and multilocus enzyme electrophoresis. *Proceedings of the National Academy of Sciences USA* **90**, 1335–1339.

Timmis, J.N., Ayliffe, M.A., Huang, C.Y., and Martin, W. (2004) Endosymbiotic gene transfer: organelle genomes forge eukaryotic chromosomes. *Nature Reviews Genetics* **5**, 123–135.

Tinti, F. and Scali, V. (1996) Androgenetics and triploids from an interacting parthenogenetic hybrid and its ancestors in stick insects. *Evolution* **50**, 1251–1258.

Torriani, S., Zapparoli, G., Malacrinò, P., et al. (2004) Rapid identification and differentiation of *Saccharomyces cerevisiae*, *Saccharomyces bayanus* and their hybrids by multiplex PCR. *Letters in Applied Microbiology* **38**, 239–244.

Tosi, A.J., Morales, J.C., and Melnick, D.J. (2000) Comparison of Y chromosome and mtDNA phylogenies leads to unique inferences of macaque evolutionary history. *Molecular Phylogenetics and Evolution* **17**, 133–144.

Tosi, A.J., Morales, J.C., and Melnick, D.J. (2002) Y-chromosome and mitochondrial markers in *Macaca fascicularis* indicate introgression with Indochinese *M. mulatta* and a biogeographic barrier in the Isthmus of Kra. *International Journal of Primatology* **23**, 161–178.

Tosi, A.J., Morales, J.C., and Melnick, D.J. (2003) Paternal, maternal, and biparental molecular markers provide unique windows onto the evolutionary history of macaque monkeys. *Evolution* **57**, 1419–1435.

Tovar, J., León-Avila, G., Sánchez, L.B., et al. (2003) Mitochondrial remnant organelles of *Giardia* function in iron-sulphur protein maturation. *Nature* **426**, 172–176.

Trachtulec, Z., Mihola, O., Vlcek, C., et al. (2005) Positional cloning of the Hybrid sterility 1 gene: fine genetic mapping and evaluation of two candidate genes. *Biological Journal of the Linnean Society* **84**, 637–641.

Trelease, W. (1917) Naming American hybrid oaks. *Proceedings of the American Philosophical Society* **56**, 44–52.

Tripet, F., Dolo, G., and Lanzaro, G.C. (2005) Multilevel analyses of genetic differentiation in *Anopheles gambiae* s.s. reveal patterns of gene flow important for malaria-fighting mosquito projects. *Genetics* **169**, 313–324.

Troy, C.S., MacHugh, D.E., Bailey, J.F., et al. (2001) Genetic evidence for Near-Eastern origins of European cattle. *Nature* **410**, 1088–1091.

Tucker, J.M. (1952) Evolution of the Californian oak *Quercus alvordiana*. *Evolution* **6**, 162–180.

Tucker, P.K., Sage, R.D., Warner, J., et al. (1992) Abrupt cline for sex chromosomes in a hybrid zone between two species of mice. *Evolution* **46**, 1146–1163.

Turgeon, J., Estoup, A., and Bernatchez, L. (1999) Species flock in the North American Great Lakes: molecular ecology of Lake Nipigon ciscoes (Teleostei: Coregonidae: *Coregonus*). *Evolution* **53**, 1857–1871.

Turner, B.J., Brett, B.-L.H., and Miller, R.R. (1980) Interspecific hybridization and the evolutionary origin of a gynogenetic fish, *Poecilia formosa*. *Evolution* **34**, 917–922.

Turner, B.L. and Horne, D. (1964) Taxonomy of *Machaeranthera* sect. *Psilactis* (Compositae—Astereae). *Brittonia* **16**, 316–331.

Turner, G.F., Seehausen, O., Knight, M.E., et al. (2001) How many species of cichlid fishes are there in African lakes? *Molecular Ecology* **10**, 793–806.

Turner, T.F., Trexler, J.C., Harris, J.L., and Haynes, J.L. (2000) Nested clade analysis indicates population fragmentation shapes genetic diversity in a freshwater mussel. *Genetics* **154**, 777–785.

Turner, T.L., Hahn, M.W., and Nuzhdin, S.V. (2005) Genomic islands of speciation in *Anopheles gambiae*. *PLoS Biology* **3**, e285.

Ungerer, M.C., Baird, S.J.E., Pan, J., and Rieseberg, L.H. (1998) Rapid hybrid speciation in wild sunflowers. *Proceedings of the National Academy of Sciences USA* **95**, 11757–11762.

Valbuena-Carabaña, M., González-Martínez, S.C., Sork, V.L., et al. (2005) Gene flow and hybridisation in a mixed oak forest (*Quercus pyrenaica* Willd and *Quercus petraea* (Matts) Liebl) in central Spain. *Heredity* **95**, 457–465.

Van den Heede, C.J., Viane, R.L.L., and Chase, M.W. (2003) Phylogenetic analysis of *Asplenium* subgenus *Ceterach* (Pteridophyta: Aspleniaceae) based on plastid and nuclear ribosomal ITS DNA sequences. *American Journal of Botany* **90**, 481–495.

van der Giezen, M., Cox, S., and Tovar, J. (2004) The iron-sulfur cluster assembly genes *iscS* and *iscU* of *Entamoeba histolytica* were acquired by horizontal gene transfer. *BMC Evolutionary Biology* **4**, 7.

Vanlerberghe, F., Dod, B., Boursot, P., *et al.* (1986) Absence of Y-chromosome introgression across the hybrid zone between *Mus musculus domesticus* and *Mus musculus musculus*. *Genetical Research* **48**, 191–197.

Vanlerberghe, F., Boursot, P., Catalan, J., *et al.* (1988) Analyse génétique de la zone d'hybridation entre les deux sous-espèces de souris *Mus musculus domesticus* et *Mus musculus musculus* en Bulgarie. *Genome* **30**, 427–437.

Van Zandt, P.A. and Mopper, S. (2002) Delayed and carryover effects of salinity on flowering in *Iris hexagona* (Iridaceae). *American Journal of Botany* **89**, 1847–1851.

Van Zandt, P.A. and Mopper, S. (2004) The effects of maternal salinity and seed environment on germination and growth in *Iris hexagona*. *Evolutionary Ecology Research* **6**, 813–832.

Van Zandt, P.A., Tobler, M.A., Mouton, E., *et al.* (2003) Positive and negative consequences of salinity stress for the growth and reproduction of the clonal plant, *Iris hexagona*. *Journal of Ecology* **91**, 837–846.

Veen, T., Borge, T., Griffith, S.C., *et al.* (2001) Hybridization and adaptive mate choice in flycatchers. *Nature* **411**, 45–50.

Verheyen, E., Salzburger, W., Snoeks, J., and Meyer, A. (2003) Origin of the superflock of cichlid fishes from Lake Victoria, East Africa. *Science* **300**, 325–329.

Verheyen, E., Salzburger, W., Snoeks, J., and Meyer, A. (2004) Origin of the superflock of cichlid fishes from Lake Victoria, East Africa. *Science* **304**, 963c.

Vilà, C., Savolainen, P., Maldonado, J.E., *et al.* (1997) Multiple and ancient origins of the domestic dog. *Science* **276**, 1687–1689.

Vilà, C., Seddon, J., and Ellegren, H. (2005) Genes of domestic mammals augmented by backcrossing with wild ancestors. *Trends in Genetics* **21**, 214–218.

Vilà, C., Walker, C., Sundqvist, A.-K., *et al.* (2003) Combined use of maternal, paternal and bi-parental genetic markers for the identification of wolf—dog hybrids. *Heredity* **90**, 17–24.

Vilà, M. and D'Antonio, C.M. (1998a) Fruit choice and seed dispersal of invasive vs. noninvasive *Carpobrotus* (Aizoaceae) in coastal California. *Ecology* **79**, 1053–1060.

Vilà, M. and D'Antonio, C.M. (1998b) Hybrid vigor for clonal growth in *Carpobrotus* (Aizoaceae) in coastal California. *Ecological Applications* **8**, 1196–1205.

Vines, T.H., Kohler, S.C., Thiel, M., *et al.* (2003) The maintenance of reproductive isolation in a mosaic hybrid zone between the fire-bellied toads (*Bombina bombina* and *B. variegata*). *Evolution* **57**, 1876–1888.

Viosca, P., Jr (1935) The irises of southeastern Louisiana—a taxonomic and ecological interpretation. *Bulletin of the American Iris Society* **57**, 3–56.

Vogel, J.P., Andrews, H.L., Wong, S.K., and Isberg, R.R. (1998) Conjugative transfer by the virulence system of *Legionella pneumophila*. *Science* **279**, 873–876.

Volkov, R.A., Borisjuk, N.V., Panchuk, I.I., *et al.* (1999) Elimination and rearrangement of paternal rDNA in the allotetraploid *Nicotiana tabacum*. *Molecular Biology and Evolution* **16**, 311–320.

Vollmer, S.V. and Palumbi, S.R. (2002) Hybridization and the evolution of reef coral diversity. *Science* **296**, 2023–2025.

Vorburger, C. and Reyer, H.-U. (2003) A genetic mechanism of species replacement in European waterfrogs? *Conservation Genetics* **4**, 141–155.

Vriesendorp, B. and Bakker, F.T. (2005) Reconstructing patterns of reticulate evolution in angiosperms: what can we do? *Taxon* **54**, 593–604.

Wagner, F.S. (1987) Evidence for the origin of the hybrid cliff fern, *Woodsia* x *abbeae* (Aspleniaceae: Athyrioideae). *Systematic Botany* **12**, 116–124.

Wagner, P.L. and Waldor, M.K. (2002) Bacteriophage control of bacterial virulence. *Infection and Immunity* **70**, 3985–3993.

Wagner, W.H., Jr (1954) Reticulate evolution in the Appalachian aspleniums. *Evolution* **8**, 103–118.

Wakeley, J. and Hey, J. (1997) Estimating ancestral population parameters. *Genetics* **145**, 847–855.

Walton, C., Handley, J.M., Tun-Lin, W., *et al.* (2000) Population structure and population history of *Anopheles dirus* mosquitoes in south-east Asia. *Molecular Biology and Evolution* **17**, 962–974.

Walton, C., Handley, J.M., Collins, F.H., *et al.* (2001) Genetic population structure and introgression in *Anopheles dirus* mosquitoes in south-east Asia. *Molecular Ecology* **10**, 569–580.

Wang, H., McArthur, E.D., Sanderson, S.C., *et al.* (1997) Narrow hybrid zone between two subspecies of Big Sagebrush (*Artemisia tridentata*: Asteraceae) IV. Reciprocal transplant experiments. *Evolution* **51**, 95–102.

Wang, H., McArthur, E.D., and Freeman, D.C. (1999) Narrow hybrid zone between two subspecies of Big Sagebrush (*Artemisia tridentata*: Asteraceae) IX: elemental uptake and niche separation. *American Journal of Botany* **86**, 1099–1107.

Wang, J., Tian, L., Madlung, A., *et al.* (2004) Stochastic and epigenetic changes of gene expression in *Arabidopsis* polyploids. *Genetics* **167**, 1961–1973.

Wang, R.-L. and Hey, J. (1996) The speciation history of *Drosophila pseudoobscura* and close relatives: inferences from DNA sequence variation at the period locus. *Genetics* **144**, 1113–1126.

Wang, R.-L., Wakeley, J., and Hey, J. (1997) Gene flow and natural selection in the origin of *Drosophila pseudoobscura* and close relatives. *Genetics* **147**, 1091–1106.

Wang, X., Shi, X., Hao, B., *et al.* (2005) Duplication and DNA segmental loss in the rice genome: implications for diploidization. *New Phytologist* **165**, 937–946.

Wang, X.-R. and Szmidt, A.E. (1994) Hybridization and chloroplast DNA variation in a *Pinus* species complex from Asia. *Evolution* **48**, 1020–1031.

Wang, X.-R., Szmidt, A.E., and Savolainen, O. (2001) Genetic composition and diploid hybrid speciation of a high mountain pine, *Pinus densata*, native to the Tibetan plateau. *Genetics* **159**, 337–346.

Wang, Y.-M., Dong, Z.-Y., Zhang, Z.-J., *et al.* (2005) Extensive *de novo* genomic variation in rice induced by introgression from wild rice (*Zizania latifolia* Griseb.). *Genetics* **170**, 1945–1956.

Watanabe, H., Fujiyama, A., Hattori, M., *et al.* (2004) DNA sequence and comparative analysis of chimpanzee chromosome 22. *Nature* **429**, 382–388.

Waterbury, J.B. and Valois, F.W. (1993) Resistance to co-occurring phages enables marine *Synechococcus* communities to coexist with cyanophages abundant in seawater. *Applied Environmental Microbiology* **59**, 3393–3399.

Webb, G.C., White, M.J.D., Contreras, N., and Cheney, J. (1978) Cytogenetics of the parthenogenetic grasshopper *Warramaba* (formerly *Moraba*) *virgo* and its bisexual relatives. IV. Chromosome banding studies. *Chromosoma* **67**, 309–339.

Webby, R.J. and Webster, R.G. (2003) Are we ready for pandemic influenza? *Science* **302**, 1519–1522.

Weber, E. and D'Antonio, C.M. (1999) Germination and growth responses of hybridizing *Carpobrotus* species (Aizoaceae) from coastal California to soil salinity. *American Journal of Botany* **86**, 1257–1263.

Weider, L.J., Hobæk, A., Hebert, P.D.N., and Crease, T.J. (1999) Holarctic phylogeography of an asexual species complex—II. Allozymic variation and clonal structure in Arctic *Daphnia*. *Molecular Ecology* **8**, 1–13.

Welch, M.E. and Rieseberg, L.H. (2002) Patterns of genetic variation suggest a single, ancient origin for the diploid hybrid species *Helianthus paradoxus*. *Evolution* **56**, 2126–2137.

Wendel, J.F. (1989) New World tetraploid cottons contain Old World cytoplasm. *Proceedings of the National Academy of Sciences USA* **86**, 4132–4136.

Wendel, J.F. (1995) Cotton—*Gossypium* (Malvaceae). In J. Smartt and N.W. Simmonds, eds. *Evolution of Crop Plants*, 2nd edn, pp. 358–366. Longman Scientific & Technical, Harlow.

Wendel, J.F., Schnabel, A., and Seelanan, T. (1995) Bidirectional interlocus concerted evolution following allopolyploid speciation in cotton (*Gossypium*). *Proceedings of the National Academy of Sciences USA* **92**, 280–284.

Werth, C.R., Guttman, S.I., and Eshbaugh, W.H. (1985) Electrophoretic evidence of reticulate evolution in the Appalachian *Asplenium* complex. *Systematic Botany* **10**, 184–192.

Wesselingh, R.A. and Arnold, M.L. (2000) Nectar production in Louisiana Iris hybrids. *International Journal of Plant Sciences* **161**, 245–251.

White, G.B. (1971) Chromosomal evidence for natural interspecific hybridization by mosquitoes of the *Anopheles gambiae* complex. *Nature* **231**, 184–185.

White, M.J.D. (1978) *Modes of Speciation*. WH Freeman and Company, San Francisco.

White, M.J.D. (1980) Meiotic mechanisms in a parthenogenetic grasshopper species and its hybrids with related bisexual species. *Genetica* **52/53**, 379–383.

White, M.J.D. and Contreras, N. (1982) Cytogenetics of the parthenogenetic grasshopper *Warramaba virgo* and its bisexual relatives. VIII. Karyotypes and C-banding patterns in the clones of *W. virgo*. *Cytogenetics and Cell Genetics* **34**, 168–177.

White, M.J.D., Cheney, J., and Key, K.H.L. (1963) A parthenogenetic species of grasshopper with complex structural heterozygosity (Orthoptera: Acridoidea). *Australian Journal of Zoology* **11**, 1–19.

White, M.J.D., Contreras, N., Cheney, J., and Webb, G.C. (1977) Cytogenetics of the parthenogenetic grasshopper *Warramaba* (formerly *Moraba*) *virgo* and its bisexual relatives. II. Hybridization studies. *Chromosoma* **61**, 127–148.

White, M.J.D., Webb, G.C., and Contreras, N. (1980) Cytogenetics of the parthenogenetic grasshopper *Warramaba* (formerly *Moraba*) *virgo* and its bisexual relatives. VI. DNA replication patterns of the chromosomes. *Chromosoma* **81**, 213–248.

White, M.J.D., Dennis, E.S., Honeycutt, R.L., *et al.* (1982) Cytogenetics of the parthenogenetic grasshopper *Warramaba virgo* and its bisexual relatives. IX. The ribosomal RNA cistrons. *Chromosoma* **85**, 181–199.

White, N.J. (2003) Melioidosis. *Lancet* **361**, 1715–1722.

Whitham, T.G. (1989) Plant hybrid zones as sinks for pests. *Science* **244**, 1490–1493.

Whitham, T.G., Morrow, P.A., and Potts, B.M. (1994) Plant hybrid zones as centers of biodiversity: the herbivore community of two endemic Tasmanian eucalypts. *Oecologia* **97**, 481–490.

Whitham, T.G., Martinsen, G.D., Floate, K.D., *et al.* (1999) Plant hybrid zones affect biodiversity: tools for a genetic-based understanding of community structure. *Ecology* **80**, 416–428.

Whitham, T.G., Young, W.P., Martinsen, G.D., *et al.* (2003) Community and ecosystem genetics: a consequence of the extended phenotype. *Ecology* **84**, 559–573.

Whittemore, A.T. and Schaal, B.A. (1991) Interspecific gene flow in sympatric oaks. *Proceedings of the National Academy of Sciences USA* **88**, 2540–2544.

Whitton, J., Wolf, D.E., Arias, D.M., *et al.* (1997) The persistence of cultivar alleles in wild populations of sunflowers five generations after hybridization. *Theoretical and Applied Genetics* **95**, 33–40.

Widmer, A. and Baltisberger, M. (1999a) Extensive intraspecific chloroplast DNA (cpDNA) variation in the alpine *Draba aizoides* L. (Brassicaceae): haplotype relationships and population structure. *Molecular Ecology* **8**, 1405–1415.

Widmer, A. and Baltisberger, M. (1999b) Molecular evidence for allopolyploid speciation and a single origin of the narrow endemic *Draba ladina* (Brassicaceae). *American Journal of Botany* **86**, 1282–1289.

Wilhelm, S.W., Brigden, S.M., and Suttle, C.A. (2002) A dilution technique for the direct measurement of viral production: a comparison in stratified and tidally mixed coastal waters. *Microbial Ecology* **43**, 168–173.

Wilkinson, J. (1944) The cytology of *Salix* in relation to its taxonomy. *Annals of Botany* **8**, 269–284.

Williams, B.A.P., Hirt, R.P., Lucocq, J.M., and Embley, T.M. (2002) A mitochondrial remnant in the microsporidian *Trachipleistophora hominis*. *Nature* **418**, 865–869.

Williams, J.H., Boecklen, W.J., and Howard, D.J. (2001) Reproductive processes in two oak (*Quercus*) contact zones with different levels of hybridization. *Heredity* **87**, 680–690.

Williams, J.H., Jr, Friedman, W.E., and Arnold, M.L. (1999) Developmental selection within the angiosperm style: using gamete DNA to visualize interspecific pollen competition. *Proceedings of the National Academy of Sciences USA* **96**, 9201–9206.

Wimp, G.M. and Whitham, T.G. (2001) Biodiversity consequences of predation and host plant hybridization on an aphid-ant mutualism. *Ecology* **82**, 440–452.

Wimp, G.M., Martinsen, G.D., Floate, K.D., *et al.* (2005) Plant genetic determinants of arthropod community structure and diversity. *Evolution* **59**, 61–69.

Witter, M.S. and Carr, G.D. (1988) Adaptive radiation and genetic differentiation in the Hawaiian Silversword alliance (Compositae: Madiinae). *Evolution* **42**, 1278–1287.

Woese, C.R. (1987) Bacterial evolution. *Microbiological Reviews* **51**, 221–271.

Wolf, P.G., Campbell, D.R., Waser, N.M., *et al.* (2001) Tests of pre- and postpollination barriers to hybridization between sympatric species of *Ipomopsis* (Polemoniaceae). *American Journal of Botany* **88**, 213–219.

Wolf, Y.I., Kondrashov, A.S., and Koonin, E.V. (2000) Interkingdom gene fusions. *Genome Biology* **1**(6), 0013.1–0013.13.

Wolfe, K.H. and Shields, D.C. (1997) Molecular evidence for an ancient duplication of the entire yeast genome. *Nature* **387**, 708–713.

Wolpoff, M.H., Hawks, J., Frayer, D.W., and Hunley, K. (2001) Modern human ancestry at the peripheries: a test of the replacement theory. *Science* **291**, 293–297.

Wommack, K.E. and Colwell, R.R. (2000) Virioplankton: viruses in aquatic ecosystems. *Microbiology and Molecular Biology Reviews* **64**, 69–114.

Won, H., and Renner, S.S. (2003) Horizontal gene transfer from flowering plants to *Gnetum*. *Proceedings of the National Academy of Sciences USA* **100**, 10824–10829.

Won, Y.-J. and Hey, J. (2005) Divergence population genetics of chimpanzees. *Molecular Biology and Evolution* **22**, 297–307.

Won, Y.-J., Sivasundar, A., Wang, Y., and Hey, J. (2005) On the origin of Lake Malawi cichlid species: A population genetic analysis of divergence. *Proceedings of the National Academy of Sciences USA* **102**, 6581–6586.

Worobey, M., Rambaut, A., Pybus, O.G., and Robertson, D.L. (2002) Questioning the evidence for genetic recombination in the 1918 'Spanish Flu' virus. *Science* **296**, 211a.

Wrigley, G. (1995) Coffee—*Coffea* spp. (Rubiaceae). In J. Smartt and N.W. Simmonds, eds. *Evolution of Crop Plants*, 2nd edn, pp. 438–443. Longman Scientific & Technical, Harlow.

Wu, C.A., and Campbell, D.R. (2005) Cytoplasmic and nuclear markers reveal contrasting patterns of spatial genetic structure in a natural *Ipomopsis* hybrid zone. *Molecular Ecology* **14**, 781–791.

Wyatt, R., Odrzykoski, I.J., Stoneburner, A., *et al.* (1988) Allopolyploidy in bryophytes: multiple origins of *Plagiomnium medium*. *Proceedings of the National Academy of Sciences USA* **85**, 5601–5604.

Wyatt, R., Odrzykoski, I.J., and Stoneburner, A. (1992) Isozyme evidence of reticulate evolution in mosses: *Plagiomnium medium* is an allopolyploid of *P. ellipticum* x *P. insigne*. *Systematic Botany* **17**, 532–550.

Xiong, L.Z., Xu, C.G., Saghai Maroof, M.A., and Zhang, Q. (1999) Patterns of cytosine methylation in an elite rice hybrid and its parental lines, detected by a methylation-sensitive amplification polymorphism technique. *Molecular and General Genetics* **261**, 439–446.

Yamada, T., Eishi, Y., Ikeda, S., *et al.* (2002) *In situ* localization of *Propionibacterium acnes* DNA in lymph nodes from sarcoidosis patients by signal amplification with catalysed reporter deposition. *Journal of Pathology* **198**, 541–547.

Yamane, K. and Kawahara, T. (2005) Intra- and interspecific phylogenetic relationships among diploid *Triticum-Aegilops* species (Poaceae) based on base-pair substitutions, indels, and microsatellites in chloroplast noncoding sequences. *American Journal of Botany* **92**, 1887–1898.

Yonekawa, H., Moriwaki, K., Gotoh, O., *et al.* (1988) Hybrid origin of Japanese mice 'Mus musculus molossinus': evidence from restriction analysis of mitochondrial DNA. *Molecular Biology and Evolution* **5**, 63–78.

Zardoya, R., Ding, X., Kitagawa, Y., and Chrispeels, M.J. (2002) Origin of plant glycerol transporters by horizontal gene transfer and functional recruitment. *Proceedings of the National Academy of Sciences USA* **99**, 14893–14896.

Zeng, L.-W. and Singh, R.S. (1993) The genetic basis of Haldane's Rule and the nature of asymmetric hybrid male sterility among *Drosophila simulans, Drosophila mauritiana* and *Drosophila sechellia*. *Genetics* **134**, 251–260.

Zerega, N.J.C., Ragone, D., and Motley, T.J. (2004) Complex origins of breadfruit (*Artocarpus altilis*, Moraceae): implications for human migrations in Oceania. *American Journal of Botany* **91**, 760–766.

Zerega, N.J.C., Ragone, D., and Motley, T.J. (2005) Systematics and species limits of breadfruit (*Artocarpus*, Moraceae). *Systematic Botany* **30**, 603–615.

Zhao, X.-P., Ji, Y., Ding, X., *et al.* (1998a) Macromolecular organization and genetic mapping of a rapidly evolving chromosome-specific tandem repeat family (B77) in cotton (*Gossypium*). *Plant Molecular Biology* **38**, 1031–1042.

Zhao, X.-P., Ji, Y., Hanson, R.E., *et al.* (1998b) Dispersed repetitive DNA has spread to new genomes since polyploid formation in cotton. *Genome Research* **8**, 479–492.

Zhaxybayeva, O. and Gogarten, J.P. (2004) Cladogenesis, coalescence and the evolution of the three domains of life. *Trends in Genetics* **20**, 182–187.

Zhaxybayeva, O., Lapierre, P., and Gogarten, J.P. (2004) Genome mosaicism and organismal lineages. *Trends in Genetics* **20**, 254–260.

Zhou, N.N., Senne, D.A., Landgraf, J.S., *et al.* (1999) Genetic reassortment of avian, swine, and human influenza A viruses in American pigs. *Journal of Virology* **73**, 8851–8856.

Zhu, G., Keithly, J.S., and Philippe, H. (2000a) What is the phylogenetic position of *Cryptosporidium*? *International Journal of Systematic and Evolutionary Microbiology* **50**, 1673–1681.

Zhu, G., Marchewka, M.J., and Philippe, H. (2000b) *Cryptosporidium parvum* appears to lack a plastid genome. *Microbiology* **146**, 315–321.

Zimmer, E.A., Martin, S.L., Beverley, S.M., *et al.* (1980) Rapid duplication and loss of genes coding for the α chains of hemoglobin. *Proceedings of the National Academy of Sciences USA* **77**, 2158–2162.

Index

Note: page numbers in *italics* refer to Figures and Tables. Page numbers in **bold** refer to Glossary entries.